Drug Delivery Across Physiological Barriers

Drug Delivery Across Physiological Barriers

edited by
Silvia Muro

PAN STANFORD PUBLISHING

Published by

Pan Stanford Publishing Pte. Ltd.
Penthouse Level, Suntec Tower 3
8 Temasek Boulevard
Singapore 038988

Email: editorial@panstanford.com
Web: www.panstanford.com

British Library Cataloguing-in-Publication Data
A catalogue record for this book is available from the British Library.

Drug Delivery Across Physiological Barriers

ISBN 978-981-4669-40-5 (Hardcover)
ISBN 978-981-4669-41-2 (eBook)

Printed in the USA

Contents

Preface

Optimal drug delivery in the body is paramount to maximizing the therapeutic efficiency of pharmaceutical compounds while minimizing their potential toxicity. In this regard, numerous advances in the last decades have rendered considerable improvement in drug delivery strategies, thereby increasing the bioavailability of therapeutic agents. These systems help solubilize pharmaceutical drugs, protect them from premature degradation, control their circulation, target them to sites of disease, and optimize their release rate. In addition, for most therapeutics, access to their targets of intervention requires penetration across body compartments, extracellular matrices, cellular linings, and/or different intracellular environments. Therefore, the design of strategies capable of improving transport of pharmaceuticals through these physiological barriers has become an imperative yet a challenging need in the quest for better therapeutics. This book aims at providing an overview of current advances in the field of drug delivery from the perspective of transport across the said physiological barriers. This is pursued by discussing fundamental knowledge pertaining to the biological function and natural mechanisms regulating these barriers, as well as by focusing on drug delivery strategies that facilitate transport of drugs and their carriers at the tissue, cell, and subcell levels.

With this in mind, the book has been divided into three independent yet complementary sections. **Section A** covers the background biological information regarding the structure, function, and regulation of constituents of the said physiological barriers, offering the reader an overview of the challenges and opportunities they pose. Within this section **Chapter 1** introduces the concept of cellular barriers or linings, which arise as a result of the physiological need to develop selective compartments within the body. They are established by epithelial or endothelial cells, which strongly adhere to each other, forming polarized linings that separate apical and basolateral compartments, such as those which coat hollow organs, glands, blood vessels, etc. These cellular linings most often act as barriers to passive transport of substances between the compartments they separate, regulating passage in a minutely controlled manner. From a drug delivery perspective, they regulate absorption or penetration of therapeutics across the inner layers of the skin, the gastrointestinal wall, passage from the bloodstream into subjacent tissues, etc., which pertains to drug access from the administration point to the intended body compartment. Subsequently, **Chapter 2** covers fundamental aspects of the cellular plasma membrane. This is important because, once in the appropriate body compartment, penetration into cells of the affected tissues requires passage through the plasmalemma that separates and permits communication between the intracellular environment and the extracellular milieu. The composition and regulation of this semipermeable barrier are described, including passive and active mechanisms of transport involving diffusion, ion and molecule channels, endocytosis, or exocytosis. Further, **Chapter 3** touches upon subcellular organization and offers a detailed overview of the biological mechanisms by which macromolecules (e.g., proteins, lipids, etc.) are naturally sorted and trafficked to their final subcellular destinations with exquisite precision. The cellular machinery, signaling cascades, and pathways employed in these events are finding valuable translational applications in the field of drug delivery. To end this section, **Chapter 4** discusses how pathogens have evolved remarkable means to overcome all these physiological barriers in order to infect their hosts. A main focus is paid to how such invasive bacteria and viruses gain access inside cells of the body, for example, by recognition and binding to particular cell-surface markers leading to endocytic uptake, disruption of the plasmalemma or the membrane of endolysosomal vesicles, hijacking the cytoskeleton and molecular elements promoting vesicular fusion or fission, etc. Mimicking such strategies and pathways established by nature is becoming a main practice in order to advance drug delivery.

With this fundamental biological background in place, Sections B and C offer an overview of recent advances on the design of vehicles capable of delivering pharmaceutical agents across said physiological barriers. **Section B** focuses on strategies to improve drug access from their administration point to their intended body compartments. Within this section, **Chapter 5** covers approaches to overcome the barrier posed by the skin, particularly focusing on the epidermis, which is the main requirement for topical administration of pharmaceuticals. The different factors influencing penetration through this barrier, natural pathways that enable it, models and tools employed to study this parameter, and strategies to modify and optimize skin penetration are discussed. Then, **Chapter 6** offers an overview on an additional element which contributes to the barrier function of epithelial linings that separate the outside milieu from the body: the mucus covering mucosal surfaces. Traversing this mucus coat is necessary for drug delivery through the intestinal, pulmonary, vaginal routes, etc. The composition and physicochemical properties of this coat, techniques used to characterize drug transport through the mucus, and strategies to surmount this barrier are offered in this chapter. **Chapter 7** subsequently focuses on drug transport across the gastrointestinal epithelium, which is the cellular lining responsible for absorption of orally dispensed drugs into the circulation and constitutes the main administration route for pharmaceutical compounds. Parameters influencing the bioavailability of oral formulations, different factors ruling drug absorption, transport mechanisms at this interface, and strategies to enhance drug delivery via this route are discussed. Next, for drugs that reach the bloodstream, whether through direct administration into the circulation or one of the already mentioned routes, passage out of blood vessels and into the underlining tissues is required for most therapeutic interventions. As such, **Chapter 8** discusses those cases where the layer of endothelial cells lining the inner wall of blood vessels constitutes a tight barrier regulating transport into the subjacent tissue. This is characteristic of areas irrigated by capillaries with continuous and non-fenestrated endothelium, such as the case of the blood–brain barrier, the pulmonary endothelium, etc. The endothelial structures and pathways that regulate drug transport across this lining, the physicochemical features that a compound must adopt in order to freely traverse it, and additional means to enhance this transport are offered. **Chapter 9** focuses instead on cases where the barrier property of this layer is compromised, such as the example of the vasculature of many solid tumors, where substances can leak into the subjacent tumor tissue through gaps between endothelial cells that line these blood vessels (the enhanced permeability and retention effect). The chapter presents key considerations about tumor pathophysiology, common challenges of delivering anticancer therapeutics, and drug delivery strategies to overcome these problems.

Finally, once a pharmaceutical compound is located at the intended tissue, most therapeutic applications require delivery within its cells. Focusing on this premise, **Section C** offers an overview of technical approaches aimed at achieving drug delivery to different cellular compartments. **Chapter 10** reviews the role exerted by lipids that compose the cell membranes, including the plasmalemma, on interactions of drugs and drugs carriers with this barrier. It discusses the lipid composition of cellular membranes, variations and changes associated with human pathologies, their biophysical characteristics, interactions with therapeutics, and models to study these aspects. **Chapter 11** further describes approaches where drug delivery is pursued by designing vehicles capable of fusing with cell membranes, so that their therapeutic content can gain access to the intracellular cytosolic milieu. The chapter comments on natural and synthetic vehicles that present such ability, either by means of physical assistance or spontaneously. **Chapter 12** focuses on drug delivery strategies that instead take advantage of endocytosis processes, where the plasmalemma of cells engulfs extracellular substances into membrane-bound vesicles that are then transported inside the cell body. Vesicles carrying the internalized materials are then sorted to different locations, most commonly the endolysosomal route, a membranous container system characterized by its degradative conditions. This provides an opportunity for the delivery of drugs needed at this particular location, intracellular elimination of drug delivery vehicles, and drug release, which are discussed in the chapter. However, certain pharmaceutical compounds may become entrapped in the endolysosomal compartment, which would preclude them from accessing additional intracellular locations. **Chapter 13** focuses on such cases and offers an overview on drug-delivery strategies aimed at overcoming this intracellular barrier, covering biological, chemical, physical, and combinatorial means to this avail. Where escape from endolysosomal vesicles is achieved, subsequent transport to other intracellular organelles may be required, and **Chapter 14** ends on this topic. Using the example of intracellular mitochondria as a main focus, a description of the mechanisms by which this can be achieved is offered, including physicochemical characteristics of agents that can intrinsically accumulate at particular intracellular locations, active transport routes, and elements, as well as molecular modifications and drug vehicles to achieve differential trafficking.

Altogether, this content aims to offer the reader a broad perspective on common paradigms of drug delivery to improve access of therapeutics across tissue, cellular, and subcellular barriers, using relevant examples that highlight recent advances in this field. Naturally, this collection is not comprehensive as new knowledge and technological advances in this arena arise continuously and rapidly, driven by tangible medical needs and fruit of a multifaceted and multidisciplinary effort. Nevertheless, the road to achieving precise, safe, and effective transport of drugs and drug delivery vehicles is still plagued with cumbersome obstacles. Some of these main challenges are discussed throughout the chapters, to ignite a critical evaluation of the achievements and limitations in this field, which shall guide the future continuation of such efforts.

Silvia Muro
Winter 2015

Section A

Physiological Barriers Controlling Penetration and Transport of Substances in the Body

Chapter 1

Structure and Function of Epithelial and Endothelial Barriers

Michael Koval

Division of Pulmonary, Allergy, Critical Care and Sleep Medicine,
Department of Medicine and Department of Cell Biology,
Emory University School of Medicine, Whitehead Biomedical Research Building
615 Michael St., Suite 205 Atlanta, GA 30322, USA
mhkoval@emory.edu

Human physiology requires selective compartmentation so that distinct physiological functions can be effectively regulated. Central to organ compartments are barriers composed of specialized polarized epithelial cells. Epithelial and endothelial cells have several features that promote barrier function, including apical-basolateral polarity and intercellular junctions called tight junctions. Cells also mediate a selective barrier through the cytoplasm of cells via transcytotic pathways. Each of these elements can be targeted to alter barrier function. Insights from pathogens that interact with cells to decrease barrier function provide clues to pharmacologic agents that can have similar effect to promote drug delivery.

Drug Delivery Across Physiological Barriers
Edited by Silvia Muro

ISBN 978-981-4669-40-5 (Hardcover), 978-981-4669-41-2 (eBook)
www.panstanford.com

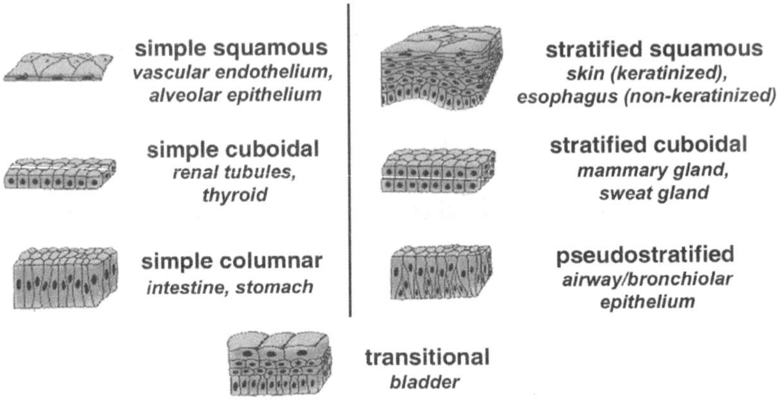

Figure 1.1 Types of epithelia. Shown are the classical types of epithelia along with examples of each type. The right-hand side of the figure shows three categories of simple epithelia; the left side shows the more complex equivalent. Most epithelia require regulated barrier function and secretory activity to ensure that they have proper moisture balance. An exception to this is skin, which has an outer keratinized layer called the stratum corneum. Cuboidal epithelia tend to contain cells with a primary role in secreting hormones or other essential products. Columnar epithelia can be absorptive, especially in the case of the gut. Pseudostratified epithelia appear to be a multilayered structure based on the distribution of nuclei throughout the tissue; however, each cell has direct contact with the basal lamina. The pseudostratified monolayer in airways contains a mixture of cells in close apposition with distinct functions ranging from mucus secretion to ciliated cells, which promote clearance of particulates from the lung. Transitional epithelium is unique in that it is primarily restricted to the bladder where it has an unusual capacity for deformation to accommodate changes in bladder fluid volume.

1.1 Introduction

Different organs have different epithelial morphology and composition, depending on organ function (Fig. 1.1). So the structure and function of skin epithelium, which is exposed to the outside world, will differ from that of the gut, which processes and absorbs nutrients, or lung, which mediates gas exchange. Moreover, vascular endothelium, which lines blood vessels, represents another specialized form of epithelium that regulates diverse functions ranging from gas exchange to inflammation. Epithelia have several

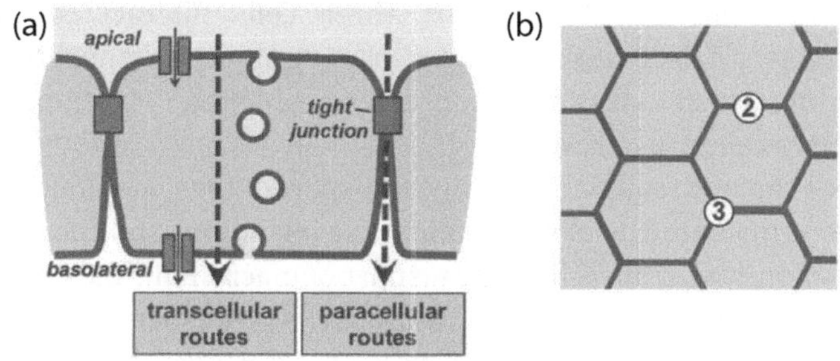

Figure 1.2 Pathways across epithelial barriers. (a) Shown are the transcellular route (through cells) and paracellular route (between cells). Transcellular transport occurs primarily through ion channels/pumps or via vesicular-mediated membrane transport. Paracellular transport is regulated by tight junctions. (b) There are predominantly two classes of cell–cell contacts in an epithelium, bicellular (2) and tricellular (3) contacts, where two or three cells, respectively, are in direct contact. This is most clear in cuboidal and columnar epithelia; other classes of epithelia (e.g., squamous epithelia) may not have clear cut distinctions between these types of contacts.

properties that create a selective permeability barrier. Clearly, a multicellular epithelium will present a more significant barrier than a monolayer. However, monolayers also have features that serve physiological barrier function.

There are two major paths that extracellular material can use to move across an epithelial monolayer: either through cells themselves (transcellular route) or between cells (paracellular route) (Fig. 1.2). Both of these transport pathways are regulated, and different materials move via different aspects of either the transcellular or paracellular routes.

Paracellular permeability between epithelial cells is regulated by structures at cell–cell contact sites known as tight junctions, which control paracellular permeability [1]. Paracellular transport of ions, metabolites, and water is mediated by passive diffusion through tight junction proteins in the claudin family, which form the equivalent of paracellular channels [2]. By contrast, paracellular transport of macromolecules requires an active process involving breaking and reformation of tight junctions [3]. Tight junctions are also categorized by whether they are sites where two cells interact (bicellular junctions) or where three cells interact (tricellular junctions). Tricellular junctions have unique composition and function, which makes them more permeable than junctions where only two cells are in contact [4].

For the transcellular pathway, transport of ions, metabolites, and water is mediated by plasma membrane ion pumps and channels [5–7]. By contrast, transcellular transport of macromolecules and large protein complexes predominantly occurs by transport of membrane vesicles, via the combined processes of endocytosis and exocytosis, called transcytosis [8, 9].

Both the paracellular and transcellular pathways represent epithelial barriers to drug delivery. This chapter mainly focuses on the molecular structure of the paracellular pathway, with particular emphasis on the role of tight junctions in control of permeability and how junction proteins can be targeted by pathogens. Some key aspects of the transcellular route are also described.

1.2 Measuring Barrier Function

A key point in considering the different aspects of paracellular epithelial barrier function is that permeability phenomena occur at different timescales. Ion permeability is measured on an almost instantaneous timescale, on the order of microseconds. On the other hand, paracellular flux of soluble molecules occurs on a much slower timescale, on the order of minutes to hours depending on the size of the solute moving across the junctional complex. This difference in timescale is critical, since changes to ion permeability do not always translate into changes to paracellular flux and *vice versa*.

Several approaches can be used to measure barrier function. When considering those approaches, it is important to keep in mind that paracellular transport does not occur in isolation. Instead, transcellular transport can also occur and confound measurements. Specifically, plasma membrane channels may contribute to net epithelial barrier permeability to ions, and transcytosis may be part of net small-molecule flux across the barrier. Using inhibitors to target specific transcellular pathways is a helpful approach to rule out roles for these processes in the measurements of paracellular flux.

Paracellular flux is measured by administering a labeled tracer molecule onto one side of an epithelial barrier and then measuring samples obtained from the other side to obtain a time course of transport. Tracers can be either fluorescently tagged or radiolabeled; Evans Blue is also used, which can be quantified by absorbance spectroscopy. This approach is amenable to *in vivo* measurement where a tracer is introduced into a compartment, such as the circulation, and then monitored by appearance in the central nervous system to assess blood–brain barrier function or into airspaces to measure lung barrier function [10]. Nondestructive methods for measuring fluid flux are also being developed; for instance live X-ray fluoroscopy to measure the extent of pulmonary edema in rodents, comparable to chest X-rays in patients [11]. Epithelial cells cultured on permeable supports (Transwells) are also highly amenable to flux measurements, where a tracer is added to the top of the chamber and appearance measured on the bottom [12].

For measuring barrier function to ions electrophysiologically, a system called an Ussing chamber can be used with cultured epithelial cells on Transwells or intact tissues such as skin or intestine [13]. In essence, samples are immersed in buffers with defined ion composition and suspended between two electrodes. The most commonly used measure of barrier function with this approach is transepithelial resistance (TER), where current across the epithelium is set to a fixed value and the resulting voltage is measured, allowing resistance to be calculated using Ohm's law. The TER of cultured epithelial cells is also readily and broadly measured with metal "chopstick" electrodes and an Ohmmeter, which is a simpler but less precise approach as compared with Ussing chamber analysis.

1.3 Epithelial Polarity

Epithelial polarity is central to the formation of a specific tissue barrier. The main determinant of polarity is cell orientation, where cells have functionally distinct top (apical) and bottom (basal) surfaces. Initiation of *de novo* polarity is induced by direct contact of cells with a structure called the basal lamina or basement membrane [14]. The basal lamina consists of a complex mixture of proteins and complex carbohydrates known collectively as the extracellular matrix (ECM). Key ECM components include proteins such as laminins, collagens, and fibronectins as well as proteoglycans and glycosaminoglycans [15, 16]. ECM composition differs extensively in a tissue and in a state-dependent manner, that is, whether the tissue is healthy, under stress, or overtly injured [17]. Cells are influenced by ECM, and they also remodel ECM, a dynamic process that synergistically affects cell function and phenotype. Mechanical stresses such as fluid flow, pressure, and direct mechanical distension can also influence cell behavior and differentiation [18].

Cell contact with the ECM provides an environmental cue that orients and defines the basal surface of the cell. This is accomplished by interactions of transmembrane proteins known as integrins, which bind to specific ECM components [19]. Active integrins are heterodimers, consisting of an alpha and beta chain, which together determine substrate specificity [20]. Thus, $\alpha 5\beta 1$ integrin binds to fibronectin, whereas $\alpha 1\beta 1$ and $\alpha 2\beta 1$ integrins bind to collagens and laminin. Integrins engaged to ligands produce outside-in signals, which then can dictate cell behavior. *In vitro* this is demonstrated by the formation of focal adhesions where integrins serve as organizing centers for signaling protein complexes, which regulate the formation and positioning of actin filaments [21]. Whether focal adhesions are formed *in vivo* is controversial; however, the principles of integrin/ECM signal transduction are likely to be comparable to observations made using cultured cell models [22].

A key step in the induction of apical/basal polarity is binding of $\beta 1$ integrin to the ECM protein laminin, which helps to define the bottom of the cell. $\beta 1$ integrin binding activates the small GTPase rac1 [23]. Rac1, in turn, stimulates actin filament formation and also enhances the secretion and deposition of laminin. Since $\beta 1$ integrin binds to laminin, rac1 helps establish a positive feedback loop, which reinforces cell polarity. Polarized actin also helps orient cells by providing tracks to promote the directional targeting of a specialized vesicle transport pathway known as the exocyst or sec6/8 pathway [24]. These vesicles transport proteins to establish the apical surface of the cell, including transmembrane proteins in the crumbs and cadherin families, which define the top and lateral aspects of the cell, respectively. Crumbs acts as an apical "signpost" that binds to cytoplasmic scaffold proteins beginning with Lin-7 (PALS1); sequential interactions between these and other scaffold and transmembrane proteins ultimately form the apical junctional complex (AJC) (Fig. 1.3) [25, 26]. Cadherins are also crucial in AJC formation, since they are capable of high affinity interactions between cells to form adherens junctions. The axis of the crumbs complex and adherens junctions ultimately determine the sites of tight junction formation and thus epithelial barrier function.

In addition to providing a reference point for the bottom of the cell, ECM composition can also have an effect on cell phenotype and behavior. Most relevant to epithelial barrier function are ECM proteins, which promote the "epithelial to mesenchyme transition" (EMT) [27]. Foremost among these is fibronectin, which is deposited into ECM as a result of tissue stress and wounding [28]. Cells binding to fibronectin induce transcription factors such as snail and slug, which cause cells to differentiate away from an epithelial phenotype toward mesenchyme (e.g., fibroblast) phenotype. This has the direct effect of decreasing the expression of key proteins required for the formation of tight junctions, especially E-cadherin and claudins, which causes a net decrease in paracellular barrier function [29]. This effect is further exacerbated in tissues under stress by deposition of transforming growth factor β (TGF-β) into the ECM; TGF-β activation by β6 integrin amplifies the effects of fibronectin on EMT [30] and also impairs barrier function by directly interfering with tight junction protein expression and assembly [31]. As a specific example, TGF-β expression in response to chronic alcohol ingestion impairs the lung epithelial barrier, which sensitizes patients to the deleterious effects of acute lung injury [25, 32].

Although EMT in and of itself is pathological and leads to fibrosis, the effects of EMT on tight junctions–controlled modulation of EMT might be a potential pharmacologic target to transiently increase paracellular permeability as a means to enhance drug delivery. It was recently found that presenting cells with a substrate that

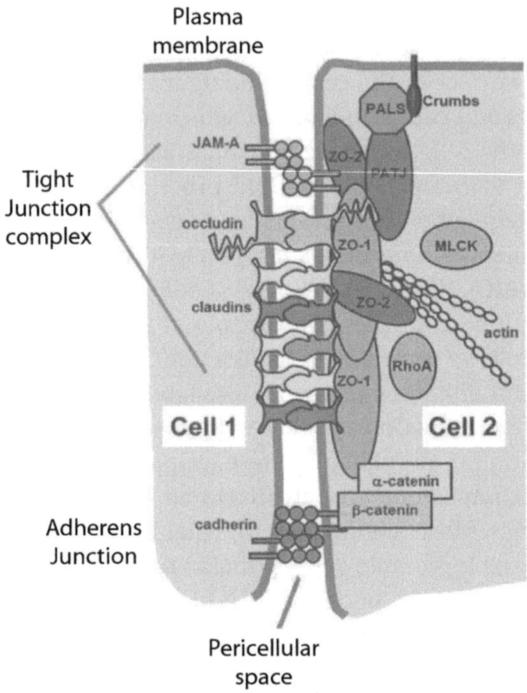

Figure 1.3 **Apical junctional complex.** Shown are the major functional components of the AJC, including polarity determinants (Crumbs, PALS, PATJ), tight junction proteins (JAM-A, occludin, claudins) and adherens junctions (E-cadherin, α-catenin, and β-catenin). The core transmembrane tight junction proteins bind scaffold proteins (ZO-1, ZO-2) and the actin cytoskeleton. Actin also binds to β-catenin and helps stabilize the AJC. Head-to-head interactions between claudins on adjacent cells form the basis for paracellular channels that restrict permeability. Homotypic interactions (between the same types of claudin) and heterotypic interactions (between different types of claudins) are depicted. Additional scaffold proteins are present, but these are not shown here for clarity. Modified with permission from Ref. 25, Copyright 2013, Annual Reviews.

mimics ECM can also provide a potential mechanism to alter barrier function. Specifically, direct contact between the apical surface of Caco-2 intestinal epithelial cells and specifically nanostructured films increases permeability to macromolecules, including IgG and etanercept [33]. This illustrates the utility of using nanostructured films to enhance drug delivery across epithelia. Although the mechanism is still being elucidated, contact of specific nanotopography with the apical surface was found to induce expression of myosin light chain kinase (MLCK) and decrease Zonula Occludens-1 (ZO-1), both of which can increase paracellular flux. Nanostructured films have also been found to influence cell phenotype depending on surface patterning [34]. Thus, it seems likely that placement of a nanotopographical cue on the apical surface has the capacity to influence epithelial barrier function by influencing epithelial phenotype and polarity; future work will determine the relative effects on these pathways.

1.4 Adherens Junctions

As a consequence of cell polarity, the plasma membrane of epithelial cells is divided into apical and basolateral domains by the AJC (Fig. 1.3). Paracellular permeability through the AJC is regulated by tight junctions, described later. Although tight junctions form the physical paracellular permeability barrier, adherens junctions are required for tight junction formation [35]. This is due, in large part, to the nature of intercellular adhesion mediated by these different classes of junctions. The strength of intercellular binding through tight junctions is mediated by a large number of low affinity binding interactions, resulting in junctional complexes that are nearly comparable to sites of membrane fusion [3]. However, small complexes lack sufficient affinity to initiate tight junction formation on their own.

Instead, tight junctions require prior formation of adherens junctions via single-pass transmembrane proteins known as cadherins [3]. There are several different classes of cadherins; foremost among these is a type I classical cadherin, Epithelial (E)-cadherin [36]. By contrast, endothelial adherens junctions are formed by intercellular binding via a type II cadherin, vascular endothelial (VE)-cadherin [37]. Although VE-cadherin is structurally similar to type I cadherins, such as E-cadherin, VE-cadherin lacks a critical extracellular His-Ala-Val motif responsible for the strong extracellular binding between E-cadherins. In other words, endothelial adherens junctions generally are weaker than epithelial adherens junctions.

There are several classes of cadherins beyond those that form adherens junctions. For instance, there are nonclassical cadherins, which form desmosomes to promote structural integrity, particularly in tissues that are subject to mechanical stress [38]. Although the topic of desmosomal and other cadherins is beyond the scope of this chapter, it is worth mentioning that desmosomes are prominent in most epithelia; however, they are rare in endothelial cells, which instead rely on the vascular smooth muscle for structural integrity [36].

In addition to mediating intercellular binding, cadherins in adherens junctions recruit filamentous actin to the AJC. This occurs via an indirect mechanism involving several cytoplasmic scaffold proteins, including β-catenin, α-catenin, and EPLIN [36, 39, 40]. Of note, α-catenin has the capacity to bind and bundle actin filaments, suggesting a direct role in linking actin to adherens junctions [41]. However, there is accumulating evidence supporting an indirect role for α-catenin, where it is first recruited to adherens junctions and then dissociates, dimerizes, and acts in concert with formins and the arp2/3 complex to nucleate actin polymerization, bundling, and branching [39, 42].

The adherens junction scaffold also controls cell signaling. The best characterized example of this is β-catenin, which has the capacity for multiple functions, including regulation of *wnt* signaling and acting as a transcription factor [43]. Thus, binding to adherens junctions serves to sequester β-catenin away from other cellular compartments as a regulatory mechanism. As an example of this, endothelial cells that are not in contact have significant nuclear β-catenin, which acts together with FOXO1 as a repressor of claudin-5 expression [44]. Adherens junctions formed by VE-cadherin make a complex with β-catenin de-repressing claudin-5 and thus providing a direct link between initiation of AJC formation, cell adhesion, and upregulation of tight junction protein expression.

1.5 Tight Junction Composition

The tight junction proteome is complex, consisting of many different components. Here the focus is on a subset of tight junction proteins, which are especially critical for barrier function.

1.5.1 *Claudins*

The physical tight junction barrier is formed by proteins in the claudin family. Claudins are tetraspan transmembrane proteins that have a short N-terminus and longer C-terminus oriented toward the cytosolic aspect of the membrane (Fig. 1.4) [32]. About two dozen human claudins have been identified. Claudins fall roughly into two different subcategories: (1) sealing claudins, which restrict paracellular permeability, and (2) pore-forming claudins, which form channels enabling paracellular diffusion of ions and small molecules.

Epithelial and endothelial cells express several claudin genes in a tissue-specific manner. By forming tight junction strands of differing composition, this enables cells to produce barriers with distinct paracellular permeability. Claudin selectivity and permeability is determined by charged amino acids in the first extracellular loop domain (for reviews, see [1, 45]). In addition, several claudins have been shown to increase barrier permeability, rather sealing tight junctions. This includes claudin-2, claudin-6, claudin-10, and claudin-15. Claudin-2, which is upregulated in response to inflammation, forms water channels, which provide a direct pathway for increasing paracellular fluid flux [46]. Whether claudin-6, claudin-10, or claudin-15 creates similar pores is not known.

Claudins are also subject to post-translational modifications, which can affect their assembly and function, such as phosphorylation (reviewed in [32]). As another example of post-translational modification, palmitoylation enhances claudin assembly into tight junctions by promoting claudin partitioning into cholesterol-enriched microdomains [47].

There are two different modes of interaction between claudins. Claudins on adjacent cells interact via the extracellular loop domains to form the junction itself; this class of interaction is called homotypic or heterotypic, depending on whether it occurs between identical or distinct claudins, respectively (Fig. 1.4). Homotypic interactions between claudins have been demonstrated by analysis of co-cultures of claudin-null cells transfected to express different claudins, using immunofluorescence co-localization or co-

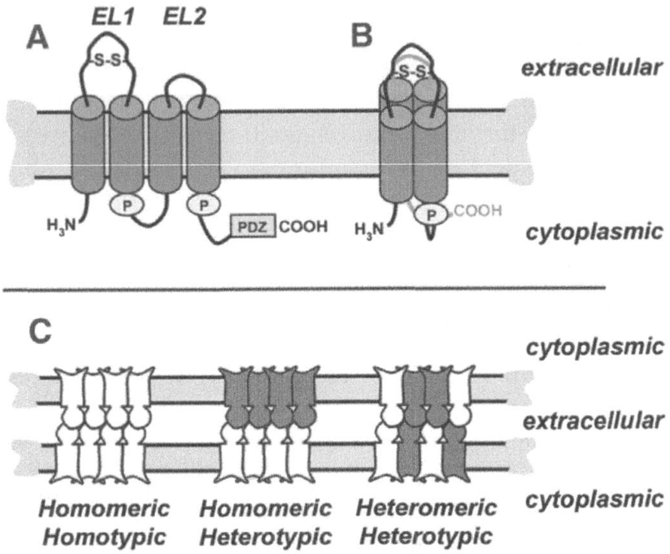

Figure 1.4 Claudin secondary structure and classes of claudin–claudin interactions. (A) Line diagram showing key features of a typical claudin in the plane of the membrane, including the two extracellular loop (EL) domains, where EL1 contains a putative disulfide bond (S–S). Cylinders represent predicted transmembrane alpha helical domains. Also illustrated are two palmitoylation motifs ("P") and the PDZ binding motif at the extreme C-terminus of the protein. (B) Proposed conformation of an individual claudin in the plane of the membrane, showing the four transmembrane alpha helical domains as a tightly packed complex. (C) Classes of claudin–claudin interactions within a tight junction strand. Depicted are tight junctions at sites where two cells in contact are composed of a single claudin (homomeric/homotypic) or multiple claudins (homomeric/heterotypic; heteromeric/heterotypic). Claudins can interact via head-to-head binding in the extracellular environment between adjacent cells (heterotypic interactions) and within the plane of the plasma membrane in the same cell (heteromeric interactions). Reprinted with permission from Ref. 32, Copyright 2011, Mary Ann Liebert, Inc.

immunoprecipitation as output variables (Table 1.1) [48–54]. By contrast, only claudin-3 has been demonstrated to heterotypically interact with other claudins [48, 49, 51]. However, several combinations of claudins have not been tested for heterotypic compatibility.

Table 1.1 Heterotypic claudin interactions

cldn	1	2	3	4	5	8	10b	15	16	19
1	Y	N	Y	N	N		N	N		
2	N	Y	Y	N		N				
3	Y	Y	Y	N	Y					
4	N	N	N	Y	N					
5	N		Y	N	Y					
8		N				Y				
10b	N						Y			
15	N							Y		
16									Y	N
19									N	Y

(Y) indicates heterotypic compatibility, (N) heterotypic incompatibility, ( ) not determined. Modified from Ref. [32] with permission.

Interestingly, claudin-3 is heterotypically incompatible with claudin-4 due to a single amino acid residue in the first extracellular loop domain [48]. Mutating this residue (asparagine 44) in claudin-3 to threonine enabled heterotypic interactions with claudin-4 but also allowed heterotypic binding to claudin-1 or claudin-5. Thus, heterotypic interactions are controlled by domains in both extracellular loops. For example, the second extracellular loop of claudin-5 has been shown to promote homotypic claudin-5 binding [52].

In addition to interactions across cell junctions, claudins expressed by the same cell interact homomerically and heteromerically in the plane of the membrane [52, 55]. Until claudins are fully assembled into tight junctions, homomeric and heteromeric interactions between them are exceptionally weak and are disrupted by most detergents [56]. Many claudins heteromerically interact despite being heterotypically incompatible (Table 1.2) [48, 51, 53, 55, 57–62]. For instance claudin-16 and claudin-19 are heteromerically compatible despite heterotypic incompatibility and interact to regulate magnesium homeostasis in the kidney [53, 58]. This suggests that claudin-16 and claudin-19 form cation-selective channels via a homotypic/heteromeric complex [63, 64]. Interestingly, other renal claudins, such as claudin-10a and claudin-18, traffic independently from claudin-16 and claudin-19 [53, 58],

Table 1.2 Heteromeric claudin interactions

cldn	1	2	3	4	5	6	7	8	9
1	F		FI	Y	FI-ny		YC		
2		F	-ny		F			-ny	
3	F	-ny	F-ny	IY	-f		I-ny	Y	
4	Y		Y	-ny			-ny	Y	
5	F-ny	F	-f		Y				
7	YC		CF-ny	-ny			Y	Y	
8		-ny	Y	Y			Y	Y	
14						-ny			-ny
15	-c								
18					I				
cldn	10a	10b	11	12	14	15	16	18	19
1						-c			
5								I	
6					-nc				
9					-nc				
10a	-y	-y					-ny	Y	Y
10b	-y	-y					-ny	-y	Y
11			Y						
12				-ny					
16	-ny	-ny					F-ny	-ny	Y
18	Y	-y					-ny	-y	
19	Y	Y					Y	Y	Y

Strong interactions shown by: Y—yeast 2 hybrid (Y2H); F—fluorescence resonance energy transfer (FRET); I—co-immunoprecipitation; C—co-recruitment. Weak interactions shown by: -y—Y2H; -f—FRET. N – no interaction shown by: -ny—Y2H; -nc—no co-recruitment.

suggesting a model where tight junction permeability is controlled by several independently regulated ensembles of claudins. In other words, the fine structure of tight junctions is relevant to modulating paracellular barrier function where different regions of cell–cell interfaces differ in claudin composition and have different permeability.

1.5.2 *Tight Junction Scaffold Proteins*

Once assembled into tight junctions, claudins are remarkably stable. This can be demonstrated biochemically as resistance to detergent solubilization [65, 66]. In intact cells, claudins are slow to diffuse in the plane of the membrane. Based on fluorescence recovery after photobleaching (FRAP) measurements, tight junction–associated claudin-1 has an immobile fraction of ~75%, which is relatively high [67]. Claudin–claudin interactions between extracellular loops certainly stabilize these proteins; however, they are also stabilized by their binding to cytoplasmic scaffold proteins. Different claudins have differing C-terminal domains ranging from ~20 to ~60 amino acids in length, which enables differential scaffold protein binding. On the other hand, nearly all claudins contain a PDZ-binding motif at the extreme C-terminus [68, 69], which binds to ZO-1 and ZO-2 [65]. ZO-1 and ZO-2 are flexible, multidomain proteins that recognize and crosslink multiple proteins into lateral complexes. ZO-1 and ZO-2 also help stabilize junction protein complexes by attaching them to the cortical actin cytoskeleton, aligned along the plane of the plasma membrane (Fig. 1.5) [70].

Other scaffold proteins help initiate and position tight junctions by functioning as links between distinct classes of transmembrane proteins. For instance, alpha-catenin, which is associated with adherens junctions, has the capacity to recruit ZO-1 to the junctional scaffold, which promotes tight junction protein assembly into a functional barrier [71]. PALS-1-associated TJ protein (PATJ) is another example of a scaffold protein that serves as part of the AJC by linking tight junctions to the apical polarity determinant crumbs [72]. This layering of multiple transmembrane proteins and scaffold proteins serves as positional information to ensure proper placement of tight junctions.

In cells where ZO-1 and ZO-2 are depleted, tight junctions still form; however, interactions of tight junction proteins with the cytoskeleton are now destabilized [73]. Apical contractility increases, much like a purse string, which decreases cell diameter at the AJC and increases the size of the apical surface. Paracellular flux significantly increases due to the increase in contractility and junction turnover; however, ion permeability and TER remain largely unaffected. As mentioned earlier, this demonstrates a key point: the electrical barrier function of tight junctions does not necessarily correlate with paracellular flux to small molecules. One way of thinking about this apparent dichotomy is to consider a revolving door, which permits efficient people to move from the

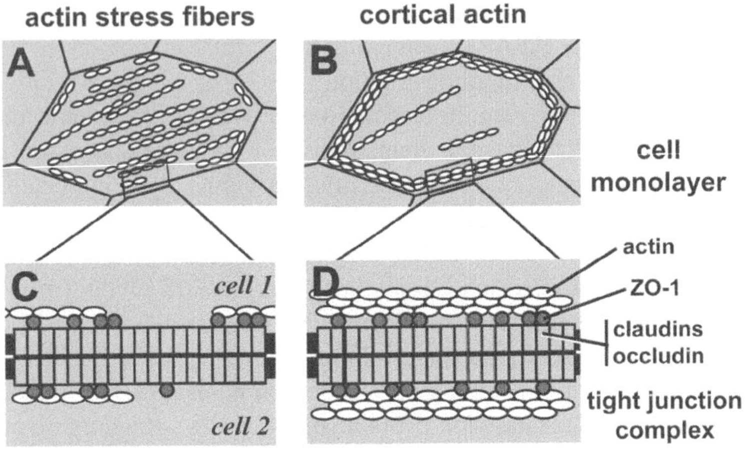

Figure 1.5 Cortical actin stabilizes junctions. A and B: diagram indicating the differential orientation of actin filaments in cells containing predominantly stress fibers (A) or predominantly peri-junctional, cortical actin (B). C and D: attachment of transmembrane tight junction proteins (claudins, occludin) to the actin cytoskeleton via the scaffold protein zonula occludens-1 (ZO-1). Increased peri-junctional bundles of actin fibers provide more ZO-1 attachment sites, which stabilize tight junction strands, promoting their barrier function. Not depicted in this diagram are other classes of scaffold and transmembrane proteins, which also crosslink tight junctions to the cytoskeleton. Adapted with permission from Ref. 70, Copyright 2010, American Physiological Society.

outside to the inside of a building, while maintaining a seal that limits air diffusion. The phenotype of ZO-1/ZO-2-depleted cells also illustrates the point that these proteins, as well as modulation of actin, are targets with the potential to be pharmacologically manipulated to alter epithelial barrier function.

ZO-1/ZO-2-depletion experiments illustrate two aspects related to tight junction stabilization: (1) crosslinking junction proteins to cortical actin and (2) limiting the ability of proteins that remodel actin to have access to junction-associated filaments. Central among these proteins are rho kinase, myosin II, and MLCK, which destabilize cell junctions [74, 75]. Rho kinase and MLCK regulate the cytoskeleton in distinct, complementary ways. Rho kinase activation of rho promotes actin filament remodeling, which has the net effect of disassembling both adherens and tight junctions, thus increasing paracellular permeability [69, 76, 77]. MLCK, on the other hand, increases the contraction of cortical actin to promote tight junction turnover and destabilization [78]. Barrier dysfunction associated with inflammatory disease and infection is frequently associated with aberrant activation of Rho kinase and/or MLCK, suggesting the utility of targeting these signaling pathways to prevent disease pathology [74, 79, 80]. Conversely, this also suggests that activating Rho kinase and MLCK may serve to promote paracellular flux as pathways for drug delivery [81].

1.5.3 *Marvel Domain Containing Proteins*

Tight junctions also contain Marvel Domain family proteins—occludin, tricellulin, and MarvelD3—which are also tetraspan proteins oriented in a manner similar to claudins [4, 82]. Instead of forming the mechanical basis for the tight junction barrier, MarvelD proteins instead regulate cell function by promoting tight junction assembly [83, 84]. For example, occludin, tricellulin, and MarvelD3 have been demonstrated to interact with several claudins and are likely to help recruit these proteins into junctional complexes [57, 85, 86]. This is due, in part, to the ability of occludin to interact with caveolin-1 and partition into cholesterol-enriched membrane microdomains, which promote tight junction formation [87]. Occludin also acts as a signal to initiate apoptosis upon disruption of junctions, consistent with a role for this protein in signaling in addition to junction formation [83, 84]. Whether tricellulin or MarvelD3 have similar roles in regulating cell function has not yet been determined.

Tricellulin was initially discovered as a tight junction protein enriched in tricellular contacts (Fig. 1.2) [88]. Depending on expression level, tricellulin can localize to bicellular contacts as well [4]. Moreover, tricellulin was found to compensate for loss of occludin, suggesting a comparable function in bicellular junctions [82]. Tricellulin localization also depends on another class of proteins, called angulins, the first of which to be identified was the lipolysis-stimulated lipoprotein receptor (LSR), a single-pass transmembrane protein with immunoglobulin fold domains [89]. The function of tricellulin and LSR at tricellular junctions is to preferentially restrict paracellular flux of aqueous molecules [4]. By contrast, tricellular junctions have only a small role in regulating ion flux, suggesting that claudin function as paracellular ion channels is restricted to bicellular junctions. The relative cross-sectional area for tricellular junctions is also minute as compared with bicellular junctions, further limiting potential contributions to ion permeability [4]. A specialized role for tricellular junctions in regulating barrier function is also suggested by the observation that leukocyte transmigration preferentially occurs through tricellular junctions [90].

1.5.4 *Junctional Adhesion Molecule A*

Junctional adhesion molecule A (JAM-A) is another immunoglobulin fold single-pass transmembrane protein that has the capacity to regulate intercellular junctions [91]. Trans-interactions promote JAM-A localization to tight junctions. Critically, JAM-A signaling is stimulated by cis-dimerization, which provides a platform for multiple proteins to cluster in close apposition [92]. In particular, JAM-A has been shown to recruit scaffold proteins, including ZO-1, ZO-2, and Par3, to tight junctions where these proteins enhance tight junction assembly [93, 94]. It was recently demonstrated that the cytoplasmic domain of JAM-A binds directly to ZO-2, which, in turn, helps recruit other scaffold proteins to tight junctions, including ZO-1 and afadin [95].

JAM-A also serves to organize the formation of signaling complexes at tight junctions. Of direct relevance to junction stability, JAM-A recruits PDZ-GTP-exchange factor 1 (PDZ-GEF1) and the small molecular weight GTPase Rap2c into a functional complex that inhibits cytoskeletal turnover [95]. A comparable JAM-A-dependent pathway has also been shown to play a role in intestinal epithelial cell migration during wound repair. However, instead of co-clustering PDZ-GEF1 and Rap2c, which regulate tight junction permeability, JAM-A controls cell migration by facilitating the formation of a PDZ-GEF2 complex that activates Rap1 [96]. A direct role for Rap1 in barrier function has been demonstrated for vascular endothelium [97, 98]. In addition, activation of Rap1 also inhibits degradation of $\beta 1$ integrin [96, 99]. Since $\beta 1$ integrin is critical for cell–matrix interactions required for epithelial polarity, this provides an additional pathway whereby JAM-A promotes epithelial barrier function.

1.6 Transcellular Transport

Contrary to the paracellular route, the transcellular route involves transport through cells, either directly through the cytoplasm, which is facilitated by plasma membrane channels, or by vesicle-mediated transport (see Fig. 1.2). These pathways are briefly mentioned here; for details refer to the cited reviews.

1.6.1 *Plasma Membrane Channels*

The plasma membrane is an effective barrier to small molecules. However, plasma membrane channels provide conduits that enable small molecules and ions to flow from the extracellular environment into the cytoplasm and *vice versa* [5–7]. Net ion transport readily occurs when ions or molecules either move along a concentration gradient, such as potassium, or ATP efflux out of cells. Movement across membranes against a concentration gradient can occur as a result of coupled transport (Na-Glucose co-transporters) or by ATP-driven pumps (e.g., Na,K-ATPase). Frequently, physiological barriers are due to the activities of multiple ion transport channels and pumps. For instance, lung airspace fluid balance is maintained by the combined activities of epithelial sodium channel, Na,K-ATPase, cystic fibrosis transmembrane conductance regulator, aquaporins, and paracellular chloride and water diffusion [5].

Large conductance channels are also found in the plasma membrane and include pannexins, connexin hemichannels, and calcium homeostasis modulator channels [100–102]. These classes of plasma membrane channels are particularly attractive targets for drug delivery, since they have broad permeability characteristics, which could potentially mediate transport of agents in the 200–800 Da range into cells. Note also that, while these classes of channels can accommodate a spectrum of molecules, they do exhibit specificity based on molecule charge and shape, which can be advantageous for the purpose of specific targeting.

1.6.2 *Vesicle-Mediated Transport*

Transcellular transport can also occur by a vesicle-mediated process known as transcytosis [8, 9]. The process is initiated by several different receptor-mediated internalization pathways, including clathrin-mediated and caveolar endocytosis, which mediate uptake of vesicles in the 100–200 nm range as well as pathways such as micropinocytosis, which can enable formation of vesicles 500 nm and larger. Whether internalized material is transcytosed is largely a function of the receptor, which contains targeting motifs that direct it along this pathway as opposed to the plasma membrane recycling or lysosomal degradative pathways [9]. Transcytosis is also dependent on cell type; it is a major function of secretory epithelia that transport significant amounts of protein across a barrier, such as secretory immunoglobulin A (sIgA), by mammary epithelia to transmit innate immunity in milk.

Transcytosis can also be used as a mode for drug delivery. Recently, it was demonstrated that CAM-mediated endocytosis, which is initiated by crosslinking of cell adhesion molecules such as ICAM-1 [103], has the capacity to enable nanoparticle transcytosis across intestinal epithelial cells [104]. This demonstrates the feasibility of targeting transcytosis as a means to promote drug delivery by tailoring a pharmacologically active ligand to a suitable receptor and cell type.

1.7 Unique Features of Endothelial Barriers

The vascular endothelium shares many features in common with epithelial barriers, including cell polarity, junction composition, and transcellular transport pathways [37]. However, the endothelium also has unique aspects related to its function. In fact, the vasculature is heterogeneous and consists of vascular beds with different architecture [105]. This results in considerably more heterogeneity in permeability than epithelia, which typically restrict paracellular diffusion of small molecules. Of the different endothelial subtypes, continuous nonfenestrated endothelium forms the tightest barriers, which includes the blood–brain barrier and pulmonary microvasculature. Other vessels are considerably more permeable. Continuous fenestrated endothelium has diaphragms that are freely permeable to water and small molecules less than 3 nm in diameter, which play important roles in fluid transfer in the kidney and gastrointestinal tract [106]. Discontinuous endothelium, as the name suggests, has large gaps between cells and essentially freely permeable, which promotes blood filtration through tissues such as liver sinusoids.

Another contributor to endothelial barrier function is the glycocalyx, which is a complex of several classes of complex plasma-membrane-associated molecules composed of protein and carbohydrate, including proteoglycans, glycoproteins, and glycolipids [107]. It is a layered highly viscous structure, which can be from 0.15 micron to over 0.5 micron in thickness, that restricts permeability of proteins and other macromolecular solutes across the endothelial barrier [108]. Thus, the glycocalyx is particularly critical for the fluid barrier in considering Starling forces that relate to fluid transport between serum, which contains high concentrations of protein, and tissues, which is driven by hydrostatic pressures that might not be otherwise maintained in the absence of a glycocalyx. Although glycocalyx is most prominent in the vasculature, some epithelia also have a glycocalyx, notably intestinal epithelium [109], and it shares some features related to the mucus layer found in airways and other epithelia [110].

Transcytosis across endothelium is most prominently via the caveolar pathway, since this is the predominant form of endocytosis in these cells [111]. In addition to the transcellular pathways mentioned above, endothelial cells have a unique, related transcellular pathway known as vesiculo–vacuolar organelles (VVOs) or transendothelial channels [112]. VVOs are related to the caveolar transcytosis pathway [9]; however, they are formed by the fusion of discrete caveolin-enriched vesicles to form a continuous opening across cells. VVO formation is stimulated by vascular endothelial growth factor, which enhances endothelial permeability during vascular remodeling. Thus, VVOs have been associated with the increases in vessel leak observed during inflammation and in tumor vascularization.

1.8 Targeting Junction Proteins

Because junction proteins are accessible from the outside of the cell, this provides a target for ligands to bind and disrupt junction integrity. This alters barrier function by directly inhibiting the incorporation of specific proteins into tight junctions, mainly by disrupting homotypic or heterotypic binding. Heteromeric complexes can also be targeted by this approach, provided that the proteins in the complex have at least one constituent with a targetable extracellular domain. Disrupting tight junctions by interfering with junction formation has already occurred in nature since several pathogens have evolved to interact directly with mammalian junction proteins to invade a host.

1.8.1 *Virus Coat Proteins*

Several classes of viruses have been found to interact with junction proteins. For example, the Ig superfamily junction protein CAR was originally defined by its capacity to act as a coxsackievirus and adenovirus receptor [113]. JAM-A also acts as a receptor for reoviruses and binds to the $\sigma 1$ coat protein [114]. Although the virus binds to JAM-A, internalization requires $\beta 1$ integrin as a cofactor, consistent with the formation of a JAM-A/$\beta 1$ integrin complex [115, 116]. An engineered soluble form of $\sigma 1$ binds to JAM-A and induces its internalization, which, in turn, disrupts junction integrity both in cultured cell models and in the intact gut *in vivo* [95]. This suggests that $\sigma 1$-derived peptides could be used to transiently increase paracellular flux as a means to promote drug delivery across epithelia.

Claudins also serve as receptors for pathogens. Hepatitis C virus binds to claudin-1 and is internalized as part of a complex with the tetraspanin CD81, scavenger receptor BI, and occludin [117–119]. The E1 and E2 viral coat proteins are required for binding to claudin-1 and formation of the complex [120]. Hepatitis C also binds to claudin-6 and claudin-9; however, these claudins are not as effective as claudin-1 at mediating virus uptake and infection [121, 122]. Although this suggests the potential for hepatitis proteins to be used in a comparable manner as the reovirus $\sigma 1$ derived proteins, intact tight junctions inhibit the ability of hepatitis C to interact with claudin-1, which may limit potential utility of hepatitis-derived peptides as a permeabilizing agent [123].

1.8.2 *Clostridium Perfringens Enterotoxin*

Clostridium perfringens enterotoxin (CPE) binds several different claudins with high affinity, chiefly claudin-3 and claudin-4, among others [124]. CPE binding to claudins ultimately leads to internalization, which in the normal life cycle of *Clostridium perfringens* is toxic to cells. However, the cytotoxic portion of CPE can be removed to create nontoxic peptides, which bind and are internalized by cells. In fact, C-terminal CPE fragments as short as 30–140 amino acids in length retain the ability to specifically bind claudins [124]. However, other motifs are likely to contribute to the claudin-binding specificity. For instance, a CPE116-319 fragment recognizes claudin-5, while CPE194-319 does not [125]. This suggests the potential to engineer CPE to change the specificity of interaction, particularly since the structure of the claudin-binding domain of CPE is known [125, 126]. The feasibility of altering CPE-binding specificity has been demonstrated by creating and selecting CPE mutants that have increased affinity for claudin-1 and claudin-2, which normally do not bind to CPE [127].

CPE has the capacity to disrupt tight junctions and thus may have utility for increasing paracellular flux in a manner comparable to virus peptides mentioned earlier [128, 129]. CPE-derived therapeutic agents have also been used to target anticancer chemotherapeutic agents, since they strongly interact with claudin-3 and claudin-4, which are highly upregulated in some forms of ovarian, breast, and pancreatic cancer [130]. Among these agents is a fusion protein consisting of CPE290-319 fused to the protein synthesis inhibitory factor (PSIF) domain of Pseudomonas exotoxin (C-CPE-PSIF), which specifically kills human breast cancer cells that express claudins-3 and claudin-4 and is effective in a mouse model of metastasis [131].

1.8.3 *Claudin Extracellular Loop Peptides*

Given the success of pathogens in targeting junction proteins as a means to alter paracellular permeability, several investigators have begun to explore the concept of using extracellular domain mimetics in a comparable fashion. Since the extracellular loop domains of claudins are required for their ability to form a tight junction barrier, one approach has been to use small peptides homologous to these domains to sterically block junction formation.

The most effective peptides that disrupt claudin assembly are homologous to the second half of the first extracellular loop domain [85, 132]. Including a portion of the hydrophobic interface region enhances efficacy [85]. By contrast, peptides corresponding to the first half of the extracellular loop or to the second extracellular domain were not effective in this context. Claudin-1 peptides decrease epithelial barrier function *in vitro* and are effective at increasing gut and perineurium permeability *in vivo* as well [85, 132]. Consistent with the ability to use claudin peptides to enhance drug delivery, claudin-1 peptides were shown to increase the efficacy of analgesic agents to the peripheral nervous system [132].

Claudin-1 peptides interact directly with claudin-1 to prevent homotypic binding and cause claudin-1 to be internalized, directly disrupting tight junction formation. Claudin-1 peptides also bind to claudin-3 and occludin, reflecting their ability to simultaneously disrupt multiple different tight junction proteins [85]. This also suggests that these peptides may be useful tools to dissect out protein interactions in tight junctions. In addition to claudin-1, other claudins are also being targeted. Most notably, claudin-5 peptides are under development, which may have particular utility in disrupting the blood–brain barrier for drug delivery [132].

Occludin extracellular loop peptides have also been used to target tight junctions. Interestingly, epithelial cells treated with an occludin extracellular loop peptides specifically increase paracellular flux without affecting TER [133]. This is reminiscent of the phenotype of cells where ZO-1/ZO-2 is depleted where barrier function is comparably affected [73]. This underscores the potential for targeting occludin as a means for selectively increasing paracellular flux as a means to drug delivery and also provides evidence for crosstalk between occludin, ZO-1, and ZO-2 in regulating tight junction assembly.

1.9 Conclusions and Perspectives

Epithelia provide barriers that are required for organ function, yet represent an obstacle to drug delivery that needs to be effectively, specifically, and reversibly breeched in order for effective administration of pharmacologic agents. Epithelial cells coordinate barrier function by orchestrating cell morphology, orientation, and the specific deposition of proteins into the AJC complex. This provides several targets with the potential to enhance drug delivery across barriers. This includes agents that modulate the cell microenvironment to alter cell phenotype, function, and organization of intercellular contacts. Epithelial permeability can also be modulated using ligands that specifically bind to individual junction proteins impairing assembly. Signaling pathways, particularly those affecting the cytoskeleton, provide another avenue for modulating junction permeability. Understanding the molecular basis for the regulation of epithelial barrier function provides principles that can be used to inform the design of strategies to specifically modulate barrier function, particularly when considering mechanisms that have evolved as a result of physiologic conditions and pathogens that specifically target tight junctions.

Acknowledgments

This work was supported by the National Institutes of Health (NIH) grants P50-AA013757 (Emory Alcohol and Lung Biology Center), R01-HL083120, R01-HL116958, Emory University Research Committee, and Emory+Children's Center of Excellence for Cystic Fibrosis Research.

References

1. Anderson, J.M., and Van Itallie, C.M. (2009). Physiology and function of the tight junction, *Cold Spring Harb. Perspect. Biol.*, **1**, pp. a002584.
2. Furuse, M., and Tsukita, S. (2006). Claudins in occluding junctions of humans and flies, *Trends Cell Biol.*, **16**, pp. 181–188.
3. Schneeberger, E.E, and Lynch, R.D. (2004). The tight juntion: A multifuctional complex, *Am. J. Physiol. Cell. Physiol.*, **286**, pp. C1213–1228.
4. Krug, S.M., Amasheh, S., Richter, J.F., Milatz, S., Gunzel, D., Westphal, J.K., Huber, O., Schulzke, J.D., and Fromm, M. (2009). Tricellulin forms a barrier to macromolecules in tricellular tight junctions without affecting ion permeability, *Mol. Biol. Cell*, **20**, pp. 3713–3724.
5. Eaton, D.C., Helms, M.N., Koval, M., Bao, H.F., and Jain, L. (2009). The contribution of epithelial sodium channels to alveolar function in health and disease, *Annu. Rev. Physiol.*, **71**, pp. 403–423.
6. Brown, D., Bouley, R., Paunescu, T.G., Breton, S., and Lu, H.A. (2012). New insights into the dynamic regulation of water and acid–base balance by renal epithelial cells, *Am. J. Physiol. Cell Physiol.*, **302**, pp. C1421–1433.
7. Verkman, A.S., and Galietta, L.J. (2009). Chloride channels as drug targets, *Nat. Rev. Drug Discov.*, **8**, pp. 153–171.
8. Muro, S., Koval, M., and Muzykantov, V. (2004). Endothelial endocytic pathways: Gates for vascular drug delivery, *Curr. Vasc. Pharmacol.*, **2**, pp. 281–299.
9. Tuma, P.L., and Hubbard, A.L. (2003). Transcytosis: Crossing cellular barriers, *Physiol. Rev.*, **83**, pp. 871–932.
10. Nitta, T., Hata, M., Gotoh, S., Seo, Y., Sasaki, H., Hashimoto, N., Furuse, M., and Tsukita, S. (2003). Size-selective loosening of the blood-brain barrier in claudin-5-deficient mice, *J. Cell Biol.*, **161**, pp. 653–660.
11. Goodson, P., Kumar, A., Jain, L., Kundu, K., Murthy, N., Koval, M., and Helms, M.N. (2012). NADPH oxidase regulates alveolar epithelial sodium channel activity and lung fluid balance *in vivo* via O_2^- signaling, *Am. J. Physiol. Lung Cell Mol. Physiol.*, **302**, pp. L410–419.
12. Shen, L., Weber, C.R., Raleigh, D.R., Yu, D., and Turner, J.R. (2011). Tight junction pore and leak pathways: A dynamic duo, *Annu. Rev. Physiol.*, **73**, pp. 283–309.

13. Clarke, L.L. (2009). A guide to Ussing chamber studies of mouse intestine, *Am. J. Physiol. Gastrointest. Liver Physiol.*, **296**, pp. G1151–1166.
14. Nejsum, L.N., and Nelson, W.J. (2009). Epithelial cell surface polarity: The early steps, *Front. Biosci. (Landmark Ed.)*, **14**, pp. 1088–1098.
15. Hynes, R.O., and Zhao, Q. (2000). The evolution of cell adhesion, *J. Cell Biol.*, **150**, pp. F89–96.
16. Hutter, H., Vogel, B.E., Plenefisch, J.D., Norris, C.R., Proenca, R.B., Spieth, J., Guo, C., Mastwal, S., Zhu, X., Scheel, J., and Hedgecock, E.M. (2000). Conservation and novelty in the evolution of cell adhesion and extracellular matrix genes, *Science*, **287**, pp. 989–994.
17. Chapman, H.A. (2004). Disorders of lung matrix remodeling, *J. Clin. Invest.*, **113**, pp. 148–157.
18. Guillot, C., and Lecuit, T. (2013). Mechanics of epithelial tissue homeostasis and morphogenesis, *Science*, **340**, pp. 1185–1189.
19. Schwarz, U.S., and Gardel, M.L. (2012). United we stand: Integrating the actin cytoskeleton and cell-matrix adhesions in cellular mechanotransduction, *J. Cell Sci.*, **125**, pp. 3051–3060.
20. Hynes, R.O. (2002). Integrins: Bidirectional, allosteric signaling machines, *Cell*, **110**, pp. 673–687.
21. Ingber, D.E. (1997). Tensegrity: The architectural basis of cellular mechanotransduction, *Annu. Rev. Physiol.*, **59**, pp. 575–599.
22. Cukierman, E., Pankov, R., and Yamada, K.M. (2002). Cell interactions with three-dimensional matrices, *Curr. Opin. Cell Biol.*, **14**, pp. 633–639.
23. Yu, W., Datta, A., Leroy, P., O'Brien, L.E., Mak, G., Jou, T.S., Matlin, K.S., Mostov, K.E., and Zegers, M.M. (2005). Beta1-integrin orients epithelial polarity via Rac1 and laminin, *Mol. Biol. Cell*, **16**, pp. 433–445.
24. Hertzog, M., and Chavrier, P. (2011). Cell polarity during motile processes: Keeping on track with the exocyst complex, *Biochem. J.*, **433**, pp. 403–409.
25. Koval, M. (2012). Claudin heterogeneity and control of lung tight junctions, *Annu. Rev. Physiol.*, **75**, pp. 551–567.
26. Margolis, B., and Borg, J.P. (2005). Apicobasal polarity complexes, *J. Cell Sci.*, **118**, pp. 5157–5159.
27. Nieto, M.A. (2011). The ins and outs of the epithelial to mesenchymal transition in health and disease, *Annu. Rev. Cell Dev. Biol.*, **27**, pp. 347–376.

28. White, E.S., and Muro, A.F. (2011). Fibronectin splice variants: Understanding their multiple roles in health and disease using engineered mouse models, *IUBMB Life*, **63**, pp. 538–546.
29. Martinez-Estrada, O.M., Culleres, A., Soriano, F.X., Peinado, H., Bolos, V., Martinez, F.O., Reina, M., Cano, A., Fabre, M., and Vilaro, S. (2006). The transcription factors Slug and Snail act as repressors of claudin-1 expression in epithelial cells, *Biochem. J.*, **394**, pp. 449–457.
30. Munger, J.S., Huang, X., Kawakatsu, H., Griffiths, M.J., Dalton, S.L., Wu, J., Pittet, J.F., Kaminski, N., Garat, C., Matthay, M.A., Rifkin, D.B., and Sheppard, D. (1999). The integrin alpha v beta 6 binds and activates latent TGF beta 1: A mechanism for regulating pulmonary inflammation and fibrosis, *Cell*, **96**, pp. 319–328.
31. Barrios-Rodiles, M., Brown, K.R., Ozdamar, B., Bose, R., Liu, Z., Donovan, R. S., Shinjo, F., Liu, Y., Dembowy, J., Taylor, I.W., Luga, V., Przulj, N., Robinson, M., Suzuki, H., Hayashizaki, Y., Jurisica, I., and Wrana, J. L. (2005). High-throughput mapping of a dynamic signaling network in mammalian cells, *Science*, **307**, pp. 1621–1625.
32. Overgaard, C.E., Daugherty, B.L., Mitchell, L.A., and Koval, M. (2011). Claudins: Control of barrier function and regulation in response to oxidant stress, *Antioxid. Redox Signal.*, **15**, pp. 1179–1193.
33. Kam, K.R., Walsh, L.A., Bock, S.M., Koval, M., Fischer, K.E., Ross, R.F., and Desai, T.A. (2013). Nanostructure-mediated transport of biologics across epithelial tissue: Enhancing permeability via nanotopography, *Nano Lett.*, **13**, pp. 164–171.
34. Kam, K.R., Walsh, L.A., Bock, S.M., Ollerenshaw, J.D., Ross, R.F., and Desai, T.A. (2013). The effect of nanotopography on modulating protein adsorption and the fibrotic response, *Tissue Eng. Part. A.*, **20**, pp. 130–138.
35. Capaldo, C.T., and Macara, I.G. (2007). Depletion of E-cadherin disrupts establishment but not maintenance of cell junctions in MDCK epithelial cells, *Mol. Biol. Cell*, **18**, pp. 189–200.
36. Gooding, J.M., Yap, K.L., and Ikura, M. (2004). The cadherin-catenin complex as a focal point of cell adhesion and signalling: New insights from three-dimensional structures, *Bioessays*, **26**, pp. 497–511.
37. Dejana, E. (2004). Endothelial cell-cell junctions: Happy together, *Nat. Rev. Mol. Cell Biol.*, **5**, pp. 261–270.
38. Kowalczyk, A.P., and Green, K. J. (2013). Structure, function, and regulation of desmosomes, *Prog. Mol. Biol. Transl. Sci.*, **116**, pp. 95–118.

39. Weis, W.I., and Nelson, W.J. (2006). Re-solving the cadherin-catenin-actin conundrum, *J. Biol. Chem.*, **281**, pp. 35593–35597.

40. Abe, K., and Takeichi, M. (2008). EPLIN mediates linkage of the cadherin–catenin complex to F-actin and stabilizes the circumferential actin belt, *Proc. Natl. Acad. Sci. U.S.A.*, **105**, pp. 13–19.

41. Noda, K., Zhang, J., Fukuhara, S., Kunimoto, S., Yoshimura, M., and Mochizuki, N. (2010). Vascular endothelial-cadherin stabilizes at cell-cell junctions by anchoring to circumferential actin bundles through alpha- and beta-catenins in cyclic AMP-Epac-Rap1 signal-activated endothelial cells, *Mol. Biol. Cell*, **21**, pp. 584–596.

42. Brieher, W.M., and Yap, A.S. (2013). Cadherin junctions and their cytoskeleton(s), *Curr. Opin. Cell Biol.*, **25**, pp. 39–46.

43. Shapiro, L. (2001). Beta-catenin and its multiple partners: Promiscuity explained, *Nat. Struct. Biol.*, **8**, pp. 484–487.

44. Taddei, A., Giampietro, C., Conti, A., Orsenigo, F., Breviario, F., Pirazzoli, V., Potente, M., Daly, C., Dimmeler, S., and Dejana, E. (2008). Endothelial adherens junctions control tight junctions by VE-cadherin-mediated upregulation of claudin-5, *Nat. Cell Biol.*, **10**, pp. 923–934.

45. Angelow, S., Ahlstrom, R., and Yu, A.S. (2008). Biology of claudins, *Am. J. Physiol. Renal Physiol.*, **295**, pp. F867–876.

46. Rosenthal, R., Milatz, S., Krug, S.M., Oelrich, B., Schulzke, J.D., Amasheh, S., Gunzel, D., and Fromm, M. (2010). Claudin-2, a component of the tight junction, forms a paracellular water channel, *J. Cell Sci.*, **123**, pp. 1913–1921.

47. Van Itallie, C.M., Gambling, T.M., Carson, J.L., and Anderson, J.M. (2005). Palmitoylation of claudins is required for efficient tight junction localization, *J. Cell Sci.*, **118**, pp. 1427–1436.

48. Daugherty, B.L., Ward, C., Smith, T., Ritzenthaler, J.D., and Koval, M. (2007). Regulation of heterotypic claudin compatibility, *J. Biol. Chem.*, **282**, pp. 30005–30013.

49. Furuse, M., Sasaki, H., and Tsukita, S. (1999). Manner of interaction of heterogeneous claudin species within and between tight junction strands, *J. Cell Biol.*, **147**, pp. 891–903.

50. Furuse, M., Furuse, K., Sasaki, H., and Tsukita, S. (2001). Conversion of zonulae occludentes from tight to leaky strand type by introducing claudin-2 into Madin–Darby canine kidney I cells, *J. Cell Biol.*, **153**, pp. 263–272.

51. Coyne, C.B., Gambling, T.M., Boucher, R.C., Carson, J.L., and Johnson, L.G. (2003). Role of claudin interactions in airway tight junctional

permeability, *Am. J. Physiol. Lung Cell. Mol. Physiol.*, **285**, pp. L1166–1178.

52. Piontek, J., Winkler, L., Wolburg, H., Muller, S.L., Zuleger, N., Piehl, C., Wiesner, B., Krause, G., and Blasig, I.E. (2008). Formation of tight junction: Determinants of homophilic interaction between classic claudins, *Faseb J.*, **22**, pp. 146–158.
53. Hou, J., Renigunta, A., Konrad, M., Gomes, A.S., Schneeberger, E.E., Paul, D.L., Waldeggerl, S., and Goodenoughl, D.A. (2008). Claudin-16 and claudin-19 interact and form a cation-selective tight junction complex, *J. Clin. Invest.*, **118**, pp. 619–628.
54. Krugl, S.M., Gunzell, D., Conradl, M.P., Leel, I.F., Amashehl, S., Fromml, M., and Yul, A.S. (2012). Charge-selective claudin channels, *Ann. N. Y. Acad. Sci.*, **1257**, pp. 20–28.
55. Blasigl, I.E., Winklerl, L., Lassowskil, B., Muellerl, S.L., Zulegerl, N., Krausel, E., Krausel, G., Gastl, K., Kolbe, M., and Piontek, J. (2006). On the self-association potential of transmembrane tight junction proteins, *Cell. Mol. Life Sci.*, **63**, pp. 505–514.
56. Mitic, L.L., Unger, V.M., and Anderson, J.M. (2003). Expression, solubilization, and biochemical characterization of the tight junction transmembrane protein claudin-4, *Protein Sci.*, **12**, pp. 218–227.
57. Cording, J., Berg, J., Kading, N., Bellmann, C., Tscheik, C., Westphal, J.K., Milatz, S., Gunzel, D., Wolburg, H., Piontek, J., Huber, O., and Blasig, I.E. (2012). Tight junctions: Claudins regulate the interactions between occludin, tricellulin and marvelD3, which, inversely, modulate claudin oligomerization, *J. Cell Sci.*, **126**, pp. 554–564.
58. Hou, J., Renigunta, A., Gomes, A.S., Hou, M., Paul, D.L., Waldegger, S., and Goodenough, D.A. (2009). Claudin-16 and claudin-19 interaction is required for their assembly into tight junctions and for renal reabsorption of magnesium, *Proc. Natl. Acad. Sci. U.S.A.*, **106**, pp. 15350–15355.
59. Hou, J., Renigunta, A., Yang, J., and Waldegger, S. (2010). Claudin-4 forms paracellular chloride channel in the kidney and requires claudin-8 for tight junction localization, *Proc. Natl. Acad. Sci. U.S.A.*, **107**, pp. 18010–18015.
60. Lu, Z., Ding, L., Hong, H., Hoggard, J., Lu, Q., and Chen, Y.H. (2011). Claudin-7 inhibits human lung cancer cell migration and invasion through ERK/MAPK signaling pathway, *Exp. Cell Res.*, **317**, pp. 1935–1946.
61. Piontek, J., Fritzsche, S., Cording, J., Richter, S., Hartwig, J., Walter, M., Yu, D., Turner, J.R., Gehring, C., Rahn, H.P., Wolburg, H., and Blasig, I.E.

(2011). Elucidating the principles of the molecular organization of heteropolymeric tight junction strands, *Cell. Mol. Life Sci.*, **68**, pp. 3903–3918.

62. Van Itallie, C.M., Mitic, L.L., and Anderson, J.M. (2011). Claudin-2 forms homodimers and is a component of a high molecular weight protein complex, *J. Biol. Chem.*, **286**, pp. 3442–3450.
63. Simon, D.B., Lu, Y., Choate, K.A., Velazquez, H., Al-Sabban, E., Praga, M., Casari, G., Bettinelli, A., Colussi, G., Rodriguez-Soriano, J., McCredie, D., Milford, D., Sanjad, S., and Lifton, R.P. (1999). Paracellin-1, a renal tight junction protein required for paracellular Mg^{2+} resorption, *Science*, **285**, pp. 103–106.
64. Konrad, M., Schaller, A., Seelow, D., Pandey, A.V., Waldegger, S., Lesslauer, A., Vitzthum, H., Suzuki, Y., Luk, J.M., Becker, C., Schlingmann, K.P., Schmid, M., Rodriguez-Soriano, J., Ariceta, G., Cano, F., Enriquez, R., Juppner, H., Bakkaloglu, S.A., Hediger, M.A., Gallati, S., Neuhauss, S.C., Nurnberg, P., and Weber, S. (2006). Mutations in the tight junction gene claudin 19 (CLDN19) are associated with renal magnesium wasting, renal failure, and severe ocular involvement, *Am. J. Hum. Genet.*, **79**, pp. 949–957.
65. Van Itallie, C.M., Fanning, A.S., Bridges, A., and Anderson, J.M. (2009). ZO-1 stabilizes the tight junction solute barrier through coupling to the perijunctional cytoskeleton, *Mol. Biol. Cell*, **20**, pp. 3930–3940.
66. Mitchell, L.A., Overgaard, C.E., Ward, C., Margulies, S S., and Koval, M. (2011). Differential effects of claudin-3 and claudin-4 on alveolar epithelial barrier function, *Am. J. Physiol. Lung Cell. Mol. Physiol.*, **301**, pp. L40–L49.
67. Shen, L., Weber, C.R., and Turner, J.R. (2008). The tight junction protein complex undergoes rapid and continuous molecular remodeling at steady state, *J. Cell. Biol.*, **181**, pp. 683–695.
68. Shen, L., and Turner, J.R. (2005). Actin depolymerization disrupts tight junctions via caveolae-mediated endocytosis, *Mol. Biol. Cell.*, **16**, pp. 3919–3936.
69. Bruewer, M., Hopkins, A.M., Hobert, M.E., Nusrat, A., and Madara, J.L. (2004). RhoA, Rac1, and Cdc42 exert distinct effects on epithelial barrier via selective structural and biochemical modulation of junctional proteins and F-actin, *Am. J. Physiol. Cell Physiol.*, **287**, pp. C327–335.
70. Koval, M. (2010). Keratinocyte growth factor improves alveolar barrier function: Keeping claudins in line, *Am. J. Physiol. Lung Cell. Mol. Physiol.*, **299**, pp. L721–L723.

71. Maiers, J.L., Peng, X., Fanning, A.S., and DeMali, K.A. (2013). ZO-1 recruitment to alpha-catenin: A novel mechanism for coupling the assembly of tight junctions to adherens junctions, *J. Cell Sci.*, **126**, pp. 3904–3915.
72. Roh, M.H., Liu, C.J., Laurinec, S., and Margolis, B. (2002). The carboxyl terminus of zona occludens-3 binds and recruits a mammalian homologue of discs lost to tight junctions, *J. Biol. Chem.*, **277**, pp. 27501–27509.
73. Fanning, A.S., Van Itallie, C.M., and Anderson, J.M. (2012). Zonula occludens-1 and -2 regulate apical cell structure and the zonula adherens cytoskeleton in polarized epithelia, *Mol. Biol. Cell*, **23**, pp. 577–590.
74. Ivanov, A.I., Parkos, C.A., and Nusrat, A. (2010). Cytoskeletal regulation of epithelial barrier function during inflammation, *Am. J. Pathol.*, **177**, pp. 512–524.
75. Shen, L., Black, E.D., Witkowski, E.D., Lencer, W.I., Guerriero, V., Schneeberger, E.E., and Turner J.R. (2006). Myosin light chain phosphorylation regulates barrier function by remodeling tight junction structure, *J. Cell Sci.*, **119**, pp. 2095–2106.
76. Samarin, S.N., Ivanov, A.I., Flatau, G., Parkos, C.A., and Nusrat, A. (2007). Rho/Rho-associated kinase-II signaling mediates disassembly of epithelial apical junctions, *Mol. Biol. Cell*, **18**, pp. 3429–3439.
77. Sahai, E., and Marshall, C.J. (2002). ROCK and Dia have opposing effects on adherens junctions downstream of Rho, *Nat. Cell Biol.*, **4**, pp. 408–415.
78. Yu, D., Marchiando, A.M., Weber, C.R., Raleigh, D.R., Wang, Y., Shen, L., and Turner, J.R. (2010). MLCK-dependent exchange and actin binding region-dependent anchoring of ZO-1 regulate tight junction barrier function, *Proc. Natl. Acad. Sci. U.S.A.*, **107**, pp. 8237–8241.
79. Guttman, J.A., and Finlay, B.B. (2009). Tight junctions as targets of infectious agents, *Biochim. Biophys. Acta.*, **1788**, pp. 832–841.
80. Sousa, S., Lecuit, M., and Cossart, P. (2005). Microbial strategies to target, cross or disrupt epithelia, *Curr. Opin. Cell Biol.*, **17**, pp. 489–498.
81. Minshall, R.D., and Malik, A.B. (2006). Transport across the endothelium: Regulation of endothelial permeability, *Handb. Exp. Pharmacol.*, **176**, pp. 107–144.
82. Raleigh, D.R., Marchiando, A.M., Zhang, Y., Shen, L., Sasaki, H., Wang, Y., Long, M., and Turner, J.R. (2010). Tight junction–associated

MARVEL proteins marveld3, tricellulin, and occludin have distinct but overlapping functions, *Mol. Biol. Cell*, **21**, pp. 1200–1213.

83. Yu, A.S., McCarthy, K.M., Francis, S.A., McCormack, J.M., Lai, J., Rogers, R. A., Lynch, R.D., and Schneeberger, E.E. (2005). Knockdown of occludin expression leads to diverse phenotypic alterations in epithelial cells, *Am. J. Physiol. Cell Physiol.*, **288**, pp. C1231–1241.
84. Beeman, N., Webb, P.G., and Baumgartner, H.K. (2012). Occludin is required for apoptosis when claudin–claudin interactions are disrupted, *Cell Death Dis.*, **3**, pp. e273.
85. Mrsny, R.J., Brown, G.T., Gerner-Smidt, K., Buret, A.G., Meddings, J.B., Quan, C., Koval, M., and Nusrat, A. (2008). A key claudin extracellular loop domain is critical for epithelial barrier integrity, *Am. J. Pathol.*, **172**, pp. 905–915.
86. Everett, R.S., Vanhook, M.K., Barozzi, N., Toth, I., and Johnson, L.G. (2006). Specific modulation of airway epithelial tight junctions by apical application of an occludin peptide. *Mol. Pharmacol.*, **69**, pp. 492–500.
87. Nusrat, A., Parkos, C.A., Verkade, P., Foley, C.S., Liang, T.W., Innis-Whitehouse, W., Eastburn, K.K., and Madara, J. L. (2000). Tight junctions are membrane microdomains, *J. Cell Sci.*, **113(Pt 10)**, pp. 1771–1781.
88. Ikenouchi, J., Furuse, M., Furuse, K., Sasaki, H., and Tsukita, S. (2005). Tricellulin constitutes a novel barrier at tricellular contacts of epithelial cells, *J. Cell Biol.*, **171**, pp. 939–945.
89. Higashi, T., Tokuda, S., Kitajiri, S., Masuda, S., Nakamura, H., Oda, Y., and Furuse, M. (2013). Analysis of the 'angulin' proteins LSR, ILDR1 and ILDR2: Tricellulin recruitment, epithelial barrier function and implication in deafness pathogenesis, *J. Cell Sci,*. **126**, pp. 966–977.
90. Burns, A.R., Bowden, R.A., MacDonell, S.D., Walker, D.C., Odebunmi, T.O., Donnachie, E.M., Simon, S.I., Entman, M.L., and Smith, C.W. (2000). Analysis of tight junctions during neutrophil transendothelial migration, *J. Cell Sci.*, **113(Pt 1)**, pp. 45–57.
91. Mandell, K.J., and Parkos, C.A. (2005). The JAM family of proteins, *Adv. Drug Deliv. Rev.*, **57**, pp. 857–867.
92. Severson, E.A., Jiang, L., Ivanov, A.I., Mandell, K.J., Nusrat, A., and Parkos, C.A. (2008). Cis-dimerization mediates function of junctional adhesion molecule A, *Mol. Biol. Cell*, **19**, pp. 1862–1872.
93. Itoh, M., Sasaki, H., Furuse, M., Ozaki, H., Kita, T., and Tsukita, S. (2001). Junctional adhesion molecule (JAM) binds to PAR-3: A possible

mechanism for the recruitment of PAR-3 to tight junctions, *J. Cell Biol.*, **154**, pp. 491–497.

94. Bazzoni, G., Martinez-Estrada, O.M., Orsenigo, F., Cordenonsi, M., Citi, S., and Dejana, E. (2000). Interaction of junctional adhesion molecule with the tight junction components ZO-1, cingulin, and occludin, *J. Biol. Chem.*, **275**, pp. 20520–20526.
95. Monteiro, A.C., Sumagin, R., Rankin, C.R., Leoni, G., Mina, M.J., Reiter, D.M., Stehle, T., Dermody, T.S., Schaefer, S.A., Hall, R.A., Nusrat, A., and Parkos, C.A. (2013). JAM-A associates with ZO-2, Afadin and PDZ-GEF1 to activate Rap2c and regulate epithelial barrier function, *Mol. Biol. Cell*, **24**, pp. 2849–2860.
96. Severson, E.A., Lee, W.Y., Capaldo, C.T., Nusrat, A., and Parkos, C.A. (2009). Junctional adhesion molecule A interacts with Afadin and PDZ-GEF2 to activate Rap1A, regulate beta1 integrin levels, and enhance cell migration, *Mol. Biol. Cell*, **20**, pp. 1916–1925.
97. Yan, J., Li, F., Ingram, D.A., and Quilliam, L.A. (2008). Rap1a is a key regulator of fibroblast growth factor 2-induced angiogenesis and together with Rap1b controls human endothelial cell functions, *Mol. Cell. Biol.*, **28**, pp. 5803–5810.
98. Birukova, A.A., Fu, P., Xing, J., and Birukov, K.G. (2009). Rap1 mediates protective effects of iloprost against ventilator induced lung injury, *J. Appl. Physiol.*, **107**, pp. 1900–1910.
99. Mandell, K.J., Babbin, B.A., Nusrat, A., and Parkos, C.A. (2005). Junctional adhesion molecule 1 regulates epithelial cell morphology through effects on beta1 integrins and Rap1 activity, *J. Biol. Chem.*, **280**, pp. 11665–11674.
100. Wang, N., De Bock, M., Decrock, E., Bol, M., Gadicherla, A., Vinken, M., Rogiers, V., Bukauskas, F.F., Bultynck, G., and Leybaert, L. (2013). Paracrine signaling through plasma membrane hemichannels, *Biochim. Biophys. Acta.*, **1828**, pp. 35–50.
101. Siebert, A.P., Ma, Z., Grevet, J.D., Demuro, A., Parker, I., and Foskett, J.K. (2013). Structural and functional similarities of Calcium Homeostasis Modulator 1 (CALHM1) ion channel with connexins, pannexins, and innexins, *J. Biol. Chem.*, **288**, pp. 6140–6153.
102. Penuela, S., Gehi, R., and Laird, D.W. (2013). The biochemistry and function of pannexin channels, *Biochim. Biophys. Acta.*, **1828**, pp. 15–22.
103. Muro, S., Wiewrodt, R., Thomas, A., Koniaris, L., Albelda, S.M., Muzykantov, V.R., and Koval, M. (2003). A novel endocytic pathway

induced by clustering endothelial ICAM-1 or PECAM-1, *J. Cell Sci.*, **116**, pp. 1599–1609.

104. Ghaffarian, R., Bhowmick, T., and Muro, S. (2012). Transport of nanocarriers across gastrointestinal epithelial cells by a new transcellular route induced by targeting ICAM-1, *J. Control Release*, **163**, pp. 25–33.
105. Aird, W.C. (2007). Phenotypic heterogeneity of the endothelium: I. Structure, function, and mechanisms, *Circ. Res.*, **100**, pp. 158–173.
106. Stan, R.V. (2007). Endothelial stomatal and fenestral diaphragms in normal vessels and angiogenesis, *J. Cell. Mol. Med.*, **11**, pp. 621–643.
107. Curry, F.E., and Adamson, R.H. (2012). Endothelial glycocalyx: Permeability barrier and mechanosensor, *Ann. Biomed. Eng.*, **40**, pp. 828–839.
108. van Haaren, P.M., VanBavel, E., Vink, H., and Spaan, J.A. (2003). Localization of the permeability barrier to solutes in isolated arteries by confocal microscopy, *Am. J. Physiol. Heart Circ. Physiol.*, **285**, pp. H2848–2856.
109. Moran, A.P., Gupta, A., and Joshi, L. (2011). Sweet-talk: Role of host glycosylation in bacterial pathogenesis of the gastrointestinal tract, *Gut*, **60**, pp. 1412–1425.
110. Hirota, J.A., and Knight, D.A. (2012). Human airway epithelial cell innate immunity: Relevance to asthma, *Curr. Opin. Immunol.*, **24**, pp. 740–746.
111. Cohen, A.W., Hnasko, R., Schubert, W., and Lisanti, M.P. (2004). Role of caveolae and caveolins in health and disease, *Physiol. Rev.*, **84**, pp. 1341–1379.
112. Dvorak, A.M., and Feng, D. (2001). The vesiculo-vacuolar organelle (VVO). A new endothelial cell permeability organelle, *J. Histochem. Cytochem.*, **49**, pp. 419–432.
113. Bergelson, J.M., Cunningham, J.A., Droguett, G., Kurt-Jones, E.A., Krithivas, A., Hong, J.S., Horwitz, M.S., Crowell, R.L., and Finberg, R.W. (1997). Isolation of a common receptor for Coxsackie B viruses and adenoviruses 2 and 5, *Science*, **275**, pp. 1320–1323.
114. Barton, E.S., Forrest, J.C., Connolly, J.L., Chappell, J.D., Liu, Y., Schnell, F.J., Nusrat, A., Parkos, C.A., and Dermody, T.S. (2001). Junction adhesion molecule is a receptor for reovirus. *Cell*, **104**, pp. 441–451.
115. Kirchner, E., Guglielmi, K.M., Strauss, H.M., Dermody, T.S., and Stehle, T. (2008). Structure of reovirus sigma1 in complex with its receptor junctional adhesion molecule-A, *PLoS Pathog.*, **4**, p. e1000235.

116. Maginnis, M.S., Forrest, J.C., Kopecky-Bromberg, S.A., Dickeson, S.K., Santoro, S.A., Zutter, M.M., Nemerow, G.R., Bergelson, J.M., and Dermody, T.S. (2006). Beta1 integrin mediates internalization of mammalian reovirus, *J. Virol.*, **80**, pp. 2760–2770.

117. Si, Y., Liu, S., Liu, X., Jacobs, J.L., Cheng, M., Niu, Y., Jin, Q., Wang, T., and Yang, W. (2012). A human claudin-1-derived peptide inhibits hepatitis C virus entry, *Hepatology*, **56**, pp. 507–515.

118. Harris, H.J., Davis, C., Mullins, J.G., Hu, K., Goodall, M., Farquhar, M.J., Mee, C.J., McCaffrey, K., Young, S., Drummer, H., Balfe, P., and McKeating, J.A. (2010). Claudin association with CD81 defines hepatitis C virus entry, *J. Biol. Chem.*, **285**, pp. 21092–21102.

119. Evans, M.J., von Hahn, T., Tscherne, D.M., Syder, A.J., Panis, M., Wolk, B., Hatziioannou, T., McKeating, J.A., Bieniasz, P.D., and Rice, C.M. (2007). Claudin-1 is a hepatitis C virus co-receptor required for a late step in entry, *Nature*, **446**, pp. 801–805.

120. Douam, F., Thi, V.L., Maurin, G., Fresquet, J., Mompelat, D., Zeisel, M.B., Baumert, T.F., Cosset, F.L., and Lavillette, D. (2013). A critical interaction between E1 and E2 glycoproteins determines binding and fusion properties of hepatitis C virus during cell entry, *Hepatology*, **59**, pp. 776–788.

121. Fofana, I., Zona, L., Thumann, C., Heydmann, L., Durand, S.C., Lupberger, J., Blum, H.E., Pessaux, P., Gondeau, C., Reynolds, G.M., McKeating, J.A., Grunert, F., Thompson, J., Zeisel, M.B., and Baumert, T.F. (2013). Functional analysis of claudin-6 and claudin-9 as entry factors for hepatitis C virus infection of human hepatocytes by using monoclonal antibodies, *J. Virol.*, **87**, pp. 10405–10410.

122. Meertens, L., Bertaux, C., Cukierman, L., Cormier, E., Lavillette, D., Cosset, F.L., and Dragic, T. (2008). The tight junction proteins claudin-1, -6, and -9 are entry cofactors for hepatitis C virus, *J. Virol.*, **82**, pp. 3555–3560.

123. Mee, C.J., Grove, J., Harris, H.J., Hu, K., Balfe, P., and McKeating, J.A. (2008). Effect of cell polarization on hepatitis C virus entry, *J. Virol.*, **82**, pp. 461–470.

124. Mitchell, L.A., and Koval, M. (2010). Specificity of interaction between Clostridium perfringens enterotoxin and claudin-family tight junction proteins, *Toxins*, **2**, pp. 1595–1611.

125. Winkler, L., Gehring, C., Wenzel, A., Muller, S.L., Piehl, C., Krause, G., Blasig, I.E., and Piontek, J. (2009). Molecular determinants of the

interaction between Clostridium perfringens enterotoxin fragments and claudin-3, *J. Biol. Chem.*, **284**, pp. 18863–18872.

126. Van Itallie, C.M., Betts, L., Smedley, J.G., 3rd, McClane, B.A., and Anderson, J.M. (2008). Structure of the claudin-binding domain of Clostridium perfringens enterotoxin, *J. Biol. Chem.*, **283**, pp. 268–274.
127. Takahashi, A., Saito, Y., Kondoh, M., Matsushita, K., Krug, S.M., Suzuki, H., Tsujino, H., Li, X., Aoyama, H., Matsuhisa, K., Uno, T., Fromm, M., Hamakubo, T., and Yagi, K. (2012). Creation and biochemical analysis of a broad-specific claudin binder, *Biomaterials*, **33**, pp. 3464–3474.
128. Kondoh, M., Masuyama, A., Takahashi, A., Asano, N., Mizuguchi, H., Koizumi, N., Fujii, M., Hayakawa, T., Horiguchi, Y., and Watanbe, Y. (2005). A novel strategy for the enhancement of drug absorption using a claudin modulator, *Mol. Pharmacol.*, **67**, pp. 749–756.
129. Wray, C., Mao, Y., Pan, J., Chandrasena, A., Piasta, F., and Frank, J.A. (2009). Claudin 4 augments alveolar epithelial barrier function and is induced in acute lung injury, *Am. J. Physiol. Lung Cell. Mol. Physiol.*, **297**, pp. L219–227.
130. Runkle, E.A., and Mu, D. (2013). Tight junction proteins: From barrier to tumorigenesis, *Cancer Lett.*, **337**, pp. 41–48.
131. Saeki, R., Kondoh, M., Kakutani, H., Matsuhisa, K., Takahashi, A., Suzuki, H., Kakamu, Y., Watari, A., and Yagi, K. (2010). A claudin-targeting molecule as an inhibitor of tumor metastasis, *J. Pharmacol. Exp. Ther.*, **334**, pp. 576–582.
132. Zwanziger, D., Hackel, D., Staat, C., Bocker, A., Brack, A., Beyermann, M., Rittner, H., and Blasig, I.E. (2012). A peptidomimetic tight junction modulator to improve regional analgesia, *Mol. Pharm.*, **9**, pp. 1785–1794.
133. Wong, V., and Gumbiner, B.M. (1997). A synthetic peptide corresponding to the extracellular domain of occludin perturbs the tight junction permeability barrier, *J. Cell Biol.*, **136**, pp. 399–409.

Chapter 2

Plasma Membrane as a Semipermeable Barrier

Guido Li Volsi

Department of Biomedical and Biotechnological Sciences, Section of Physiology, University of Catania, Viale Andrea Doria, 695125 Catania, Italy

glivolsi@unict.it

This chapter reviews some basic physiological and biophysical aspects of the plasma membrane. Seen as a semipermeable barrier, this particular *wall* is able to allow diffusional flows of ions and molecules through passive and active transport. Because of its character of semipermeable barrier, it gives rise to the separation of charges (cations and anions) along its entire thickness and molecules at the intracellular and extracellular side. This generates electrochemical gradients used by the cell as a driving force. The chapter also gives emphasis on transmembrane proteins (ion channels) and mechanisms involved in both release (exocytosis) and uptake of molecules (endocytosis). These mechanisms are useful for cellular communication and internalization of endogenous and exogenous molecules, as well as particles present in the extracellular space.

Drug Delivery Across Physiological Barriers

Edited by Silvia Muro

ISBN 978-981-4669-40-5 (Hardcover), 978-981-4669-41-2 (eBook)

www.panstanford.com

2.1 Introduction

The plasma membrane can be considered the barrier of an open thermodynamic system (the cell) able to carry out structural functions, separate intracellular and extracellular environment, and act as support and interface. The plasma membrane plays a primary role in the communication between cells.

The term plasma membrane was coined by Nageli (*Plasmamembran*) [1] to describe the aggregation (a film) formed when cells that are no longer intact come in contact with water. This observation appears to be a sort of precipitation reaction, described as such by Heilbrunn [2]. A better understanding of the cell membrane comes from the study of red blood cells. Its lipid nature was proposed over a hundred years ago. The idea that the plasma membrane was a purely lipid layer was subsequently abandoned, following the results of calculations made to measure surface tension. Danielli proposed the presence of two-protein films, associated with polar head groups in each of the sides of the lipid bilayer. This model was defined as the "paucimolecular membrane" of Davson–Danielli [3]. Some years later (1972), Singer and Nicolson [4, 5] proposed a model in which the lipids give rise to a fluid matrix in which proteins are immersed or floating (Fig. 2.1).

According to this model, lipids and proteins would move on a two-dimensional plane [6–8], providing the membrane

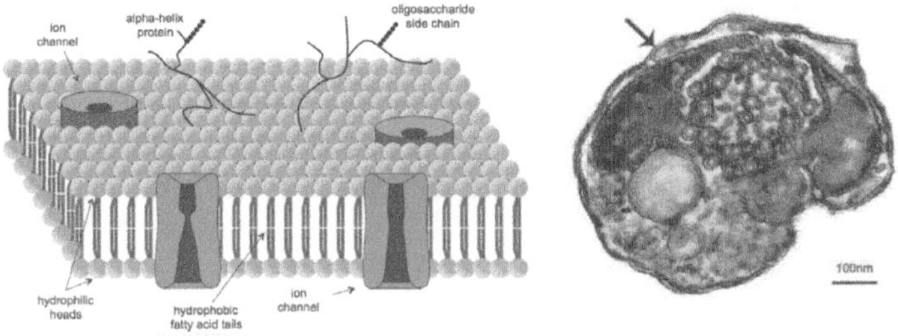

Figure 2.1 (Left) Schematic drawing of plasma membrane. (Right) Thin-section electron micrograph of a sperm flagellum of aphid. Courtesy of Prof. Renata Viscuso and Danilo Vitale, Histology Lab., Department of Biological, Geological and Environmental Sciences, University of Catania, Italy.

Table 2.1 Concentrations of ions and proteins in different body compartments

Compartment	Na^+ (mOsm/LH$_2$O)	K^+ (mOsm/LH$_2$O)	Cl^- (mOsm/ LH$_2$O)	Proteins (mOsm/LH$_2$O)
Intracellular	14	140	4	4
Extracellular	139	4.0	108	0.2
Plasmatic	142	4.2	108	1.2

with the characteristic of a fluid mosaic whose composition is heterogeneous.

Variations in the plasma membrane lipid composition across species appear to respond to evolutionary adaptations able to maintain an appropriate fluidity under specific environmental conditions [9, 10]. The different habitats of living organisms (from the marine to terrestrial ones) gave rise to a quite surprising evolutionary driving force. However, some general principles regarding the cellular architecture were preserved. In this context, the plasma membrane has played the key role as an element of separation between the extracellular and intracellular environment. For instance, this is the case with regard to the concentrations of ions and proteins. Table 2.1 shows a few examples illustrating that ions and proteins are not equally distributed inside and outside the cell, and this differential location is maintained by various mechanisms that counteract the free diffusion and balance across the plasma membrane [11].

2.2 Membrane Permeability and Transport

Although many substances move freely within an organic compartment, exchanges between intracellular and extracellular compartments are limited by the plasma membrane. The entry of a substance in a cell depends on the properties of the membrane and the substance (ion or molecule) itself [12, 13]. From here, a fundamental concept emerges: plasma membranes are selectively permeable. In other words, the lipoprotein composition of a cell membrane determines which ions/molecules will enter and exit a

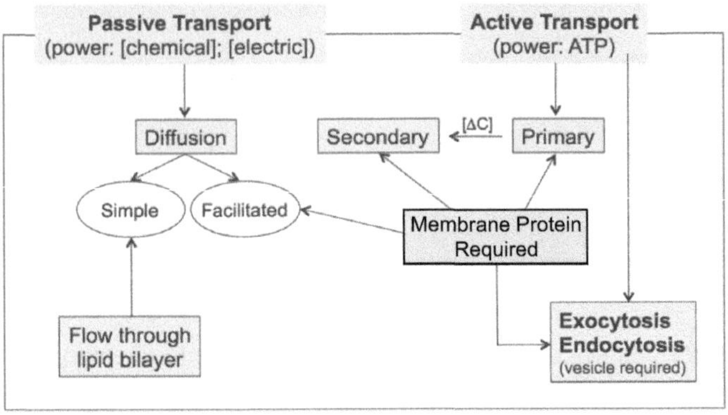

Figure 2.2 Chemical and physical forces rule the transport of ions and molecules across the plasma membrane. Simple diffusion and facilitated diffusion are considered equilibrating transport, while the primary and secondary active transports are considered nonequilibrating. The driving force (ΔC) of the secondary active transport comes from the concentration gradient generated by primary active transport.

cell. If a membrane allows a substance to cross it, the membrane is said to be permeable to that substance. Conversely, if a membrane does not allow a substance to cross the membrane, it is impermeable to the said substance.

More generally, the permeability of a membrane is variable and can be modified by changing lipid and/or protein composition. Water, gas (oxygen, carbon dioxide), and some lipids move without any difficulty. Conversely, ions and polar molecules will not move freely. Finally, the proteins may cross a plasma membrane with extreme difficulty, or not cross it at all. The movement of ions and molecules through a membrane is defined as transport, a process influenced by two parameters: the size of the substance and its solubility in lipids. Small or some fat-soluble substances can pass through the phospholipid bilayer. In contrast, ions and molecules of moderate size or not soluble in lipids need transport systems such as those provided by a class of transmembrane proteins that form ion channels. Finally, large molecules and polar ones are transported via vesicles (Fig. 2.2).

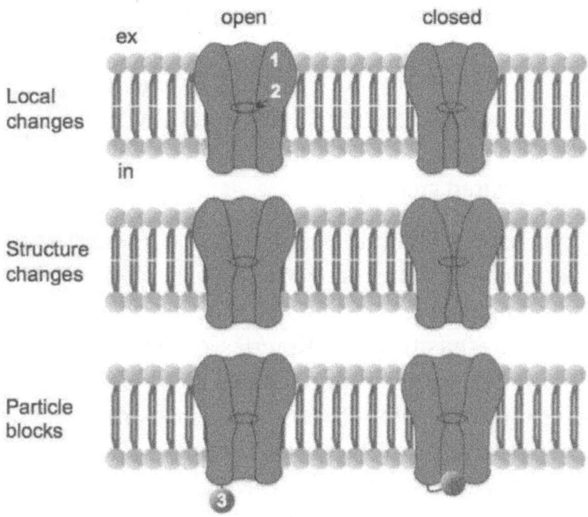

Figure 2.3 Schematic drawing of an ion channel: (1) channel domains; (2) selectivity filter; (3) blocking particle; in: intracellular side; ex: extracellular side.

2.3 Ion and Molecule Channels

Protein channels ensure transport of ions and small molecules [14, 15] (Fig. 2.3). Ion channels play an essential role keeping the vital functions of cells. They are composed of proteins, which assemble into structures that transverse the bilayer, with an inner pore through which ions flow in two directions (from inside to outside the cell or *vice versa*). The driving force is given by their electrochemical gradient. Ion channels show selectivity, which is developed on several levels. They can be either always open (passive channels), or they can be operated (i.e., open or closed) by events of chemical, electrical, mechanical, and electromagnetic nature. These channels can be classified referring to the element/agent responsible (directly or indirectly) for their operational activities (*gating*). According to these general criteria, ion channels can be divided into:

- Ligand-gated, if operated by extracellular or intracellular ligands (molecules, ions, cyclic nucleotides);

- Voltage-gated, if operated by changes in voltage membrane (Vm);
- Mechanosensitive, if operated by a mechanical stimulus;
- Phosphorylation-gated, if operated by transfer of the high-energy phosphate, P_i; and
- Light-gated channels.

Each type of ion channel has its own characteristics that relate to the level of conductance, selectivity, permeability, and gating. Ion channels are also subject to control processes such as modulation, blocking, and desensitization. Kinetics of activation are estimated to be in the range of miliseconds, while conductance ranges from a few to some hundreds picosiemens and have values that may change with the concentration of the permeant ion. After the opening phase of the channel, a current (I) flows through, whose value is

$$I = \gamma(V_m - E_i) \tag{2.1}$$

where γ is the conductance, V_m is the transmembrane potential, and E_i is the equilibrium potential of the considered ion. The selective permeability depends in part by the architecture of the pore.

2.3.1 *Ligand-Gated Ion Channels*

The activation process of ligand-gated ion channels [16–18] takes place by interaction of a *ligand* (ion, molecule) with a *receptor*, which can be part of the channel or located remotely. These channels are operated by extracellular or intracellular ligands. Following the interaction of the receptors with these chemicals, the receptor modulates ionic fluxes through that channel. The currents generated by these flows can be depolarizing (V_m becomes less negative) or hyperpolarizing (V_m becomes more negative) at post- and/or presynaptic level. These channels are also subject to desensitization (loss of response by excess neurotransmitter) and/or blockage (the presence of ions or molecules that act as plug).

The literature distinguishes this kind of receptors between ionotropic [19–21] and metabotropic [22–25]. Ionotropic receptors have constitutive subunits enclosing a pore through which the ion flow occurs. In the absence of ligand–receptor interaction, the structural conformation prevents the entrance of the ion. After the ligand binds with the receptor, the conformational structure changes with consequent ion transit. By contrast, activation of metabotropic receptors triggers a series of reactions mediated by an intracellular cascade (second messengers are involved), which provides the signal transduction. Ligand-gated ion channels play a central role in cellular communication (synapses, secretory processes, control of cellular function by ion and molecules). This key function has two consequences: first, these receptors are major targets for drug discovery because of their potential involvement in numerous brain diseases; second, they are often found to be the target of plant and animal toxins. This makes toxin/receptor interactions important for drug discovery [26]. Ligand-gated channels are less selective than voltage-gated ones.

2.3.2 *Voltage-Gated Ion Channels* ($i_V^{\pm}$)

Voltage-gated ion channels are operated by changes in the level of polarization of the plasma membrane. These channels, in addition to the state of opening and closing, can present a condition of inactivation. In addition, they show a reversible transition from one state to the other. Each domain of this type of channel has transmembrane segments, including S4 (the S4 domain is the voltage-sensing helix), which contains several basic residues—arginine or lysine. These residues are responsible for sensitivity to voltage, being provided with electrical charges. Following depolarization, positive charges are pushed outward. This translocation and the reorientation of the charges provoke the opening of the pore. Na^+, Ca^{2+}, K^+, and Cl^-, if not passively transported, utilize also voltage-gated ion channels for their transport across the plasma membrane [25–28]. This occurs when the membrane potential assumes a *certain* value (the operating voltage of the *given* ion channel). The higher is the difference between the membrane potential and the working potential of a voltage-gated channel, the greater its probability to be in the open state.

Voltage-gated ion channels are targets of numerous toxins [29]. For example, most of the Na_V^+ channels are blocked by nanomolar concentrations of saxitoxin (STX). This toxin was found in dinoflagellates and has no effect on Ca_V^{2+} and K_V^+ channels.

Tetrodotoxin, produced by the puffer fish, induces the same effects of STX. The effect may relate to activation and inactivation, permeation, and block.

2.3.3 *Mechanosensitive Ion Channels*

Various mechanical forces affect the functioning of animal cells, and probably mechanosensitivity represents one of the oldest transduction processes [30]. A molecular system of mechano-electric type is able to transduce mechanical forces exerted on the plasma membrane into electrical signals suitable to trigger or regulate some biochemical or physiological processes. Mechanosensitive channels are already present in prokaryotes (bacteria and archaea) [31]. More generally, these channels respond to forces (internal or external) applied along the plane of the plasma membrane. The final effect, with some exceptions, is the generation of a cationic current (selective or nonselective) directed from the extracellular to the intracellular side [32]. Touch, hearing, proprioception, osmotic gradients, geotropism, and cell turgor are some examples of the use of mechanosensitive ion channels. These channels are anchored to cationic aqueous pore. Mechanosensitive ion channels give rise to disorders of neurons [33, 34] and muscle [35]. Malfunctions of mechanosensitive ion channels in the kidney give rise to polycystic kidney disease [36] due to an alteration of the permeability to Ca^{2+}, while at the cardiac level, malfunctions may give rise to arrhythmias [37], atrial fibrillation, and hypertension [38].

2.3.4 *Phosphorylation-Gated Ion Channels*

Some ligands can activate cellular signaling cascades, which can operate the channel through a phosphorylation process. Experimental evidences suggest that intracellular second messengers can modulate the activity of ion channels and that protein phosphorylation by the different protein kinases is a frequent intermediary in these modulatory effects. The conclusion that ion channel proteins are indeed substrates for phosphorylation has been verified in numerous biochemical studies [39–42]. Channel phosphorylation involves a change in channel-open probability and, in the case of voltage-sensitive ion channels, a shift in the voltage dependence of channel activation.

Phosphorylation-gated channels are opened by the binding of a phosphate to the channel protein on the cytoplasmic side. Within neurons, these types of channels are opened and closed by metabotropic processes that determine their internal chemical state.

2.3.5 *Light-Gated Ion Channels*

Light-gated ion channels are a group of transmembrane proteins that form ion channels [43]. These channels have a pore whose conductance reversibly changes in response to irradiation with UV or visible light. Light-gated ion channels are useful for laboratory studies; one naturally occurring example is channelrhodopsin.

The functioning of light-gated ion channels is similar to other gated ion channels. When exposed to a given stimulus, a conformational change occurs in the transmembrane proteins of the pore. This change opens or closes the ion channel, allowing for the flow of ions according to their electrochemical gradient. In the specific case of light-gated ion channels, the transmembrane proteins are usually coupled with a molecule that acts as a photo-transducer. Retinal is a good example of a molecular photo-transducer and is found in the naturally occurring channelrhodopsins. The photo-transducer absorbs a specific photon and modifies its conformation, which, in turn, modifies the conformation of the transmembrane proteins opening or closing the pore through which ions flow.

2.4 Equilibrium and Diffusion Potentials

Ionic species present on both sides of a membrane in unequal concentrations and with different coefficients of permeability will give rise to a diffusion potential.

Fick's first law states that the amount of solute that passes through a membrane by diffusion (diffusional flow) depends on the area of the exchange surface, the number of moles passing through the membrane in a given time, and the diffusion properties of the substance. As in steady-state conditions, the concentration gradient is equal to the difference of the concentrations at the two sides of the membrane.

In a thin membrane, the concentration gradient is considered to be linear [44].

Diffusion has the following properties [45]:

- It is a passive process.
- Ions and molecules move from a higher concentration to a lower concentration environment.
- Diffusional flow proceeds until the concentration of a particular substance in the two environments does not assume the same value. (It is fast for short distances, but it can become very slow for longer distances.)
- It is directly related to temperature and is inversely related to the molecular size.
- It can occur in the presence of a semipermeable barrier (the plasma membrane) in which the possibility of transit exists, although with difficulty.

This kind of potential, by its nature, is subject to decay and, thus, cannot be maintained for an indefinite time.

Otherwise, if a membrane is permeable to an ion species and impermeable to another present at the two sides of the membrane at unequal concentration, an *equilibrium* potential is generated. The equilibrium potentials have a defined value that essentially depends on the concentrations of ion species considered at the two sides of the membrane.

A plasma membrane shows both these properties. Normally, animal cells contain protein anions (A^-) not diffusible, a high concentration of K^+, and a low concentration of Na^+, Cl^-, and Ca^{2+}. Conversely, their extracellular environment is rich in Na^+, Cl^-, and Ca^{2+}. If we consider an ion capable of moving through the plasma membrane, we must take into account the forces involved in this transport. The driving forces are both chemical (W_C) and electrical (W_E), and we can quantify these works.

$$W_C = RT\,\{\ln[i_e] - \ln[i_i]\} \tag{2.2}$$

where R is the gas constant (8.3 J/mol/K), T is the absolute temperature in kelvin (K = 273 + °C), $[i_e]$ and $[i_i]$ are ion concentrations at the extracellular (e) and intracellular (i) levels, and ln is the natural log.

$$W_E = z_i F V_{m(i)} \tag{2.3}$$

where F is the Faraday constant (96,500 C/eq), z is the valence (±), and $V_{m(i)}$ is the voltage membrane due to the ion.

This ion can reach a state of equilibrium, that is, its net flow through the membrane will be equal to zero if the chemical work done to move it from one side is equal to the electrical work done to move it from the opposite side. Since $W_c = W_E$,

$$RT\,\{\ln[i_e] - \ln[i_i]\} = z_i F V_{m(i)} \tag{2.4}$$

At this point, we can calculate the value of $V_{m(i)}$, the value of the potential assumed by the membrane because of the concerned ion (E_i). The final equilibrium potential is expressed by the Nernst equation [46], and we can use it to calculate the equilibrium potential for each of ions discussed above:

$$E_i = \frac{RT}{zF}\ln\frac{[i]_e}{[i]_i} \text{ or } E_i = \frac{2.303RT}{zF}\log\frac{[i]_e}{[i]_i} \tag{2.5}$$

where 2.303 is the factor for conversion of natural log (ln), or log to base e (2.717), to log to the base 10. By assuming an environmental temperature of 20°C ($T = 293$) and a valence of 1 ($z = 1$), Eq. 2.5 becomes

$$E_i = 0.058V\log\frac{[i]_e}{[i]_i} \tag{2.6}$$

This reached equilibrium does not means that the ion is equally distributed to both sides of the membrane. On the basis of the distributions of the intracellular and extracellular permeant ions, the Nernst equation allows, thus, to determine the relative equilibrium potentials. For a mammalian neuron, these values are

$$E_{Na} = +56\,\text{mV};\ E_{Ca} = +125\,\text{mV};\ E_K = -102\,\text{mV};\ E_{Cl} = -76\,\text{mV}$$

2.5 Membrane and Action Potentials

2.5.1 *Membrane Potential*

The structure of the plasma membrane, the presence of permeant ions, and their equilibrium potentials allow to calculate the value of the potential (V_m) of that membrane. The recording of membrane potential in individual neurons, myocytes, or red blood cells indicates that each value does not match any of the equilibrium potentials of permeant ions. A membrane potential of −80 mV approaches the equilibrium potential of Cl^- or K^+ but is much lower than that of Na^+ or Ca^{2+}. This is a steady-state condition in which the ion concentrations on the two sides of the membrane remain constant over time, even in the presence of net flows different from zero. One such operating condition is possible if there is a system of transmembrane transport able to maintain permeant ions away from their equilibrium potential. A transport system against electrochemical gradient can ensure this type of balance. To this purpose, plasma membranes are equipped with $Na^+/K^+{}_{\mathrm{ATPase}}$ pumps able to ensure a type of transport called *antiporter*.

In the 1940s, starting from the observations of Nernst, Goldman [47] of the Columbia University suggested the theory of "constant field." According to this theory, one could define the transmembrane potential in steady-state conditions, taking into account the ionic concentrations of permeant ions present on both sides of the membrane and of the membrane permeability to these ions. In terms of current flows, the membrane conductance (g) can be a good measure of how quickly an ion passes through the membrane. In addition, another valid measure is the permeability (P) of the membrane to a particular ion, expressed in terms of speed (cm/s). This concept can be quantified by the Goldman equation [48]:

$$V_{\mathrm{m}} = \frac{RT}{F} \ln \frac{P_{\mathrm{K}}[K^+]_{\mathrm{e}} + P_{\mathrm{Na}}[Na^+]_{\mathrm{e}} + P_{\mathrm{Cl}}[\mathrm{Cl}^-]_{\mathrm{i}}}{P_{\mathrm{K}}[K^+]_{\mathrm{i}} + P_{\mathrm{Na}}[Na^+]_{\mathrm{i}} + P_{\mathrm{Cl}}[\mathrm{Cl}^-]_{\mathrm{e}}} \tag{2.7}$$

This equation is known as the Goldman–Hogkin–Katz (GHK) equation and allows to obtain the value of the membrane potential as a function of the concentrations and the permeability of the considered ions. Furthermore, the same equation shows that the importance of an ion in determining the membrane potential increases with an increase in its concentration and the permeability of membranes toward that ion [48].

The GHK equation represents, therefore, a very general quantification that includes the Nernst equation. Normally, cell membranes are more permeable to K^+ than to other ions. Because of this, K^+

plays a key role in setting the membrane potential of many cell types to a given level. The value of the membrane potential of neurons has a strong dependence on the concentration of K^+ present at the extracellular level. Glial cells, present in their proximity, are able to control the concentration of this ion very finely and thus indirectly control the excitability of neurons [49].

2.5.2 *Action Potential*

The permeability of the plasma membrane to different ions does not have an absolute value and is not unmodifiable [50]. This is because all animal cells behave as rechargeable batteries. They can be considered devices tending to maintain a stable potential in the face of disturbances that may modify it. This potential has a value (*negative*) ranging from a few mV (in developing cells) up to 90 mV (adult neurons). Plasma membranes react to depolarizing events, generating cationic outward currents or, if possible, anionic inward currents [51]. Conversely, if they are hyperpolarized, they give rise to cationic inward currents or to anionic outward currents.

However, cell types such as excitable cells [15] (neurons and smooth, skeletal, or cardiac cells) are able to modify the amplitude of their membrane potential (usually called resting potential) for a few to several hundred milliseconds, following "conditions" that we can define as "new."

The opening of voltage-gated ion channels generates these new conditions. In other words, ions such as Na^+ and Ca^{2+}, powered by their electrochemical driving forces, are able to cross the plasma membrane also by means of channels (voltage-gated ones) that are normally closed. The trigger for this new temporary condition is an ion transport across the membrane able to depolarize it at the opening value of the channel (operating voltage for that channel). The trigger can be a ligand or the architecture of the cell membrane (presence of voltage-dependent ion channels) capable of self-depolarization (autorhythmic activity).

From a biophysical point of view, the action potential can be considered the result of a condition of the membrane voltage (threshold value) at which the cationic inward current exceeds the outward one. During this temporary condition, the membrane

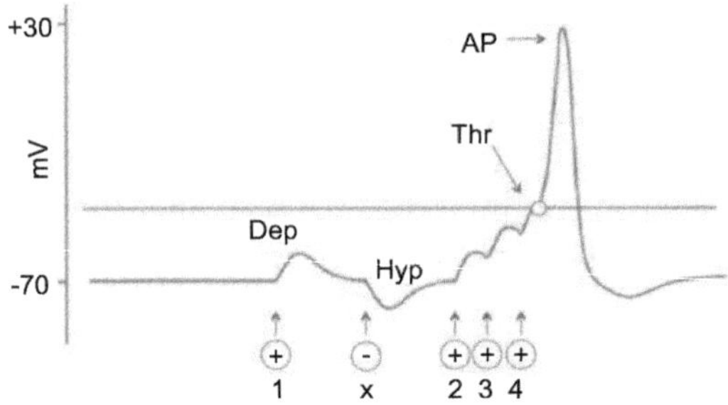

Figure 2.4 Schematic drawing of subthreshold depolarizing potentials (1 to 4) and action potential (5) of a hypothetic neuron. The arrows represent a subthreshold stimulus unable to evoke an action potential. If a stimulus (see line below) is provided before the end of effects evoked by subthreshold one, the neuron will give rise to an action potential. The stimulus *x* generates a hyperpolarization due to inward anionic current. Dep: depolarization; Thr: threshold; Hyp: hyperpolarization.

potential will assume a positive mean value of 55 mV that, broadly, will express the equilibrium potential of the ion (usually Na^+) that caused it. Subsequently, voltage-gated K^+ outward current will provide the potential to restore to its resting value (Fig. 2.4). The pumps Na^+/K^+_{ATPase} will restore Na^+ inside and K^+ outside.

2.6 Endocytosis and Exocytosis

Cells are dynamic units that can give rise to a multiplicity of activities. Some of these are used to adjust their content. Materials can be transported either from the outside toward the inside of the cell, yet not across the plasma membrane (endocytosis), or from the inside toward the outside of the cell (exocytosis).

2.6.1 *Endocytosis*

Endocytosis is a process by which cells internalize molecules (such as proteins) by engulfing them. All cells of the body use it because most substances important to them are large polar molecules that cannot pass through the hydrophobic plasma or cell membrane.

During endocytosis, the membrane surface invaginates and a vesicle forms. Endocytosis is, by its nature, constitutive, that is, it is an essential function that always takes place [52]. Endocytosis is an active process that requires ATP as an energy source. Also, it may be of nonselective type, allowing the extracellular fluid to enter into the cell (pinocytosis), or of the selective type, allowing entry only to specific molecules.

If solid particles, it is phagocytosis; if soluble, then it is pinocytosis. Pinocytosis can occur via large vesicles (macropinocytosis) or small ones (micropinocytosis). Micropinocytosis can occur via clathrin-mediated endocytosis, caveolae-mediated endocytosis, or non-clathrin non-caveolar types (this is cell adhesion molecule or CAM-mediated endocytosis, flotilin pathway, Arf6 pathway, or cdc42 pathway). Clathrin, caveolae, CAM, and other pathways are dynamin dependent, while others are independent. Macropinocytosis and some caveolar pathways can be non-receptor mediated, while the rest are mediated by receptors.

Two types of endocytosis require that a ligand binds to a membrane receptor: receptor-mediated endocytosis and potocytosis.

2.6.1.1 Receptor-mediated endocytosis

Receptor-mediated endocytosis takes place in different regions of the cell membrane. They are known as clathrin[a]-coated vesicles. They are recesses in which the cytoplasmic side of the membrane has a high concentration of this protein. The first stage of

---

[a]Clathrin is a protein containing a three-legged structure termed triskelion. The triskelion consists of three heavy chain polypeptides, each bound to a light chain (LC) of which there are two different types, LCa and LCb. The clathrin heavy chain (CHC) is known to provide the backbone for the clathrin lattice. CHC is composed of a terminal globular domain, a distal segment, and a proximal segment containing an LC-binding site. Clathrin LCs contain regulatory domains that influence the recruitment, assembly, and disassembly of clathrin within the cell. Clathrin was first isolated and named by Barbara Pearse in 1975. It forms a triskelion shape composed of three CHCs and three LCs. When the triskelia interact, they form a polyhedral lattice that surrounds the vesicle. Coat proteins, like clathrin, are used to build small vesicles to safely transport molecules within and between cells. The endocytosis and exocytosis of vesicles allow cells to transfer nutrients to import signaling receptors to mediate an immune response after sampling the extracellular world, and to clean up the cell debris left by tissue inflammation. On occasion, this mechanism also provides a pathway for raiding pathogens or toxins.

the endocytic process involves the binding of the substrates to extracellular membrane receptors. The binding of the ligand with the receptor triggers a signal that induces the cell response. In many cases, a protein called ubiquitin [53] tags the receptor. This means that the receptor has been marked for internalization. The receptor–ligand complex induces the formation of a clathrin-coated vesicle (Fig. 2.5). When the receptor–ligand complex is located within the dimple, the membrane withdraws inward and invaginates. Then, the invagination detaches from the cell membrane and becomes a cytoplasmic vesicle. Thus, the clathrin molecules are released and return to the membrane. In vesicles, the receptor and the ligand are often separated, leaving the ligand in the endosome, a membranous organelle designed to sort between receptors that can be recycled back to the membrane and those that need to be degraded or inactivated in lysosomes. If the receptor is to be destroyed, it is placed within the endosome. A protein complex named endosomal sorting complex required for transport (ESCRT) is responsible for recognizing a receptor to be destroyed and forming the vesicle around it. Conversely, receptors that are to be recycled are kept out of these internal vesicles. If the ligand must be destroyed, the endosome is transferred to a lysosome (another membranous organelle that contains enzymes to destroy the receptor), whereas if the ligand must be treated, it is transferred to the Golgi apparatus.

These receptors can be used in a process known as "recycling of membrane." In this case, the vesicle containing receptors migrates to the plasma membrane and merges with it. Thus, the membrane of the vesicle is incorporated again in the cell membrane by exocytosis.

Receptor-mediated endocytosis allows the entrance within the cell to different substances: growth factors, proteins, hormones, and plasma proteins that act as carriers for Fe^{2+} and cholesterol (Fig. 2.5).

Potocytosis or, more generally, caveolae-mediated endocytosis [54] represents the mechanism by which small and large molecules as well as macromolecular complexes are sequestered and transported by caveolae [55, 56]. Caveolae are flask-shaped plasma membrane specializations characterized by a filamentous coat consisting of caveolins, which decorates cytosolic surface of

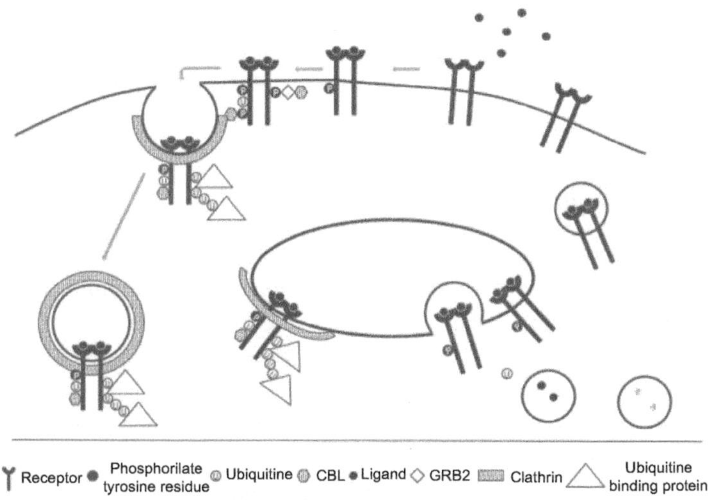

Figure 2.5 Schematic drawing of endocytosis-mediated receptors. See text for details.

each caveolar membrane. This pathway has endocytotic functions different from the clathrin-coated pit pathway. Ligands bound to receptors that are internalized by caveolae can be delivered to four different locations in the cell bypassing the lysosome. At least four different caveolae membrane traffic patterns during potocytosis can be distinguished. Hence, cells have two endocytic machines and each is designed to accomplish different tasks.

2.6.1.2 Non-clathrin and non-caveolar endocytosis

The recent development of new techniques has provided new insights into clathrin and caveolae-independent internalization (CLIC). The literature describes several CLIC pathways. One of these is dynamin-independent CLIC; it internalizes large molecular weight ligands and mediates the uptake of glycosylphosphatidylinositol-linked proteins. A CLIC dynamin dependent (CLIC-d) was described for the IL-2β receptor immune cells. In endothelial cells, CLIC allows the internalization of proteoglycans and bound factors such as FGF-2. Muro et al. described a novel endocytosis pathway in endothelial cells [57]. They called it cell adhesion molecule-mediated endocytosis pathway (CAM-ME). It is induced by crosslinking of either adhesion molecules ICAM-1 (CD34) or PECAM-1 (CD31). CLIC can deliver molecules to various intracellular compartments (Golgi apparatus, the ER as well as to classic endocytic compartments).

2.6.2 *Exocytosis*

Exocytosis is the energy-consuming process by which a cell directs the contents of secretory vesicles out of the cell and into the extracellular space. The cellular secretion requires a transient vesicle fusion with the cell membrane, dumping its contents out of the cell's environment. The secretory vesicle can be recycled for subsequent rounds of exo- and endocytosis, until completely empty of its contents.

In multicellular organisms, there are two types of exocytosis [58]:

1. Non-Ca^{2+}-triggered constitutive
2. Ca^{2+}-triggered non-constitutive

Constitutive exocytosis is performed by all cells and serves the release of components of the extracellular matrix, hormones, and neurotransmitters, or just delivery of newly synthesized membrane proteins and lipids that are incorporated in the plasma membrane after the fusion of the transport vesicle. Regulated exocytosis, on the other hand, requires an external signal, a specific sorting signal on the vesicles, a clathrin coat, as well as an increase in intracellular calcium.

Secretory vesicles transiently dock at the cell plasma membrane, preceding the formation of a tight t-/v-SNARE [an acronym derived from "SNAP (Soluble NSF Attachment *Protein*) Receptor"] complex, leading to priming and the establishment of continuity between the opposing bilayers [59].

In neurons, exocytosis includes all of the molecular rearrangements and ATP-dependent protein and lipid modifications. These events take place after initial docking of a synaptic vesicle but before exocytosis, such that the influx of Ca^{2+} through Ca_V^{2+} is all. It triggers nearly instantaneous neurotransmitter release. In other cell types, whose secretion is constitutive (i.e., continuous, calcium ion independent, non-triggered), there is no priming.

Transient vesicle fusion is driven by SNARE proteins, resulting in the release of vesicle contents into the extracellular space (or in the case of neurons in the synaptic cleft).

The merging of the membranes (donor and acceptor) accomplishes three tasks:

1. The surface of the plasma membrane increases.
2. The substances within the vesicle are released into the extracellular side. These might be neurotransmitters during synaptic transmission, hormones, waste products, or toxins.
3. The proteins embedded in the vesicle membrane are now part of the plasma membrane. The side of the protein before facing the inside of the vesicle now faces extracellular side.

Non-constitutive exocytosis and subsequent endocytosis are highly energy-expending processes and are thus dependent on mitochondria [60, 61].

2.7 Conclusions and Perspectives

In conclusion, the plasma membrane allows the delimitation of environments (intracellular space), the key to the identity of a single cell. It is also the interface for communication between cells, which carries out this task with great precision and without a lot of noise. Many drugs need to pass through one or more cell membranes to reach their site of action. Chemical and electrical properties of a plasma membrane can act as obstacles or gates. Investigators, membrane engineers, and factories need to evaluate these properties when building ideas, methods, and molecules for the health of living organisms.

References

1. Nageli, K. W. (1887). *The Microscope in Theory and Practice* (Swan Sonnenschein, Lowrey, London, UK).
2. Heilbrunn, L. V. (1956). *The Dynamics of Living Protoplasm* (Academic Press, New York, USA).

3. Davson, H., and Danielli, J. F. (1943). *The Permeability of Natural Membranes* (Hafner Publishing Company, New York, USA).
4. Singer, S. J., and Nicolson, G. L. (1972). The fluid mosaic model of the structure of cell membrane, *Science*, **175**, pp. 720–731.
5. Yeagle, P. L. (2005). *The Structure of Biological Membranes*, 2nd Ed. (Boca Raton, CRC Press, FL, USA).
6. Sonnino, S., and Prinetti, A. (2013). Membrane domains and the "lipid raft" concept, *Curr. Med. Chem.*, **20**, pp. 4–21.
7. Vaz, W. L. (1994). Diffusion and chemical reactions in phase-separated membranes, *Biophys. Chem.*, **50**, pp. 139–145.
8. Sheikh, K. H., and Jarvis, S. P. (2011). Crystalline hydration structure at the membrane–fluid interface of model lipid raft indicates a highly reactive boundary region, *J. Am. Chem. Soc.*, **133**, pp. 18296–18303.
9. Cossis, A. R., and Prosser, C. L. (1978). Evolutionary adaptation of membranes to temperature, *Proc. Nat. Acad. Sci.*, **75**, pp. 2040–2043.
10. Reece, J. B., Urry, L. A., Cain, M. L., Wasserman, S. A., Minorsky, P. V., and Jackson, R. B. (2010). *Campbell Biology*, 9th Ed. (Benjamin Cummings, USA).
11. Guyton and Hall. (2011). *Textbook of Medical Physiology*, 12th Ed. Constituents of Extracellular and Intracellular Fluids (Saunders, Elsevier).
12. Niikimura, K., Nambara, K., Okajiama, T., Matsuo, Y., and Ijiro, K. (2010). Influence of hydrophobic structures on the plasma membrane permeability of lipid like molecules, *Langmuir*, **26**, pp. 9170–9175.
13. Laude, A. J., and Prior, I. A. (2004). Plasma membrane microdomains: Organization, function and trafficking, *Mol. Membr. Biol.*, **21**, pp. 193–205.
14. Andersen, O. S., and Koeppe, R. E. (1992). Molecular determinants of ion channel function, *Physiol. Rev.*, **72**, pp. S89–S158.
15. Koester, J., and Siegelbaum, S. A. (2000), Chapter 6: Membrane Potential, In *Principles of Neural Sciences*, 4th Ed., Kandel, E. R., Schwartz, J. H., and Jessel, T. M., eds. (McGraw-Hill, USA), pp. 125–139.
16. Hucho, F., and Weise, C. (2001). Ligand-gated ion channels, *Angewandte Chemie Internat. Ed.*, **40**, pp. 3100–3116.
17. Collingridge, G. L., Olsen, R. W., Peters, J., and Spedding, M. (2009). A nomenclature for ligand-gated ion channels, *Neuropharmacology*, **56**, pp. 2–5.
18. Trovar, R., and Westbrook, G. L. (2012). Chapter 31: Ligand-Gated Ion Channels, In *Cell Physiology Source Book: Essentials of Membrane*

Biophysics, 4th Ed., Sperelakis, N., ed. (Academic Press, New York, USA), pp. 549–562.

19. Maksay, G. (2005). Activation of ionotropic receptors and thermodynamics of binding, *Neurochem. Int.*, **46**, pp. 281–291.
20. Barrera, N. P., and Henderson, R. M., Edwardson, J. M. (2008). Determination of the architecture of ionotropic receptors using AFM imaging, *Pfluger Arch.*, **456**, pp. 199–209.
21. Barrera, N. P., and Edwardson, J. M. (2008). The subunit arrangement and assembly of ionotropic receptors, *Trends Neurosci.*, **31**, pp. 569–576.
22. Lledo, P. M., Vernier, P., and Vincent, J. D. (1993). Coupling of metabotropic receptors to ionic channels. Role of G-Proteins, *C. R. Seances Soc. Biol. Fil.*, **187**, pp. 7–13 (in French).
23. Conn, P. J., and Pin, J. P. (1007). Pharmacology and function of metabotropic glutamate receptors, *Ann. Rev. Pharmacol., Toxicol.*, **37**, pp. 205–237.
24. Pin, J. P., and Duvoisin, R. (1995). The metabotropic glutamate receptors: Structure and function, *Neuropharmacology*, **34**, pp. 1–26.
25. Gilsbach, R., and Hein, L. (2008). Presynaptic metabotropic receptors for acetylcholine and adrenaline/noradrenaline, *Handb. Exp. Pharmacol.*, **184**, pp. 261–288.
26. Nasiripourdori, A., Taly, V., Grutter, T., and Taly, A. (2011). From toxins targeting ligand-gated ion channels to therapeutic molecules, *Toxins*, **3**, pp. 260–293.
27. Bezanilla, F. (2005). Voltage-gated ion channels, *IEEE Trans. Nanobiosci.*, **4**, pp. 34–48.
28. Bezanilla, F. (2007). Chapter 3: Voltage-Gated Ion Channels, In *Biological Membrane Ion Channels: Dynamics, Structures, and Applications*, Chung, S.-O., Andersen, O. S., and Krishnamurthy, V., eds. (eBook, Springer), pp. 81–118.
29. Catterall, W. A. (1977). Activation of the action potential Na-ionophore by neurotoxin: An allosteric model, *J. Biol. Chem.*, **252**, pp. 8669–8676.
30. Martinac, B. (2012). Mechanosensitive ion channels: An evolutionary and scientific tour de force in mechanobiology, *Channels*, **6**, pp. 211–213.
31. Martinac, B. (2004). Mechanosensitive ion channels: Molecules of mechanotransduction, *J. Cell. Sci.*, **117**, pp. 2449–2460.
32. Blount, P., Li, Y., Moe, P., and Iscla, I. (2008). Chapter 3: Mechanosensitive Channels Gated by Membrane Tension, In *Mechanosensitive Ion Channels*, Kamkin, A., and Kiseleva, I., eds., Springer, New York, pp. 71–102.

33. Driscoll, M., and Chalfie, M. (1991). The MEC-4 gene is a member of a family of Caernorabditis elegans genes that can mutate to induce neuronal degeneration, *Nature*, **349**, pp. 588–593.
34. Hong, K., and Driscoll, M. (1994). A transmembrane domain of the putative channel subunit MEC-4 influences mechanotransduction and neurodegeneration in *C. elegans*, *Nature*, **367**, pp. 470–473.
35. Franco, A., and Lansman, J. B. (1990). Calcium entry through stretch-inactivated channels in mdx myotubes, *Nature*, **344**, pp. 670–673.
36. Chen, X. Z., Vassilev, P. M., Basora, N., Peng, J. B., Nomura, H., Segal, Y., Brown, E. M., Reeders, S. T., Hediger, M. A., and Zhou, J. (1999). Polycystin-L is a calcium-regulated cation channel permeable to calcium ions, *Nature*, **401**, pp. 383–386.
37. Franz, M. R., Cima, R., Wang, D., Profitt, D., and Kurz, R. (1992). Electrophysiological effects of myocardial stretch and mechanical determinants of stretch-activated arrhythmias, *Circulation*, **86**, pp. 968–978.
38. Kohler, R., Distler, A., and Hoyer, J. (1999). Increased mechanosensitive currents in aortic endothelial cells from genetically hypertensive rats, *J. Hypertension*, **17**, pp. 365–371.
39. Ismailov, I. I., and Benos, D. J. (1995). Effects of phosphorylation on ion channel function, *Kidney Internat.*, **48**, pp. 1167–1179.
40. Davis, M. J. D., Wu, X., Nurkiewicz, T. M., Kawasaki, J., Gui, P., Hill, M. A., and Wilson, E. (2001). Regulation of ion channels by protein tyrosine phosphorylation, *Am. J. Physiol.*, **281**, pp. H1835–H1862.
41. Kilfoil, P. J., Tipparaju, S. M., Barski, O. A., and Bhatnagar, A. (2013). Regulation of ion channels by pyridine nucleotides, *Circ. Res.*, **112**, pp. 721–741.
42. Kaupp, U. B., and Seifert, R. (2002). Cyclic nucleotide-gated ion channels, *Physiol. Rev.*, **82**, pp. 769–824.
43. Banghart, M. R., Volgraf, M., and Traumer, D. (2006). Engineering light-gated ion channels, *Biochemistry*, **45**, pp. 15130–15141.
44. Sperelakis, N., and Freedman J. C. (2012), Chapter 8: Diffusion and Permeability, In *Cell Physiology Source Book: Essentials of Membrane Biophysics*, 4th Ed., Nicholas Sperelakis, ed., Academic Press, New York, USA, pp. 113–120.
45. Silverthon, D. U. (2013). *Human Physiology: An Integrated Approach*, 4th Ed., Pearson Education Inc., USA.
46. Wahl, D. (2005). A short history of electrochemistry, *Galvanotechtnik*, **96**, pp. 1820–1828.

47. Goldman, D. E. (1943). Potential, impedance, and rectification in membranes, *J. Gen. Physiol.*, 27, pp. 37–60.
48. Koester, J., and Siegelbaum, S. A. (2000), Chapter 7: Membrane potential, In Kandel, E. R., Schwartz, J. H., and Jessel, T. M., eds., *Principles of Neural Sciences*, 4th Ed., McGraw-Hill, USA, pp. 125–139.
49. Wu, X. X., and Shuai, J. W. (2013). Multistability in a neuron model with extracellular potassium dynamics, *Phys. Rev. E. Stat. Nonlun. Soft Matter Phys.*, **85**, pp. 061911.
50. LaMantia, G., McNamara, O., and Williams, S. M. (2001) *Ion Channels of Excitable Membranes*, 3rd Ed., Sinauer Associates Inc., Sunderland, Massachusetts, USA.
51. Hodgkin, A. L. (1951). The ionic basis of electrical activity in nerve and muscle, *Biol. Rev.*, **26**, pp. 339–409.
52. Gordon, S. (1999). *Phagocytosis: The Host*, 1st Ed., Elsevier Science.
53. Urbé, S. (2005). Ubiquitin and endocytic protein sorting, *Essay Biochem.*, **41**, pp: 81–98.
54. Widmaier, P. E., Hershel, R., and Strang, K. T. (2008). *Vander's Human Physiology*, 11th Ed., McGraw-Hill.
55. Mineo, C., and Anderson, R. G. (2001). Potocytosis. Robert Feulgen Lecture, *Histochem. Cell Biol.*, **116**, pp. 109–118.
56. Haglund, K., and Dikic, I. (2012). The role of ubiquitylation in receptor endocytosis and endosomal sorting, *J. Cell Sci.*, **125**, pp. 266–275.
57. Stan, R. V. (2006). Endocytosis pathways in endothelium: How many? *Am. J. Physiol. Lung Cell Mol. Physiol.*, **290**, pp. L806–L808.
58. Keller, S., Sanderson, M. P., Stoeck, A., and Altevogt, P. (2006). Exosomes: From biogenesis and secretion to biological function, *Immunol. Lett.*, **107**, pp: 102–108.
59. Pelham, H. R. B. (1999). Snare and secretory pathway, *Exp. Res. Cell*, **247**, pp: 1–8.
60. Boron, W. F., and Boulpaep, E. L. (2009). *Medical Physiology: A Cellular and Molecular Approach*, Saunders/Elsevier, Philadelphia, USA.
61. Ivannikov, M., Sugimori, M., and Llinás, R. R. (2013). Synaptic vesicle exocytosis in hippocampal synaptosomes correlates directly with total mitochondrial volume, *J. Mol. Neurosci.*, **49**, pp. 223–230.

Chapter 3

Biology and Regulation of Protein Sorting and Vesicular Transport

Ian H. Mather
Department of Animal and Avian Sciences, University of Maryland, College Park, MD 20740, USA
imather@umd.edu

Eukaryotic cells maintain a complex internal structure in the face of continuous synthesis and turnover of endogenous constituents, uptake of nutrients, and secretion of diverse products. Such dynamic and interrelated activities require spatially separated compartments containing distinct complements of proteins and lipids, which are essential for their unique functions. Within compartments, many components are organized in functional complexes and interface with both cytoplasmic[a] and exoplasmic spaces. Copies of most membrane proteins have an identical topology, phospholipids are selectively enriched in one leaflet of the bilayer, and most glycolipids

[a]Throughout this chapter, the term "cytoplasmic" is used to denote the spaces and membrane surfaces within and facing the cytoplasm, and the term "exoplasmic" refers to all extracellular spaces and contiguous membrane surfaces, including the luminae of the endoplasmic reticulum, Golgi complex, transport and secretory vesicles, endosomes, and lysosomes.

Drug Delivery Across Physiological Barriers
Edited by Silvia Muro

ISBN 978-981-4669-40-5 (Hardcover), 978-981-4669-41-2 (eBook)
www.panstanford.com

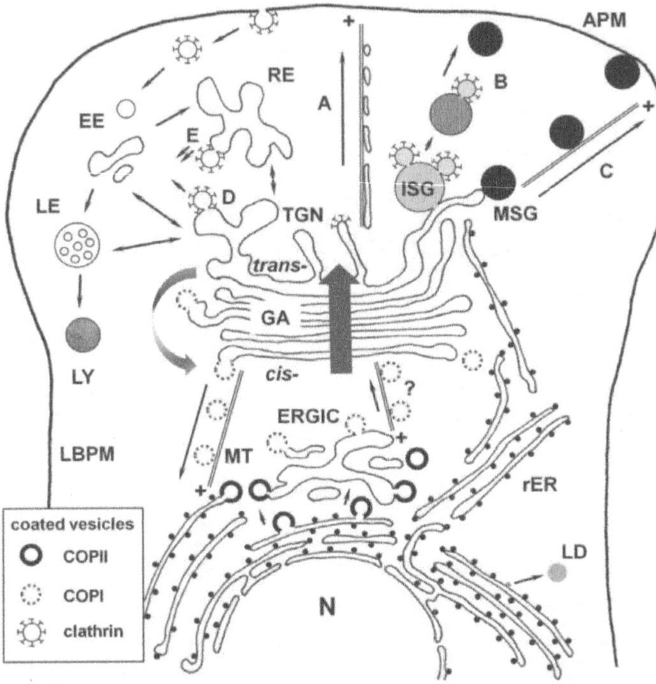

Figure 3.1 Major trafficking pathways. Secretory pathway—rER: rough endoplasmic reticulum; ERGIC: ER/Golgi intermediate compartment; GA: Golgi apparatus; TGN: trans-Golgi network; A: constitutive secretion (vesicles and tubules); B: secretory vesicles formed by post-TGN maturation; ISG: immature secretory granule; C: secretory vesicles formed in the TGN; MSG: mature secretory granule; APM: apical plasma membrane; LBPM: lateral/basal plasma membrane. **Endocytic pathway**—EE: early endosome; LE: late endosome; RE: recycling endosome; LY: lysosome; D: sorting of lysosomal proteins to endosomes (single arrow); E: sorting of PM proteins to LBPM (double arrow); LD: lipid droplet; N: nucleus; MT: microtubule.

face the exoplasmic space. Thus, the cell faces prodigious sorting and trafficking problems.

3.1 Major Traffic Routes

Notable traffic routes (Fig. 3.1) include pathways for the secretion of proteins—synthesized in the rough endoplasmic reticulum (ER), processed through the Golgi complex, and packaged into secretory vesicles (the "classical secretory pathway" [1])—and for the uptake, processing, recycling, and degradation of materials from the plasma membrane and extracellular milieu through endosomal compartments (the endocytic pathway [2]). These two major pathways are interconnected, for example by the transport of proteins through the secretory pathway to lysosomes via endosomes, or the bidirectional traffic of proteins and lipids from the trans-Golgi network (TGN) to early and late endosomes [3]. In epithelial and endothelial cells, materials may be transported across the cells in either direction by transcytosis. Other major pathways include bidirectional transport to and from the nucleus (Chapter 18) and mitochondria (Chapter 19).

In this brief overview, trafficking mechanisms, especially the formation and targeting of vesicles, will be discussed and the secretory pathway highlighted to provide examples of the conserved sorting machineries currently recognized. Within the context of this book, an understanding of the mechanisms underlying membrane traffic is essential if we are to develop optimal systems for the delivery of drugs to specific targets within eukaryotic cells.

3.2 Formation and Targeting of Vesicles

Several key steps in vesicular transport can be delineated (Fig. 3.2): budding and formation of a cargo-filled coated vesicle from a "donor" compartment; transport and targeting toward an "acceptor" compartment; tethering of the vesicle on the acceptor compartment; followed by uncoating, membrane fusion, and release of vesicle contents into the exoplasmic spaces of the acceptor compartment.

3.2.1 *Formation of Cargo-Filled Transport Vesicles*

Three classes of transport vesicles have been extensively characterized in eukaryotic cells: COPI and COPII vesicles, which function in traffic between the ER and Golgi complex, and clathrin-coated vesicles, which play multiple roles between the TGN, plasma membrane, and endosomes [4, 5]. In each case, vesicle formation

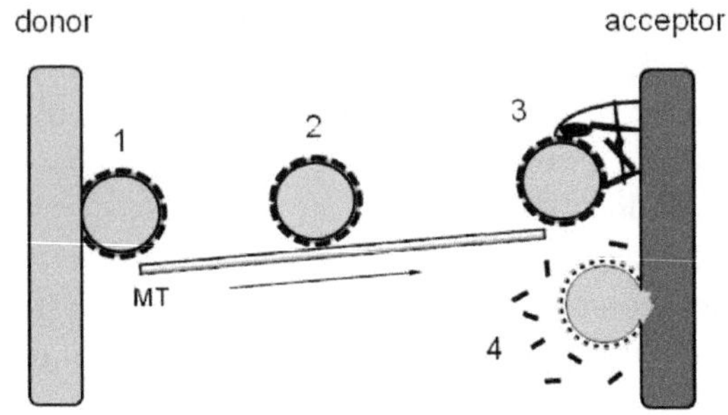

Figure 3.2 Vesicular transport: (1) formation of coated vesicle; (2) transport on microtubule (MT); (3) tethering; (4) uncoating and fusion.

requires initiating factors, usually GTPases, which recruit adaptor and coat proteins that impart curvature to the membrane to form a nascent bud. As the vesicle begins to form, the adaptors recruit cargo and receptors, which coassemble as an internal layer of protein linking the membrane to the outer coat (Fig. 3.2, step 1). Final release requires scission of the neck linking the vesicle bud to the donor membrane and removal of the coat proteins and adaptors.

3.2.2 *Vesicle Transport and Tethering*

Vesicle transport from the donor to the acceptor compartment typically requires physical guidance on elements of the cytoskeleton (Fig. 3.2, step 2). In most cases, transport is on microtubules and driven by motors in the kinesin and dynein families [6]. Kinesins drive vesicles away from the centrosome toward the periphery of the cell and the plus-end of microtubules. Traffic toward the center of the cell and the minus-end is driven by dynein. Also, vesicles may be transported by myosin motors on short actin filaments in the cell cortex [6], or in some cells for long distances through the cytoplasm, as in mouse oocytes [7]. Cytoskeletal-independent transport is possible, especially over short distances, most likely through the interaction of vesicles with matrix proteins and tethering factors to ensure targeting specificity.

Upon arrival at the acceptor compartment, vesicles are anchored to the membrane by scaffold proteins before the final docking and fusion steps (Fig. 3.2, step 3) [8]. Such so-called tethering factors comprise a large class of proteins that are recruited either singly or in hetero-oligomeric complexes to specific membranes by rab GTPases and other factors. At least eight multimeric complexes have been characterized from yeast and mammalian sources, including TRAPP1 (between the rough ER and Golgi complex) and exocyst (between secretory vesicles and the plasma membrane) [8]. Each complex assembles into a structural scaffold of globular and rod-shaped proteins, which may help to bind the vesicle to the acceptor compartment. Besides providing structural support, individual subunits may interact with regulatory proteins, such as rab GTPases, or themselves function *in situ* to regulate docking and fusion.

3.2.3 *Vesicle Fusion at the Acceptor Compartment*

Final fusion of the transport vesicle to the membrane of an acceptor compartment (Fig. 3.2, step 4) is effected by a complex of proteins in the SNAP receptor (SNARE) family [9], which are distributed on membranes and vesicles throughout the cell [10]. All SNAREs in the active state have extensive, unfolded cytoplasmic domains, each of which contains a so-called SNARE motif comprising heptad repeats of approximately 60–70 amino acids that form coiled coils [10]. Engagement of one SNARE motif on the vesicle with three SNARE motifs on the acceptor membrane triggers zippering from the N-termini inward to form a four-helix bundle. The core of the bundle comprises a series of 15 hydrophobic residues on either side of a central ionic layer, which is made up of three Q and one R residue, one each from the four component SNARE motifs. SNAREs are thus classified as R- or Q-SNAREs, with the Q-SNAREs further split into three subfamilies known as Qa, Qb, and Qc based on conserved sequences [10]. A cognate four-helix bundle is thus formed from one R-SNARE motif on the vesicle and one each Qa, Qb, and Qc SNARE motifs on the acceptor membrane. The folding reactions necessary to form the SNARE complexes provide the energy required to overcome the repulsive forces keeping the membranes apart. About 35 kBT of energy accumulates during the formation of one four-helix SNARE bundle, an amount that is similar to the 40–50 kBT required to form hemi-fused bilayers [11]. Therefore, the formation of two or three SNARE bundles should be sufficient to effect total fusion.

In liposome-based assays, many non-cognate SNARE combinations will form stable complexes, that is, the interactions are promiscuous. *In vivo*, the fidelity of R- and Q-SNARE recognition is modulated by the action of rab GTPases, tethering factors, and regulatory proteins in the Sec 1/Munc 18 (SM) family. These latter proteins are essential and function in temporal fashion as first inhibitors and then facilitators of fusion, but their exact mode of action remains unclear [12, 13].

Following membrane fusion, the SNARE complexes are disentangled and the R-SNAREs recycled to the donor compartment to enable further rounds of transport. Failure to do this rapidly leads to a traffic jam, as the number of available active R- and Q-SNAREs is depleted. Disentanglement of SNARE complexes is effected by the combined actions of N-ethylmaleimide-sensitive factor (NSF) [14] and soluble NSF attachment proteins (SNAPs) [15]. NSF is in the AAA+ family of ATPases, which typically form hexameric doughnut-shaped rings and remodel macromolecules through the hydrolysis of ATP. The mechanical force required for NSF action is transmitted by SNAPs, which serve as molecular ratchets that bind NSF to the entangled SNAREs and disassemble the SNARE bundles through ATP-stimulated changes in the conformation of NSF [15, 16].

The discovery of SNAREs established the molecular basis for the action of botulinum and tetanus toxins, which were shown to cleave specific SNARE sequences and thus block neurotransmission.

3.2.4 *Traffic Control by Rab GTPases*

The rab GTPases deserve special mention, as they are key regulators of membrane traffic [17]. Rabs are small (20–29 kDa) GTPases that associate with specific membranes and cytoskeletal elements throughout the cell and cycle between active and inactive GTP-/GDP-bound states [8, 17]. A cohort of accessory proteins regulate the function of each rab: notably guanine nucleotide exchange factors (GEFs), which exchange bound GDP for GTP, and guanine nucleotide activating proteins (GAPs), which stimulate GTP hydrolysis. Two domains (switch regions I and II) interact with the accessory proteins and downstream target proteins, which are known as rab "effectors."

Rabs, working through effector proteins, have been implicated in numerous traffic-related functions, including recycling of receptors, modulation of cytoskeletal-associated motors, assembly and function of tethering factors, and regulation of membrane fusion. On a more global scale, rabs may regulate the maturation of compartments by forming "rab domains" in which they are held in an active GTP-bound state by positive feedback loops. Such domains accumulate effectors, which recruit additional rabs that initiate maturation into a downstream compartment, for example as in the conversion of early to late endosomes by the recruitment of rab 5 followed by rab 7 [18].

3.2.5 *Vesicles or Tubules?*

The characterization of COPI-, COPII-, and clathrin-coated vesicles led to an initial consensus that membrane traffic is mediated by vesicles shuttling between specific compartments throughout the cell. Therefore, it was somewhat unexpected when the first live-cell imaging studies based on the trafficking of GFP/VSV-G protein chimeras revealed the pleomorphic character of traffic intermediates in both pre- and post-Golgi compartments. Abundant vesicles and vesiculo-tubular clusters that extended out as tubules were observed moving on microtubules from the ER to the Golgi complex [19, 20]. These pre-Golgi intermediates were labeled in temporal fashion with antibodies to COPII and COPI [19] in agreement with a sequential transport role for coated vesicle proteins. However, the pleomorphic nature of the carriers suggested other coat-independent mechanisms. Furthermore, transport from the Golgi complex to the plasma membrane was largely mediated by tubular structures, some of which broke into more spherical units on microtubular tracks between the TGN and the cell surface (Fig. 3.1A) [21, 22]. These transport intermediates did not appear to be regulated by any of the known coat proteins. Thus, membrane-bounded carriers come in many shapes and sizes and they are only partly regulated by currently identified coat structures.

The recognition that soluble and membrane-bound cargo is selectively sorted into specific subsets of coated vesicles supported the contention that there is little membrane mixing between intracellular compartments. However, the identification of tubular carriers, which are apparently uncoated, implied a less regulated nonselective transport of membrane and soluble constituents. So how does a compartment maintain a distinct complement of lipid and protein in the face of rapid bidirectional transport of constituents to other destinations? Are compartments stable entities or kinetically transient? These issues have been central questions in the trafficking field for decades [1] and are reiterated by examples at sorting and trafficking stages throughout the cell. The current consensus favors the concept of kinetically labile compartments. Many "resident" proteins and lipids turn over rapidly with the flux of membrane traffic. However, at steady state, relatively constant intra-compartmental concentrations are maintained, thus giving rise to the illusion that compartments are discrete and stable, with specific complements of proteins and lipids (i.e., "membrane markers").

3.3 Secretory Pathway

Most cells secrete proteins and other materials via the "classical secretory pathway" [1, 4]. Proteins destined for export from the cell are synthesized and processed in the rough ER, transported through the Golgi complex for further modification, packaged into secretory vesicles, and secreted by exocytosis from the plasma membrane. This pathway plays a central role in the turnover and maintenance of most cellular membranes and organelles and is discussed here to provide specific examples of sorting mechanisms. Furthermore, many trafficking mechanisms in the secretory pathway are reiterated in pathways encompassing endocytosis and transcytosis [19] and are, therefore, relevant to issues surrounding the delivery of drugs from external sources to intracellular targets.

3.3.1 *Endoplasmic Reticulum*

The ER is the principal site for the segregation of proteins between cytoplasmic and exoplasmic spaces and the synthesis and integration of proteins within and on either side of the membrane bilayer [23]. Nearly all transmembrane proteins originate in the ER with topologies that are maintained throughout their lifetimes and regardless of their final destinations. Also, the ER is the principal site for the synthesis of membrane lipids, neutral lipid droplets, and lipid precursors [24]. In actively secreting cells, the ER may constitute up to 60% of the total membrane; thus, it is, *ipso facto*, the primary membrane-generating organelle in the cell (Fig. 3.3).

In most instances, proteins are segregated between the cytoplasm and the ER when their respective mRNAs are translated in the cytoplasm. Proteins destined for the ER (i.e., most membrane and all exoplasmic proteins) are synthesized with an N-terminal leader sequence, which contains an essential central stretch of approximately 7–13 contiguous hydrophobic residues [25, 26]. Sorting is initiated when the nascent leader peptide emerges from the ribosome exit site and binds to the signal recognition particle (SRP) [27]. Binding slows translation and allows engagement of the mRNA-ribosome-SRP complex with the SRP receptor (SR) on the ER surface. Following release from the SRP and SR, the m-RNA-ribosome complex engages with Sec 61 or "translocon" (Fig. 3.3), a hetero-trimeric protein complex with a central pore between the cytoplasm and the ER exoplasm [28]. Translation then continues, and the protein is translocated through the pore to the ER lumen. Some proteins may be translocated after synthesis is completed in the cytoplasm [23, 27]. Such post-translational mechanisms require the concerted action of chaperones to maintain the protein in an unfolded state in the cytoplasm and the luminal chaperone, BiP, which pulls the completed peptide through the pore by an ATP-dependent ratchet-like mechanism. In the case of most membrane proteins, the hydrophobic anchor segments are incorporated into the ER bilayer via a lateral gate in the Sec 61 complex (Fig. 3.3, horizontal double arrows) [29]. Tail-anchored type II proteins ($N_{cyto}.C_{exo}$) with extensive cytoplasmic domains, such as SNAREs, are post-translationally targeted by a Sec 61-independent mechanism [30].

At some point during Sec 61-dependent translocation, the leader sequence is cleaved by the signal peptidase complex, a heteropentamer associated with the translocon [31]. Thus, proteins are targeted to the ER via cleavable leader sequences, which are inherent to the structure of the proteins. Proteins destined for the cytoplasm lack such sequences and, therefore, they do not bind to the SRP and translation of their respective mRNAs is completed in the cytoplasm. Leader sequences were a seminal discovery that presaged the recognition of many other signal-mediated targeting mechanisms at sites throughout the cell [25].

Apart from protein sorting and translocation, the ER is the first site for the covalent modification and processing of proteins. Asparagine-linked glycosylation is mediated by oligosaccharyl transferase, a complex of seven or eight proteins closely associated with the translocon [32]. Sugars are transferred en bloc as a 14-mer glycan ($Glc_3Man_9GlcNAc_2$) from a dolichol intermediate to N-residues in the nascent proteins, within the sequence context NXS/T (Fig. 3.3, right side, curved arrow). The initial covalently linked $Glc_3Man_9GlcNAc_2$ glycan is subsequently processed by a complex series of reactions in the ER and Golgi complex to generate the final high mannose, complex, or hybrid glycans associated with mature

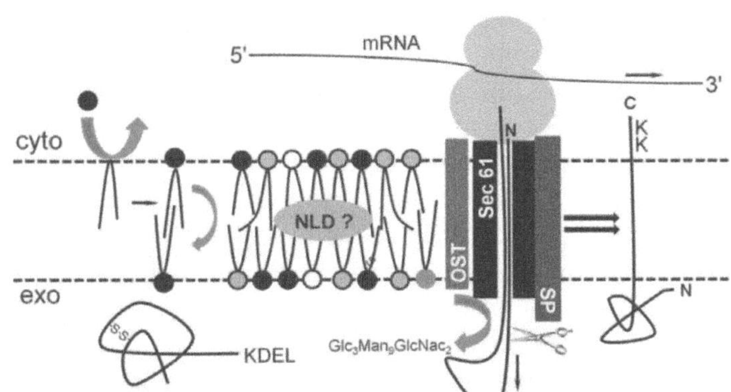

Figure 3.3 The endoplasmic reticulum is a very busy place. Left to right: Phospholipids are synthesized on the cytoplasmic side, and lipids equilibrate such that the bilayer is mostly symmetrical; nascent lipid droplets (NLDs) may form in the hydrophobic core of the bilayer; Sec 61 translocon with associated oligosaccharyl transferase (OST) and signal peptidase (SP) complex. Note that nascent proteins insert into the channel as a hairpin with their N-termini on the cytoplasmic side; ER residents with retention signals at the C-termini for a soluble protein (KDEL) and a type 1 integral protein (di-K) are shown.

N-linked glycoproteins [33]. The ER is also the initiation site for O-linked glycosylation, in which sugars are added in a stepwise fashion from sugar nucleotides to S- or T-residues.

As a consequence of the luminal disposition of the glycosylation machineries, only secretory proteins and the exoplasmic domains of membrane proteins synthesized in the ER contain N- or O-linked glycans. Since the topology of integral proteins is maintained after synthesis, their associated glycans remain exoplasmic irrespective of cellular location, including the plasma membrane. Similarly, the glycosyl phosphatidylinositol (GPI) anchors of "glypiated" proteins maintain an exoplasmic orientation after synthesis in the ER. O-linked glycoproteins containing GlcNAc residues may be synthesized in the cytoplasm by a separate mechanism [34].

During and after translocation, proteins begin to fold into their final conformations in the ER lumen with the assistance of many enzymes and chaperones [35]. Two families of enzymes are of special significance: (1) prolyl peptidyl cis-/trans-isomerases (PPTs) [35] and (2) thiol oxido-reductases (TORs) [35–38]. PPTs modulate cis-/trans-configurations around P-residues. This reaction is a rate-limiting step for proteins to adopt their native configurations, at least *in vitro*. TORs promote either the reshuffling of disulfide bonds by isomerization reactions [the protein disulfide isomerases (PDIs) [36]] or their net insertion [either PDIs in concert with the flavoprotein Ero1 [37] or the sulfhydryl oxidases [38]]. Disulfide bond formation requires an oxidizing environment, which in the ER lumen is manifested by a much higher ratio of oxidized-to-reduced glutathione than is encountered in the cytoplasm. High levels of reduced glutathione in the cytoplasm prevent the formation of disulfide bonds in nearly all cytoplasmic proteins [39]. How these redox differences across the ER membrane are actively maintained in mammalian cells is unclear. In yeast, oxidized glutathione in the cytoplasm is either rapidly reduced by glutathione reductase or actively sequestered in the vacuole by an ABC-C transporter [39].

Protein folding in both cytoplasmic and exoplasmic spaces is also dependent on a large number of chaperones. Chaperones in the ER lumen include proteins in the hsp 40, hsp 70 (including BiP), and hsp 90 families [35]. Many are ATPases with low turnover rates that bind to exposed hydrophobic regions. Folding is modulated by successive rounds of binding and release in the presence of ATP and activating and nucleotide exchange factors. A separate class of ER-resident chaperones, the calnexins and calreticulins, function as lectins, which bind to partially processed N-linked glycans containing one outer glucose residue ($GlcMan_9GlcNAc_2$). These lectins mediate the folding of nascent N-linked glycoproteins through repeated binding cycles, modulated by deglucosylation and reglucosylation reactions [35].

Under stress conditions, the quantity of chaperones and other molecules can be rapidly increased and the protein load in the ER decreased by three signaling pathways initiated in the ER, collectively known as the unfolded protein response [40]. Thus, misfolded proteins may be rescued through additional chaperone-mediated refolding reactions. However, if the native conformation is not achieved within a certain timeframe (under either unstressed or stressed conditions), the partially folded proteins are removed by retrotranslocation through ER channels to the cytoplasm. The nature of these channels has not been entirely resolved and several possibilities have been suggested based on either Sec 61 components, integral proteins in the derlin family, or E3 ubiquitin ligases [41]. Following translocation to the cytoplasm, the misfolded proteins are degraded in the proteosome (the ER-associated degradation pathway).

Besides the sorting and integration of proteins, the ER is also the principal generator of the phospholipid bilayer, through the synthesis of glycerophospholipids, cholesterol, ceramide, and, in some cells, galactosylceramide [24]. In molar terms, lipids are the most abundant membrane constituent, accounting for approximately 50 molecules to one "average" 60,000 Da protein. The most abundant ER lipid is phosphatidylcholine (PC), which is synthesized *in situ* by the transfer of the phosphocholine head group to diacylglycerol (Fig. 3.3). Other glycerophospholipids are made by analogous reactions or by base exchange. How the bilayer forms is an intriguing question, because glycerophospholipids and ceramide are synthesized on the cytoplasmic leaflet of the ER and the half-life for spontaneous movement from the cytoplasmic to exoplasmic leaflets for PC is several hours [42]. Transfer appears to require nonspecific poorly characterized transporters that "flip"

PC and other glycerophospholipids across the membrane, such that lipids in the ER are equally disposed between the cytoplasmic and exoplasmic leaflets (Fig. 3.3, left side, curved arrow) [24, 42]. Asymmetry between PC, PS, and PE is initiated in the Golgi complex and maintained at the plasma membrane by ATP-requiring aminophospholipid translocators, which transfer PS and PE from the exoplasmic to cytoplasmic leaflets [24, 42].

In addition to the generation of membrane lipids, the ER is the initial site for synthesis of neutral lipid droplets, which are believed to form within the hydrophobic core of the membrane and bud into the cytoplasm, coated with a monolayer of phospholipid derived from the cytoplasmic half of the ER bilayer (Fig. 3.3) [43].

3.3.2 *ER-Golgi Interface*

Materials exported from the ER include the proteins and lipids of most other compartments with the exception of some components of mitochondria and chloroplasts. Against this background of extensive anterograde traffic, sorting mechanisms ensure that the ER and Golgi complex maintain their unique complements of resident constituents.

The outward flow of materials to the Golgi complex is partly mediated by COPII vesicles, which form in specialized regions of the ER known as "ER exit sites." Such sites are marked by Sec 16, which serves as a scaffold protein that binds to all COPII subunits and may regulate the GAP activity associated with the coat [44]. Formation of COPII coats on the ER membrane is initiated by Sec 12, which is a resident GEF that activates the GTPase Sar 1 in the cytoplasm. Binding of GTP to Sar 1 exposes an amphipathic α-helix in the N-terminus, which inserts into the ER membrane [45]. Accommodation of this helix into the cytolasmic leaflet of the bilayer introduces a geometrical distortion and potentiates membrane curvature in concert with other proteins [46]. Sar 1 subsequently recruits the Sec 23-24 complex, which may bind to the membrane through a positively charged concave surface [4, 5, 45, 47]. At this point, cargo is recruited by Sec 24 (see below), and the coat is completed when the heterotetramer Sec 13-31 binds to the Sec 23-24-cargo complex to form an outer cage and

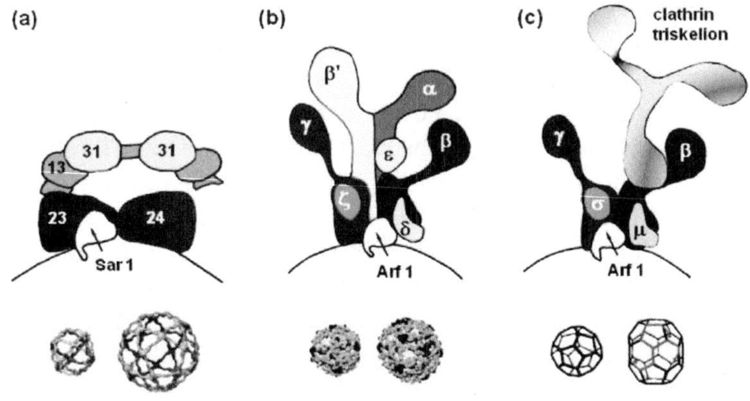

Figure 3.4 Structure of coated vesicles: (a) COPII, (b) COPI, (c) clathrin/AP1. Possible vesicles of different sizes are shown under each figure. Reprinted from Ref. 5, Copyright 2013, with permission from Elsevier.

imparts additional curvature to the membrane (Fig. 3.4a) [4, 5, 47, 48]. The assembled coat forms striking cuboctahedral and larger icosidodecahedral structures, which are flexible enough to accommodate regular sized cargo in 60–100 nm vesicles and much larger cargo, such as procollagen fibrils and prechylomicrons in mega-vesicles or tubules [5, 48].

Coat stability is partly regulated by maintaining Sar 1 in the membrane-associated GTP-bound form [49]. Hydrolysis of the bound GTP causes Sar 1 to disengage from the membrane, and thus the coat loses its initial membrane anchor. Activated Sar 1 is maintained on the membrane, both by the continued presence of Sec 12, which replaces GDP with GTP after each hydrolytic cycle and by the recruitment of cargo, which tends to slow down GTP hydrolysis [45, 50]. On the other hand, Sec 23 serves as a coat-associated GAP, which is further stimulated by Sec 31 and thus shortens Sar 1 association with the membrane [49]. An additional checkpoint is provided by Sec 16, which in association with Sec 24 inhibits GAP activity [44]. In the event that misfolded or incorrectly assembled cargo is incorporated into nascent COPII buds, GTP hydrolysis is stimulated by Sec 23 and the pre-budding complex of Sar 1, coat proteins and cargo disassociates. However, if correctly assembled cargo is recruited into the nascent coat complex, vesicle formation is presumed to continue to completion. Thus, Sar 1 not only initiates vesicle formation but also may serve as a timer of cargo recruitment that regulates formation of bona fide cargo-filled vesicles [50, 51]. At some point, coat proteins persist on the membrane, even in the absence of Sar 1 [50, 52], presumably through interactions that develop between Sec 23-24 and both the membrane bilayer and cargo as the vesicles mature.

Cargo is directly recruited to the forming vesicles by binding through short linear motifs, or folded epitopes, to one of at least four sites on the surface of Sec 24 [4, 53]. Such motifs are diverse, including, for example, di-acidic sequences in the cytoplasmic domains of integral proteins and di-hydrophobic or mixed hydrophobic/acidic motifs in many proteins, including SNAREs. Cargo may be indirectly sorted by binding to receptors that interact with Sec 24, such as the mannose-binding lectin ERGIC53, which functions in the export of some glycoproteins [54], p24/23 receptor complexes that are linked to the trafficking of GPI-anchored proteins [55], and Erv family members, which bind to both membrane-bound and soluble cargo [53]. Other proteins serve accessory roles, such as TANGO1/cTAG5, which facilitate the packaging of procollagen fibrils into enlarged COPII vesicles [56], and the ubiquitin ligase CUL3-KLHL12, which regulates the formation of such "mega-vesicles" through the ubiquitination of Sec 31 [57]. Formation of COPII vesicles is the first sorting step at which cargo is *concentrated* in the secretory pathway [53]. However, this does not exclude the possibility that proteins are incorporated into COPII vesicles by default [53, 58], especially soluble proteins that are secreted in large quantities from specialized exocrine cells.

The creation of different types of COPII vesicle is made possible by the expression of multiple paralogues of Sec 16 and several COPII components in different tissues. At least 64 kinds of COPII vesicles are theoretically feasible, since there are two paralogues each of Sar 1, Sec 16, Sec 23, and Sec 31, and four Sec 24 proteins in humans [45]. Tissue-specific roles for several such paralogues have been suggested, as in, for example, the role of Sec 23A in collagen secretion from chondrocytes [45].

Final release of completed vesicles from the ER requires scission of the neck of apposing membranes that forms between the emerging bud and ER membrane. Several mechanisms have been proposed, based on the ability of the amphipathic helix of Sar 1 to potentiate membrane curvature and thus raise the energy state in the curved neck of adjacent membranes. Subsequently, this heightened energy state is dissipated as the charged phospholipid bilayers fuse together and separate [5, 46]. GTP hydrolysis may [45] or may not [5, 59] be a requirement for the terminal scission step.

Cargo in COPII vesicles is first transported to the "ER-Golgi-intermediate compartment" (ERGIC), a complex of vesicles and tubules, which serves as the first post-ER sorting station in mammalian cells (Fig. 3.1). How this initial transport step is regulated is uncertain as microtubules may not be required [45]. Given the close proximity of the ER to the ERGIC, tethering factors and matrix proteins may be sufficient to serve as directional guides. At some point, the vesicles have to lose their coats to allow fusion with the acceptor compartment. Uncoating was originally thought to occur during or soon after vesicle release from the donor compartment [4, 8]. However, coat proteins are now known to bind to tethering factors associated with *acceptor* membranes [8, 45, 60]. COPII vesicles retain some coat proteins until they reach the ERGIC in mammalian cells or the cis-Golgi in yeast, at which point Sec 23 in the COPII coat binds to the tethering factor TRAPP1 [45, 60]. How COPII coats are subsequently dismantled after tethering has not been resolved.

Fusion of uncoated COPII vesicles with the ERGIC in mammalian cells allows initial sorting of cargo and constituent vesicle proteins. How the membranes of ERGIC are maintained and turned over is a challenging question. Membrane may be replenished by SNARE-mediated homotypic fusion of COPII vesicles from the ER and by heterotypic fusion of COPI vesicles from cis-Golgi [61]. As discussed in Section 3.2.5, forward transport from the ERGIC to the cis-Golgi is mediated by pleomorphic carriers, which may be coated with COPI and driven by dynein/dynactin motors on microtubules (Fig. 3.1) [19, 20, 62].

All of this forward traffic is balanced by the retrieval of ER-resident proteins and essential components of the trafficking machinery from multiple sites, including the ERGIC and cis-Golgi cisternae. Such retrograde transport is mediated by COPI vesicles on microtubules and is driven by kinesin motors [63] (Fig. 3.1). COPI-coated vesicles assemble on membranes by broadly similar mechanisms to COPII vesicles at ER exit sites [5]. Vesicle formation is initiated when ARF1, a GTPase in the ARF family, is activated by a GEF in the cytoplasm [64]. Like Sar 1, activation of ARF1 exposes an amphipathic helix in the N-terminus and thus promotes association with and initial curvature of target membranes [5, 65]. ARF1 initiates vesicle formation by binding to subunits of the COPI coat, a complex of seven proteins—α-, β-, β'-, γ-, δ-, ε-, and ζ—which are collectively known as coatomer (Fig. 3.4b). Unlike COPII coat subunits, coatomer is preassembled in the cytoplasm as a 600 kDa oligomer that associates en bloc with activated ARF1 [5]. The assembled coat comprises two subcomplexes: the F-subcomplex composed of the β-, γ-, δ-, and ζ-subunits and the B-subcomplex composed of the α-, β'-, and ε-subunits (Fig. 3.4b) [5, 66]. Components of the F-subcomplex share similarities with the clathrin AP adaptors (Fig. 3.4b,c), whereas the B-subcomplex, clathrin, and Sec 13-31 of COPII vesicles comprise α-solenoid and β-propeller domains in different configurations, which assemble into cages of distinctly different structure (Fig. 3.4) [5, 66, 67].

Formation of coatomer-coated buds leads to the incorporation of cargo (see below) and the recruitment of a GAP (ARFGAP1) from the cytoplasm. Unlike COPII coats, none of the coatomer subunits are GAPs or have GAP-stimulating activity. The activity of ARFGAP1 is stimulated when a central unstructured domain in the protein folds into an amphipathic helix and inserts into deformed (curved) membrane surfaces [68]. GAP activity increases as membrane curvature increases. Thus, ARFGAP1 serves as a sensor of membrane curvature and times the hydrolysis of GTP bound to ARF1 as the membrane bends to form a vesicle and the outer protein coat becomes established. Final release of the completed COPI vesicle is presumably potentiated by the membrane-embedded ARF1 helix and coatomer subunits, in direct analogy with Sar 1 and COPII coat proteins [5]. Also like COPII vesicles, it is not clear whether GTP hydrolysis is required to elicit membrane scission [5, 59]. How COPI coats disassemble has not been resolved.

Cargo incorporated into COPI vesicles includes soluble and membrane-bound proteins that are to be recycled to the ER from the ERGIC and Golgi complex. Soluble proteins are retrieved by a receptor KDEL that binds to the sequence, which is conserved at the C-terminus of nearly all luminal ER proteins in mammals [69]. Binding to the KDEL receptor thus ensures the retrieval of soluble ER-resident proteins that have "escaped" to later compartments. ER type I membrane proteins ($N_{exo}.C_{cyto}$) are retrieved through the interaction of di-K residues in their cytoplasmic termini with the α- or β'-subunits of the COPI coat [70, 71] (type II proteins are rare in the ER). The retrieval of multi-spanning type III proteins depends on which ends are available for interaction with COPI components in the cytoplasm. In the case of the KDEL receptor, which has seven membrane spans, an atypical di-K motif and phosphorylation of an S-residue in the C-terminus are required [72].

In an alternative mechanism, proteins may be retrieved through interaction of their membrane anchors with Rer1p, a multi-spanning integral protein that retains Sec 12 (the Sar 1 GEF), among other proteins, in the ER [53, 73]. Besides retrograde retrieval, some proteins may never leave the ER, that is, they are actively excluded from anterograde COPII vesicles by poorly defined mechanisms.

The dynamic relationship between the ER, ERGIC, and Golgi complex was experimentally shown by using the antibiotic brefeldin A to inhibit ARF1 GEF [74], thus preventing the formation of COPI vesicles [63]. In the absence of COPI vesicles, Golgi membranes extended as uncoated tubulo-vesicular processes along the length of the microtubules between the Golgi complex and ER [63]. Within minutes, cisternae from the cis- and medial-Golgi complex were subsumed into the ER to form a mixed ER/Golgi compartment. Remarkably, the Golgi complex was rapidly reconstituted on wash out of the inhibitor. Thus, the entire molecular machinery for construction of the Golgi complex can be sorted from ER components within minutes of reestablishing bidirectional traffic between the two compartments. Most significantly, these experiments revealed the kinetically transient nature of the Golgi complex; a compartment previously assumed to be discrete and permanent.

3.3.3 *Traffic through Golgi Complex*

The Golgi complex typically comprises a stack of four to seven closely apposed membrane-bounded cisternae and vesicles that function as a central processing station in most eukaryotic cells [75]. Proteins and lipids from the ER and ERGIC enter the cis-most cisterna and are progressively sorted and modified during transit through the stack. N-linked glycans are trimmed of outer-branch mannose residues and processed into high mannose, hybrid, and complex forms [33]; O-linked glycans are extended, and sphingolipids and glycosphingolipids are completed from ceramide precursors [24]. Enzymes required for these activities are distributed in overlapping gradients across the stack that are kinetically maintained in the face of massive anterograde secretory traffic.

How the Golgi complex retains its structure has proved to be a challenging question. In the two most widely promoted models, the cisternae either remain as distinct stable compartments [76] ("stable cisternal model") or they mature and progress through the stack ("cisternal progression model") [77]. Forward moving cargo in the stable cisternal model is mediated by COPI vesicles, direct connections between cisternae, or other means and in the cisternal progression model by the maturing cisternae themselves. In either case, the anterograde traffic of vesicles or cisternae is compensated by retrograde retrieval of membrane in the form of COPI vesicles or other intermediates (Fig. 3.1, curved arrow). The cisternal progression model has been favored in recent years because it can be directly monitored in yeast by live-cell imaging [77] and it provides a mechanism for the transport of large cargo molecules, such as procollagen fibrils, which can be accommodated within the cisternae regardless of size. However, definitive live-cell imaging is not feasible with mammalian Golgi complexes and so doubts remain, despite the recent application of some elegant experimental approaches [78].

Retention of membrane-bound enzymes in the Golgi complex is partly regulated by the length of their hydrophobic anchors. Proteins in the ER and Golgi complex have transmembrane domains that are four amino-acid residues shorter on average than integral proteins in the TGN and plasma membrane [79]. These differences in anchor length correlate with the relative dimensions of membranes in the secretory pathway, which increase in bilayer thickness from the ER through the Golgi complex and reach a maximum in the plasma membrane because of increasing cholesterol and sphingolipid concentrations [79, 80]. TGN and plasma membrane proteins are more readily accommodated in thicker cholesterol/sphingolipid-rich post-Golgi membranes and are sorted away from Golgi proteins. Thus, proteins tend to segregate into their thermodynamically most favorable lipid environments.

3.3.4 *Sorting in TGN*

Proteins and lipids processed through the Golgi stack enter the TGN [81], a pleomorphic network of vesicles and tubules, which arise from the last two to three trans-cisternae of the Golgi stack and may directly interface with elements of the rough ER (Fig. 3.1) [3, 22, 75]. TGN capacity expands or contracts in volume depending on the amount of material arriving from the Golgi complex or consumed by anterograde traffic to post-TGN destinations. Cells with extensive endosomal pathways tend to have an abundance of TGN membranes, whereas cells dedicated to regulated secretion may have very little [3].

The trans-Golgi and TGN are sites for the terminal addition of sialic acid and fucose to N- and O-linked proteins, proteolytic cleavage of prohormones, and sulfation of glycosaminoglycans [3]. Sphingomyelin and glycosphingolipids are completed within the Golgi/TGN lumen [24] and segregate with cholesterol through hydrogen bonding. In addition, the compact planar cholesterol molecules undergo phase separation with the relatively long, straight cis-unsaturated sphingosine bases associated with sphingolipids and with lipids containing straight saturated acyl chains. As discussed below, such "lipid rafts" may function to sort cargo for export from the TGN [80].

The TGN stands at the crossroads of the secretory, endocytic, and lysosomal pathways and serves as the central sorting station for post-Golgi compartments. As discussed in Section 3.2.5, much post-TGN cargo is initially sorted into tubular-like structures, which extend out into the cytoplasm attached to elements of the cytoskeleton (Fig. 3.1) [3, 21, 22, 82]. Tubulation may be induced by multiple mechanisms, including the modification of membrane lipids by lipases, or by ARF1-mediated membrane bending, followed by nucleation of adaptor proteins and formation of clathrin coats. F-actin and specific myosins may impart tension to the initial bud, which is then extended into a tubule through the pulling action of specific motors on microtubules [3, 82]. Subsequent severing of the membranes is completed by a number of protein complexes, including the dynamins or BAR domain proteins. Clathrin-coated vesicles play a major role in sorting and are seen to bud from multiple sites in the TGN, including the tubular extensions. Clathrin consists of heavy and light chains that assemble into polyhedral lattices of hexagons and pentagons (Fig. 3.4c) [83]. Three heavy and three light chains form a triskelion, which functions as a preassembled coat precursor in the cytoplasm. As in the case of COPI and COPII vesicles, formation of clathrin coats is initiated by nucleation factors (ARF1 or phosphoinositides) and adaptor and accessory proteins that coassemble with cargo and coat proteins.

The first adaptors to be characterized were the structurally related heterotetrameric adaptor protein (AP) complexes, some of which bind to clathrin and some that function independently. A total of five human AP complexes have been characterized (AP1-5), which are distributed between specific post-Golgi compartments [67, 84]. Each adaptor complex comprises two 100 kDa subunits (one, denoted either, α-, γ-, δ-, ε-, or ζ-, and the other β-1-5), one 50 kDa subunit (μ1-5), and one 20 kDa subunit (σ1-5) (AP1 example, Fig. 3.4c). Several isoforms of individual subunits have been identified (denoted A, B, etc) that increase the possible number of AP subtypes. Both AP1 and AP2 (and possibly AP3) bind to clathrin and function as linker proteins between nucleation sites on the membrane surface and the outer clathrin coat. In the case of AP1, AP3, and AP4, coat formation is initiated by ARF1, which recruits the AP complex and cargo as well as imparts membrane curvature in an analogous fashion to the formation of COPI vesicles [5]. AP2 is recruited to the plasma membrane by phosphatidylinositol-4,5-bisphosphate and functions in endocytosis [67, 84]. Neither AP4 nor AP5 binds clathrin, but AP4 at least appears to form bona fide cargo vesicles [67, 85]. Other well-characterized clathrin adaptors comprise the monomeric GGA proteins, which are recruited to membranes by activated ARF [86].

AP1 and GGAs, together with the clathrin-independent AP4, mediate vesicular traffic between the TGN and various endosomal compartments (for an updated map see Ref. [67]). Cargo molecules are recruited through Y- or di-L motifs in their cytoplasmic tails, either of which binds to the AP-μ subunits or, in the case of di-L signals, to the VHS domain of the GGAs [84, 86, 87]. In some cases, traffic is indirect, as in the transport of lysosomal proteins in AP1 or GGA vesicles to endosomes, followed by resorting for delivery to lysosomes (Fig. 3.1D, single arrow) [88]. Also, some proteins from the TGN and destined for the basal/lateral plasma membrane are resorted into AP-μ1B vesicles in endosomes (Fig. 3.1E, double arrow) for delivery to the cell surface [3, 89].

At least 20 different accessory proteins bind to clathrin, AP complexes, or GGAs and recruit specific sets of cargo molecule. Examples include epsin 1, which recruits the EGF receptor at the plasma membrane [90], and CALM, which retrieves VAMP R-SNAREs from post-Golgi compartments [91]. This impressive diversity of adaptors and accessory proteins enables great flexibility for the formation of tailor-made vesicles (with or without clathrin) for the transport of selected cargo molecules to multiple destinations.

Sorting in the Golgi complex and TGN is also mediated by the association of proteins with lipid rafts [80]. Integral glycoproteins and GPI-anchored proteins may be incorporated into separate vesicles enriched in sphingolipids and cholesterol for delivery to the apical surface [3, 80, 92]. In addition, glycoproteins and glycolipids may be sorted into apically directed vesicles by binding through their associated glycans to lectins in the galectin family [93, 94].

Regulated secretory proteins are first sorted by aggregation into clusters induced by a number of means, including specific sorting motifs in the structures of the proteins, the acidic and calcium-rich milieu of the TGN exoplasm, interaction with membrane-bound forms of the sorted secretory proteins, or by proteolytic processing [95, 96]. Vesicle formation may be completed in the TGN and the granule released as a mature secretory vesicle (Fig. 3.1C) [95]. Alternatively, immature vesicles may be released into the cytoplasm for further processing (Fig. 3.1B) and become larger by SNARE-induced homotypic fusions [95, 96]. Final conversion to a mature secretory granule is effected by the budding of AP1-containing clathrin-coated vesicles, which remove excess membrane and soluble proteins (Fig. 3.1B) [97]. Thus, cargo is concentrated and the granule eventually reduced in size. In many cells, the contents of the clathrin-coated vesicles are secreted separately or sorted to endosomes by a so-called "constitutive-like secretion"[95]. Regardless of the site of maturation, regulated granules are stored in the cytoplasm until a physiological signal stimulates release of the vesicle contents by exocytosis.

3.3.5 *Exocytosis*

Exocytosis of the contents of either regulated or constitutive secretory vesicles at the plasma membrane is the final step in the pathway for the release of proteins and other molecules into the extracellular milieu. The mechanism is fundamentally the same regardless of species and cell type and is based on the fusion of two bilayer membranes through the formation of SNARE complexes, as discussed in Section 3.2.3. Fusion is regulated by rabs and SM proteins and is often triggered by increases in calcium or cAMP, and phosphorylation events [12, 98, 99]. The R-SNAREs on the vesicle membrane are typically type II proteins in the VAMP family, and the Q-SNAREs on the plasma membrane comprise a type II syntaxin and an acylated SNAP containing two SNARE motifs.

Although the core mechanism is universal, there are many differences in detail, depending on cell type and physiological requirements. Most strikingly, the rate of exocytosis can vary widely from less than 1 ms in the synaptic cleft to 10–100 s in gastric parietal cells [99]. Multiple factors contribute to these differences, including the nature of external signals, the position of the vesicles in relation to the plasma membrane, the nature of the cytoplasmic matrix surrounding the vesicle, the size of the vesicle, and the specific SNAREs and cohort of regulatory proteins required [98,

99]. The folding state of the SNARE complex is a major critical factor [99]. In the most open configuration, the vesicle is not bound to the membrane, the R-SNARE and SNAP Q-SNAREs are unfolded, and the syntaxin Q-SNARE is bound to an SM protein (exocytosis in 10 to >100 s). In the most tightly folded state, all three SNAREs and SM protein are bound in a complex and the vesicle is docked at the plasma membrane (exocytosis in less than 1 ms). Such rapid secretion is best exemplified by synaptic vesicles [13]. Final membrane fusion is prevented by complexin, a soluble protein, which binds between the interfaces of the R- and Q-SNAREs. On stimulation of the nerve, rising calcium levels activate the integral protein, synaptotagmin, which frees complexin from the complex, thus promoting final SNARE assembly, membrane fusion, and neurotransmitter release [13].

The fate of vesicle membrane after the release of vesicle contents varies widely depending on the secretory system. Membrane may be completely incorporated into the cell surface, retrieved by endocytosis after some membrane mixing, or recovered by the so-called "kiss-and-run" mechanism, in which the entire vesicle is recycled into the cell [98].

3.4 Drug Delivery and Membrane Traffic

Attempts to deliver drugs to specific intracellular targets have to take into account the sheer complexity of membrane traffic, the number of regulatory proteins and lipids required, and the sequential nature of specific pathways. A single block in transport within or between compartments can have multiple long-range consequences, as exemplified by the collapse of Golgi structure when ARF1 GEF is inhibited by brefeldin A [63]. Furthermore, the delivery of many drugs to inappropriate sites may initiate the unfolded protein response leading to the overproduction of chaperones and, in extremis, apoptosis [40]. Thus, the development of optimal drug delivery systems will require careful attention to cellular entry routes, the underlying sorting machineries, and potential secondary complications.

References

1. Palade, G. E. (1975). Intracellular aspects of the process of protein secretion, *Science*, **189**, pp. 347–358.
2. Grant, B. D., and Donaldson, J. G. (2009). Pathways and mechanisms of endocytic recycling, *Nat. Rev. Mol. Cell Biol.*, **10**, pp. 507–608.
3. De Matties, M. A., and Luini, A. (2008). Exiting the Golgi complex, *Nat. Rev. Mol. Cell Biol.*, **9**, pp. 273–284.
4. Bonifacino, J. S., and Glick, B. S. (2004). The mechanism of vesicle budding and fusion, *Cell*, **116**, pp. 153–166.
5. Faini, M., Beck, R., Wieland, F. T., and Briggs, J. A. G. (2013). Vesicle coats: Structure, function, and general principles of assembly, *Trends Cell Biol.*, **23**, pp. 279–288.
6. Ross, J. L., Ali, M. Y., and Warshaw, D. M. (2008). Cargo transport: Molecular motors navigate a complex cytoskeleton, *Curr. Opin. Cell Biol.*, **20**, pp. 41–47.
7. Schuh, M. (2011). An actin-dependent mechanism for long-range vesicle transport, *Nat. Cell Biol.*, **13**, pp. 1431–1436.
8. Cai, H., Reinisch, K., and Ferro-Novick, S. (2007). Coats, tethers, rabs, and SNAREs work together to mediate the intracellular destination of a transport vesicle, *Develop. Cell*, **12**, pp. 671–682.
9. Sollner, T., Whiteheart, S. W., Brunner, M., Erdjument-Bromage, H., Geromanos, S., Tempst, P., and Rothman, J. E. (1993). SNAP receptors implicated in vesicle targeting and fusion, *Nature*, **362**, pp. 318–324.
10. Kloepper, T. H., Kienle, C. N., and Fasshauer, D. (2007). An elaborate classification of SNARE proteins sheds light on the conservation of the eukaryotic endomembrane system, *Mol. Biol. Cell*, **18**, pp. 3463–3471.
11. Li, F., Pincet, F., Perez, E., Eng, W. S., Melia, T. J., Rothman, J. E., and Tareste, D. (2007). Energetics and dynamics of SNAREpin folding across lipid bilayers, *Nat. Struct. Mol. Biol.*, **14**, pp. 890–896.
12. Rizo, J., and Sudhof, T. C. (2012). The membrane fusion enigma: SNAREs, Sec1/Munc18 proteins, and their accomplices: guilty as charged? *Annu. Rev. Cell Develop. Biol.*, **28**, pp. 279–308.
13. Sudhof, T. C., and Rothman, J. E. (2009). Membrane fusion: Grappling with SNARE and SM proteins, *Science*, **323**, pp. 474–477.
14. Lenzen, C. U., Steinmann, D., Whiteheart, S. W., and Weis, W. I. (1998). Crystal structure of the hexamerization domain of N-ethylmaleimide-sensitive fusion protein, *Cell*, **94**, pp. 525–536.

15. Rice, L. M., and Brunger, A. T. (1999). Crystal structure of the vesicular transport protein Sec17: Implications for SNAP function in SNARE complex disassembly, *Mol. Cell*, **4**, pp. 85–95.
16. Hanson, P. I., Roth, R., Morisaki, H., Jahn, R., and Heuser, J. E. (1997). Structure and conformational changes in NSF and its membrane receptor complexes visualized by quick-freeze/deep-etch electron microscopy, *Cell*, **90**, pp. 523–535.
17. Hutagalung, A. H., and Novick, P. J. (2011). Role of rab GTPases in membrane traffic and cell physiology, *Physiol. Rev.*, **91**, pp. 119–149.
18. Rink, J., Ghigo, E., Kalaidzidis, Y., and Zerial, M. (2005). Rab conversion as a mechanism of progression from early to late endosomes, *Cell*, **122**, pp. 735–749.
19. Scales, S. J., Pepperkok, R., and Kreis, T. E. (1997). Visualization of ER-to-Golgi transport in living cells reveals a sequential mode of action for COPII and COPI, *Cell*, **90**, pp. 1137–1148.
20. Presley, J. F., Cole, N. B., Schroer, T. A., Hirschberg, K., Zaal, K. J. M., and Lippincott-Schwartz, J. (1997). ER-to-Golgi transport visualized in living cells, *Nature*, **389**, pp. 81–85.
21. Hirschberg, K., Miller, C. M., Ellenberg, J., Presley, J. F., Siggia, E. D., Phair, R. D., and Lippincott-Schwartz, J. (1998). Kinetic analysis of secretory protein traffic and characterization of Golgi to plasma membrane transport intermediates in living cells, *J. Cell Biol.*, **143**, pp. 1485–1503.
22. Polishchuk, R. S., Polishchuk, E. V., Marra, P., Alberti, S., Buccione, R., Luini, A., and Mironov, A. A. (2000). Correlative light-electron microscopy reveals the tubular-saccular ultrastructure of carriers operating between Golgi apparatus and plasma membrane, *J. Cell Biol.*, **148**, pp. 45–58.
23. Park, E., and Rapoport, T. A. (2012). Mechanisms of Sec61/SecY-mediated protein translocation across membranes, *Annu. Rev. Biophys.*, **41**, pp. 21–40.
24. van Meer, G., Voelker, D. R., and Feigenson, G. W. (2008). Membrane lipids: Where they are and how they behave, *Nat. Rev. Mol. Cell Biol.*, **9**, pp. 112–124.
25. Blobel, G., and Dobberstein, B. (1975). Transfer of proteins across membranes. 1. Presence of proteolytically processed and unprocessed nascent immunoglobulin light-chains on membrane-bound ribosomes of murine myeloma, *J. Cell Biol.*, **67**, pp. 835–851.
26. Hegde, R. S., and Bernstein, H. D. (2006). The surprising complexity of signal sequences, *Trends Biochem. Sci.*, **31**, pp. 563–571.

27. Mandon, E. C., Trueman, S. F., and Gilmore, R. (2013). Protein translocation across the rough endoplasmic reticulum, *Cold Spring Harb. Perspect. Biol.*, pp. a013342.
28. van den Berg, B., Clemons, W. M., Collinson, I., Modis, Y., Hartmann, E., Harrison, S. C., and Rapoport, T. A. (2004). X-ray structure of a protein-conducting channel, *Nature*, **427**, pp. 36–44.
29. Frauenfeld, J., Gumbart, J., van der Sluis, E. O., Funes, S., Gartmann, M., Beatrix, B., Mielke, T., Berninghausen, O., Becker, T., Schulten, K., and Beckmann, R. (2011). Cryo-EM structure of the ribosome-SecYE complex in the membrane environment, *Nat. Struct. Mol. Biol.*, **18**, pp. 614–621.
30. Hegde, R. S., and Keenan, R. J. (2011). Tail-anchored membrane protein insertion into the endoplasmic reticulum, *Nat. Rev. Mol. Cell Biol.*, **12**, pp. 787–798.
31. Kalies, K. U., Rapoport, T. A., and Hartmann, E. (1998). The beta subunit of the Sec61 complex facilitates cotranslational protein transport and interacts with the signal peptidase during translocation, *J. Cell Biol.*, **141**, pp. 887–894.
32. Kelleher, D. J., and Gilmore, R. (2006). An evolving view of the eukaryotic oligosaccharyltransferase, *Glycobiology*, **16**, pp. 47R–62R.
33. Kornfeld, R., and Kornfeld, S. (1985). Assembly of asparagine-linked oligosaccharides, *Annu. Rev. Biochem.*, **54**, pp. 631–664.
34. Wells, L., and Hart, G. W. (2003). O-GlcNAc turns twenty: Functional implications for post-translational modification of nuclear and cytosolic proteins with a sugar, *FEBS Lett.*, **546**, pp. 154–158.
35. Hebert, D. N., and Molinari, M. (2007). In and out of the ER: Protein folding, quality control, degradation, and related human diseases, *Physiol. Rev.*, **87**, pp. 1377–1408.
36. Hatahet, F., and Ruddock, L. W. (2009). Protein disulfide isomerase: A critical evaluation of its function in disulfide bond formation, *Antioxid. Redox Signal.*, **11**, pp. 2807–2850.
37. Riemer, J., Bulleid, N., and Herrmann, J. M. (2009). Disulfide formation in the ER and mitochondria: Two solutions to a common process, *Science*, **324**, pp. 1284–1287.
38. Kodali, V. K., and Thorpe, C. (2010). Oxidative protein folding and the quiescin-sulfhydryl oxidase family of flavoproteins, *Antioxid. Redox Signal.*, **13**, pp. 1217–1230.
39. Morgan, B., Ezerina, D., Amoako, T. N. E., Riemer, J., Seedorf, M., and Dick, T. P. (2013). Multiple glutathione disulfide removal pathways mediate cytosolic redox homeostasis, *Nat. Chem. Biol.*, **9**, pp. 119–125.

40. Walter, P., and Ron, D. (2011). The unfolded protein response: From stress pathway to homeostatic regulation, *Science*, **334**, pp. 1081–1086.
41. Hampton, R. Y., and Sommer, T. (2012). Finding the will and the way of ERAD substrate retrotranslocation, *Curr. Opin. Cell Biol.*, **24**, pp. 460–466.
42. Kol, M. A., de Kroon, A. I. P. M., Killian, J. A., and de Kruijff, B. (2004). Transbilayer movement of phospholipids in biogenic membranes, *Biochemistry*, **43**, pp. 2673–2681.
43. Martin, S., and Parton, R. G. (2006). Lipid droplets: A unified view of a dynamic organelle, *Nat. Rev. Mol. Cell Biol.*, **7**, pp. 373–378.
44. Kung, L. F., Pagant, S., Futai, E., D'Arcangelo, J. G., Buchanan, R., Dittmar, J. C., Reid, R. J. D., Rothstein, R., Hamamoto, S., Snapp, E. L., Schekman, R., and Miller, E. A. (2012). Sec24p and Sec16p cooperate to regulate the GTP cycle of the COPII coat, *EMBO J.*, **31**, pp. 1014–1027.
45. Zanetti, G., Pahuja, K. B., Studer, S., Shim, S., and Schekman, R. (2012). COPII and the regulation of protein sorting in mammals, *Nat. Cell Biol.*, **14**, pp. 20–28.
46. Lee, M. C. S., Orci, L., Hamamoto, S., Futai, E., Ravazzola, M., and Schekman, R. (2005). Sar1p N-terminal helix initiates membrane curvature and completes the fission of a COPII vesicle, *Cell*, **122**, pp. 605–617.
47. Lederkremer, G. Z., Cheng, Y., Petre, B. M., Vogan, E., Springer, S., Schekman, R., Walz, T., and Kirchhausen, T. (2001). Structure of the Sec23p/24p and Sec13p/31p complexes of COPII, *Proc. Natl. Acad. Sci. U.S.A.*, **98**, pp. 10704–10709.
48. Stagg, S. M., LaPointe, P., Razvi, A., Gurkan, C., Potter, C. S., Carragher, B., and Balch, W. E. (2008). Structural basis for cargo regulation of COPII coat assembly, *Cell*, **134**, pp. 474–484.
49. Antonny, B., Madden, D., Hamamoto, S., Orci, L., and Schekman, R. (2001). Dynamics of the COPII coat with GTP and stable analogues, *Nat. Cell Biol.*, **3**, pp. 531–537.
50. Sato, K., and Nakano, A. (2005). Dissection of COPII subunit-cargo assembly and disassembly kinetics during Sar1p-GTP hydrolysis, *Nat. Struct. Mol. Biol.*, **12**, pp. 167–174.
51. Sato, K., and Nakano, A. (2004). Reconstitution of coat protein complex II (COPII) vesicle formation from cargo-reconstituted proteoliposomes reveals the potential role of GTP hydrolysis by Sar1p in protein sorting, *J. Biol. Chem.*, **279**, pp. 1330–1335.

52. Forster, R., Weiss, M., Zimmermann, T., Reynaud, E. G., Verissimo, F., Stephens, D. J., and Pepperkok, R. (2006). Secretory cargo regulates the turnover of COPII subunits at single ER exit sites, *Curr. Biol.*, **16**, pp. 173–179.
53. Dancourt, J., and Barlowe, C. (2010). Protein sorting receptors in the early secretory pathway, *Annu. Rev. Biochem.*, **79**, pp. 777–802.
54. Appenzeller, C., Andersson, H., Kappeler, F., and Hauri, H. P. (1999). The lectin ERGIC-53 is a cargo transport receptor for glycoproteins, *Nat. Cell Biol.*, **1**, pp. 330–334.
55. Bonnon, C., Wendeler, M. W., Paccaud, J.-P., and Hauri, H.-P. (2010). Selective export of human GPI-anchored proteins from the endoplasmic reticulum, *J. Cell Sci.*, **123**, pp. 1705–1715.
56. Saito, K., Yamashiro, K., Ichikawa, Y., Erlmann, P., Kontani, K., Malhotra, V., and Katada, T. (2011). cTAGE5 mediates collagen secretion through interaction with TANGO1 at endoplasmic reticulum exit sites, *Mol. Biol. Cell*, **22**, pp. 2301–2308.
57. Jin, L. Y., Pahuja, K. B., Wickliffe, K. E., Gorur, A., Baumgartel, C., Schekman, R., and Rape, M. (2012). Ubiquitin-dependent regulation of COPII coat size and function, *Nature*, **482**, pp. 495–500.
58. Thor, F., Gautschi, M., Geiger, R., and Helenius, A. (2009). Bulk flow revisited: Transport of a soluble protein in the secretory pathway, *Traffic*, **10**, pp. 1819–1830.
59. Adolf, F., Herrmann, A., Hellwig, A., Beck, R., Brugger, B., and Wieland, F. T. (2013). Scission of COPI and COPII vesicles is independent of GTP hydrolysis, *Traffic*, **14**, pp. 922–932.
60. Lord, C., Bhandari, D., Menon, S., Ghassemian, M., Nycz, D., Hay, J., Ghosh, P., and Ferro-Novick, S. (2011). Sequential interactions with Sec23 control the direction of vesicle traffic, *Nature*, **473**, pp. 181–186.
61. Xu, D. L., and Hay, J. C. (2004). Reconstitution of COPII vesicle fusion to generate a pre-Golgi intermediate compartment, *J. Cell Biol.*, **167**, pp. 997–1003.
62. Shima, D. T., Scales, S. J., Kreis, T. E., and Pepperkok, R. (1999). Segregation of COPI-rich and anterograde-cargo-rich domains in endoplasmic-reticulum-to-Golgi transport complexes, *Curr. Biol.*, **9**, pp. 821–824.
63. Lippincott-Schwartz, J., Donaldson, J. G., Schweizer, A., Berger, E. G., Hauri, H.-P., Yuan, L. C., and Klausner, R. D. (1990). Microtubule-dependent retrograde transport of proteins into the ER in the presence of brefeldin a suggests an ER recycling pathway, *Cell*, **60**, pp. 821–836.

64. Jackson, C. L., and Casanova, J. E. (2000). Turning on ARF: The Sec7 family of GEFs, *Trends Cell Biol.*, **10**, pp. 60–67.
65. Antonny, B., Beraud-Dufour, S., Chardin, P., and Chabre, M. (1997). N-terminal hydrophobic residues of the G-protein ADP-ribosylation factor-1 insert into membrane phospholipids upon GDP to GTP exchange, *Biochemistry*, **36**, pp. 4675–4684.
66. Lee, C., and Goldberg, J. (2010). Structure of coatomer cage proteins and the relationship among COPI, COPII, and clathrin vesicle coats, *Cell*, **142**, pp. 123–132.
67. Hirst, J., Barlow, L. D., Francisco, G. C., Sahlender, D. A., Seaman, M. N. J., Dacks, J. B., and Robinson, M. S. (2011). The fifth adaptor protein complex, *PLoS Biol.*, **9**, pp. e1001170.
68. Bigay, J., Casella, J. F., Drin, G., Mesmin, B., and Antonny, B. (2005). ArfGAP1 responds to membrane curvature through the folding of a lipid packing sensor motif, *EMBO J.*, **24**, pp. 2244–2253.
69. Lewis, M. J., and Pelham, H. R. B. (1990). A human homolog of the yeast HDEL receptor, *Nature*, **348**, pp. 162–163.
70. Letourneur, F., Gaynor, E. C., Hennecke, S., Demolliere, C., Duden, R., Emr, S. D., Riezman, H., and Cosson, P. (1994). Coatomer is essential for retrieval of dilysine-tagged proteins to the endoplasmic reticulum, *Cell*, **79**, pp. 1199–1207.
71. Jackson, L. P., Lewis, M., Kent, H. M., Edeling, M. A., Evans, P. R., Duden, R., and Owen, D. J. (2012). Molecular basis for recognition of dilysine trafficking motifs by COPI, *Develop. Cell*, **23**, pp. 1255–1262.
72. Cabrera, M., Muniz, M., Hidalgo, J., Vega, L., Martin, M. E., and Velasco, A. (2003). The retrieval function of the KDEL receptor requires PKA phosphorylation of its C-terminus, *Mol. Biol. Cell*, **14**, pp. 4114–4125.
73. Sato, K., Sato, M., and Nakano, A. (1997). Rer1p as common machinery for the endoplasmic reticulum localization of membrane proteins, *Proc. Natl. Acad. Sci. U.S.A.*, **94**, pp. 9693–9698.
74. Helms, J. B., and Rothman, J. E. (1992). Inhibition by brefeldin A of a Golgi membrane enzyme that catalyses exchange of guanine nucleotide bound to ARF, *Nature*, **360**, pp. 352–354.
75. Ladinsky, M. S., Mastronarde, D. N., McIntosh, J. R., Howell, K. E., and Staehelin, L. A. (1999). Golgi structure in three dimensions: Functional insights from the normal rat kidney cell, *J. Cell Biol.*, **144**, pp. 1135–1149.
76. Rothman, J. E., and Wieland, F. T. (1996). Protein sorting by transport vesicles, *Science*, **272**, pp. 227–234.

77. Losev, E., Reinke, C. A., Jellen, J., Strongin, D. E., Bevis, B. J., and Glick, B. S. (2006). Golgi maturation visualized in living yeast, *Nature*, **441**, pp. 1002–1006.
78. Morriswood, B., and Warren, G. (2013). Stalemate in the Golgi battle, *Science*, **341**, pp. 1465–1466.
79. Sharpe, H. J., Stevens, T. J., and Munro, S. (2010). A comprehensive comparison of transmembrane domains reveals organelle-specific properties, *Cell*, **142**, pp. 158–169.
80. Lingwood, D., and Simons, K. (2010). Lipid rafts as a membrane-organizing principle, *Science*, **327**, pp. 46–50.
81. Griffiths, G., and Simons, K. (1986). The trans-Golgi network: Sorting at the exit site of the Golgi complex, *Science*, **234**, pp. 438–443.
82. Anitei, M., and Hoflack, B. (2012). Bridging membrane and cytoskeleton dynamics in the secretory and endocytic pathways, *Nature Cell Biol.*, **14**, pp. 11–19.
83. Kirchhausen, T. (2000). Clathrin, *Annu. Rev. Biochem.*, **69**, pp. 699–727.
84. Robinson, M. S. (2004). Adaptable adaptors for coated vesicles, *Trends Cell Biol.*, **14**, pp. 167–174.
85. Hirst, J., Irving, C., and Borner, G. H. H. (2013). Adaptor protein complexes AP-4 and AP-5: New players in endosomal trafficking and progressive spastic paraplegia, *Traffic*, **14**, pp. 153–164.
86. Bonifacino, J. S. (2004). The GGA proteins: Adaptors on the move, *Nat. Rev. Mol. Cell Biol.*, **5**, pp. 23–32.
87. Bonifacino, J. S., and Traub, L. M. (2003). Signals for sorting of transmembrane proteins to endosomes and lysosomes, *Annu. Rev. Biochem.*, **72**, pp. 395–447.
88. Hirst, J., Sahlender, D. A., Choma, M., Sinka, R., Harbour, M. E., Parkinson, M., and Robinson, M. S. (2009). Spatial and functional relationship of GGAs and AP-1 in Drosophila and HeLa cells, *Traffic*, **10**, pp. 1696–1710.
89. Gravotta, D., Deora, A., Perret, E., Oyanadel, C., Soza, A., Schreiner, R., Gonzalez, A., and Rodriguez-Boulan, E. (2007). AP1B sorts basolateral proteins in recycling and biosynthetic routes of MDCK cells, *Proc. Natl. Acad. Sci. U.S.A*, **104**, pp. 1564–1569.
90. Kazazic, M., Bertelsen, V., Pedersen, K. W., Vuong, T. T., Grandal, M. V., Redland, M. S., Traub, L. M., Stang, E., and Madshus, I. H. (2009). Epsin 1 is involved in recruitment of ubiquitinated EGF receptors into clathrin-coated pits, *Traffic*, **10**, pp. 235–245.

91. Sahlender, D. A., Kozik, P., Miller, S. E., Peden, A. A., and Robinson, M. S. (2013). Uncoupling the functions of CALM in VAMP sorting and clathrin-coated pit formation, *PLoS ONE.*, **8**, pp. e64514.
92. Paladino, S., Sarnataro, D., Pillich, R., Tivodar, S., Nitsch, L., and Zurzolo, C. (2004) Protein oligomerization modulates raft partitioning and apical sorting of GPI-anchored proteins, *J. Cell Biol.*, **167**, pp. 699–709.
93. Delacour, D., Greb, C., Koch, A., Salomonsson, E., Leffler, H., Le Bivic, A., and Jacob, R. (2007). Apical sorting by galectin-3-dependent glycoprotein clustering, *Traffic*, **8**, pp. 379–388.
94. Mishra, R., Grzybek, M., Niki, T., Hirashima, M., and Simons, K. (2010). Galectin-9 trafficking regulates apical-basal polarity in Madin-Darby canine kidney epithelial cells, *Proc. Natl. Acad. Sci. U.S.A*, **107**, pp. 17633–17638.
95. Arvan, P., and Castle, D. (1998). Sorting and storage during secretory granule biogenesis: Looking backward and looking forward, *Biochem. J.*, **332**, pp. 593–610.
96. Tooze, S. A., Martens, G. J. M., and Huttner, W. B. (2001). Secretory granule biogenesis: Rafting to the SNARE, *Trends Cell Biol.*, **11**, pp. 116–122.
97. Austin, C., Hinners, I., and Tooze, S. A. (2000). Direct and GTP-dependent interaction of ADP-ribosylation factor 1 with clathrin adaptor protein AP-1 on immature secretory granules, *J. Biol. Chem.*, **275**, pp. 21862–21869.
98. Burgoyne, R. D., and Morgan, A. (2003). Secretory granule exocytosis, *Physiol. Rev.*, **83**, pp. 581–632.
99. Kasai, H., Takahashi, N., and Tokumaru, H. (2012). Distinct initial SNARE configurations underlying the diversity of exocytosis, *Physiol. Rev.*, **92**, pp. 1915–1964.

Chapter 4

Intracellular Transport: Learning from Pathogens

Lali K. Medina-Kauwe

Department of Biomedical Sciences, Cedars-Sinai Medical Center and UCLA Geffen School of Medicine, 8700 Beverly Blvd., Los Angeles, CA 90048, USA

medinal@cshs.org

Therapeutic molecules can refer to just about any molecular agent that produces a therapeutic effect, including small-molecule drugs currently used in the clinic, as well as antibodies, peptides, proteins, toxins, and nucleic acids. Studies related to the delivery of any of these agents in recent years have reflected a better appreciation of endocytic processes facilitating drug efficacy, in particular, because of their ability to avoid drug efflux and degradative pathways. Invasive bacteria and viruses have evolved means to evade such cell defense pathways; hence it helps to examine the intracellular transport mechanisms used by these pathogens that may contribute to enhanced drug delivery.

4.1 Introduction

Membrane-permeable drugs can easily enter cells by diffusion through cell membranes (Fig. 4.1A) but, at the same time, can

Drug Delivery Across Physiological Barriers
Edited by Silvia Muro

ISBN 978-981-4669-40-5 (Hardcover), 978-981-4669-41-2 (eBook)
www.panstanford.com

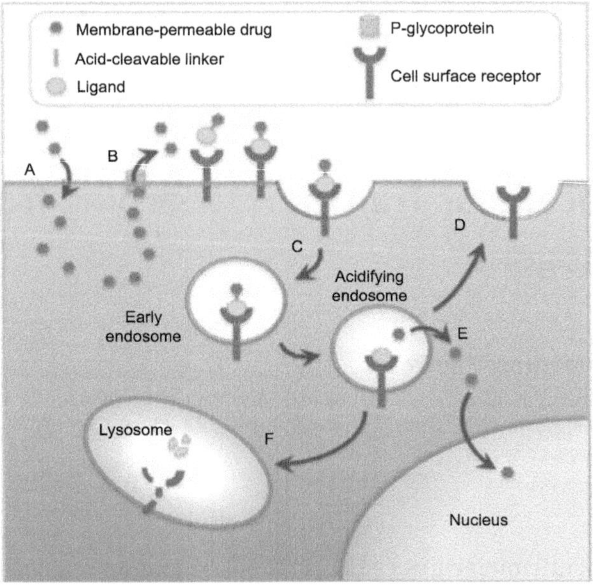

Figure 4.1 **Uptake and trafficking of targeted drugs**. Membrane-permeable drugs can easily enter cells by diffusion through cell membranes (**A**) but can be pumped out of cells via multidrug transporters (**B**). Targeted drugs bind to cell-specific surface receptors, triggering internalization into endosomes (**C**) and avoiding exposure to p-glycoprotein. If the drug is normally membrane permeable but covalently linked to a targeting ligand, endosomal escape can be facilitated by using an acid-cleavable linker to conjugate the drug and ligand. Acidification of the maturing endosome induces cleavage of the linker and release of the drug in the endosome, which can then diffuse into the cytoplasm through the endosomal membrane (**E**). Endosomal escape avoids recycling back out of the cell (**D**) and allows drug access to intracellular targets (most likely residing in the cytoplasm or nucleus). Vesicle escape also avoids delivery to the lysosomal compartment (**F**), which poses a particular barrier to the delivery of therapeutic nucleic acids, proteins, or peptides because these molecules are substrates for degradation by lysosomal enzymes.

be pumped out of cells via multidrug transporters, thus reducing efficacy (Fig. 4.1B) [1]. Targeted drugs tend to evade drug efflux, at least during the early stages of cell entry due to the endocytic transport mediating uptake [2, 3]. Drug targeting typically entails linking a drug to a ligand or antibody that recognizes a cell-specific surface molecule for homing the drug to selective cells. Targeting the drug to a cell-surface marker re-routes the drug to the internalization pathway used by that marker (Fig. 4.1C). Commonly, the marker protein undergoes endocytosis and enclosure in an endocytic vesicle, which would avoid exposure to p-glycoprotein and similar transport systems [4]. Endosomal escape would be necessary to avoid endosomal recycling back out of the cell (Fig. 4.1D) [5] and allow drug access to intracellular targets (most likely residing in the cytoplasm or nucleus). If the drug is normally membrane permeable but covalently linked to a targeting ligand, endosomal escape can be facilitated by using an acid-cleavable linker to conjugate the drug and ligand [6]. The acidification of the maturing endosome induces cleavage of the linker and release of the drug in the endosome, which can then diffuse into the cytoplasm through the endosomal membrane (Fig. 4.1E). It is believed that at this stage of intracellular trafficking, while cytosolic entry still has the possibility of exposing the drug to p-glycoprotein transport, this possibility is less likely if the drug is released at a sufficient distance from the plasma membrane where efflux proteins reside [7].

For drugs that are not membrane permeable but trapped in endosomes, the inclusion of a membrane-lytic agent is necessary to facilitate endosomal escape and entry into the cytoplasm and avoid recycling to the cell surface. Such agents can utilize a variety of strategies to accomplish this, including osmotic swelling, pore formation, membrane flipping, or fragmentation, that may or may not occur in response to the acidifying environment of the endosomal lumen [8, 9]. Vesicle escape also avoids delivery to the lysosomal compartment, which poses a particular barrier to the delivery of therapeutic nucleic acids, proteins, or peptides because these molecules are substrates for degradation by lysosomal enzymes (Fig. 4.1F) [10]. With regard to the delivery of exogenous therapeutic genes, additional barriers are encountered in the cytoplasm: the crowded cytosolic milieu can prevent rapid motility to the nucleus [11], while cytosolic nucleases can degrade the DNA cargo [12].

Bacteria and viruses have evolved strategies to hijack the cell's own machinery for invasion and cargo delivery. Hence, an examination of the different intracellular transport pathways exploited by pathogens may be beneficial for improving drug delivery.

Endocytic uptake can occur through a variety of pathways, including clathrin-mediated endocytosis, macropinocytosis, caveolae-mediated internalization, and phagocytosis. While these pathways have the potential to converge with one another at certain stages of uptake, there are distinct features of each pathway that distinguish one from another. **Clathrin-mediated endocytosis** is characterized by the assembly of clathrin protein cages on the cytoplasmic side of vesicles budding from the plasma membrane [13]. Sorting motifs on clathrin-associated adaptor protein (AP) complexes facilitate navigation of the maturing vesicle (which eventually sheds its clathrin coat) to different intracellular trafficking routes that include recycling back to the cell surface or delivering the cargo to lysosomes (discussed in more detail later) [14]. **Caveolae** are flask-shaped membrane invaginations rich in cholesterol and sphingolipids and contain the caveolin integral membrane proteins [15]. Caveolae can be sites of membrane signaling due to the association of caveolins with signaling molecules [16]. Additionally, caveolae can pinch off into caveolin-coated vesicles with the potential to merge with other trafficking pathways or with neutral pH structures known as caveosomes (discussed in more detail later) [17]. **Macropinocytosis** is characterized by the uptake of large volumes of extracellular fluid facilitated by ruffling of the cell membrane, thus creating large non-coated endosomes, or macropinosomes, resulting from the fusion of undulating regions of the plasma membrane [18]. **Phagocytosis** describes the engulfment of foreign bodies by the plasma membrane and tends to be a predominant endocytic feature of immune cells, such as macrophages [19]. Still yet, pathways that do not bear any of the distinguishing features that would classify them in the abovementioned routes have been vaguely categorized as **clathrin- and caveolae-independent endocytosis** pathways [20], though this classification is still undergoing an evolving identity. Although flotillin has been indicated as a marker for clathrin- and caveolae-independent endocytosis [21], the intercellular adhesion molecule 1 (ICAM-1)-mediated cell entry used by major group rhinoviruses lacks overlap with clathrin- and caveolae-dependent routes and does not colocalize with flotillin [22]. Additionally, ICAM-1 and platelet-endothelial cell adhesion molecule 1 (PECAM-1)-mediated clustering and internalization of immunoconjugates share some, but not all, properties of macropinocytosis [23]. Hence, this last category of endocytosis may encompass a growing set of pathways that have yet to be fully identified.

Here we highlight a few examples of cell uptake pathways that by no means sum up the totality of different approaches pathogens have evolved to invade eukaryotic cells. The following will discuss how adenoviruses and bacteria utilize specialized mechanisms associated with clathrin-mediated endocytosis to gain cell entry and spread; how bacterial and plant toxins use a retrograde route of trafficking to bypass degradative compartments; and how studies probing SV40 infection have revealed a novel caveolar route of entry. These examples highlight some strategies of interest that are currently being explored and exploited for improving drug delivery.

4.2 Pathogens Clash with the Classic Route

Binding to cell-surface receptors initiates the cell uptake of targeted drugs as well as a majority of pathogens and can route receptor–ligand pairs to several possible internalization pathways. A common uptake pathway is clathrin-mediated endocytosis, understood as the classic route of uptake. In this pathway, receptor binding recruits clathrin on the cytoplasmic surface of the plasma membrane at the receptor-binding sites, causing localized membrane invaginations known as clathrin-coated pits (Fig. 4.2A). With the assistance of a number of different effector proteins [24], the dynamin GTPase [25, 26] pinches off these pits into clathrin-coated vesicles (Fig. 4.2B), which then begin to shed their clathrin coats through an enzymatic process [27]. These "naked" vesicles deliver their cargo to early (or sorting) endosomes (Fig. 4.2C) that undergo "maturation," which is characterized by luminal acidification via ATP-dependent proton pumps in the vesicle membrane [28]. These endosomes can traffic their cargo through either the recycling pathway (Fig. 4.2D), where the vesicle delivers the cargo back to the cell surface, or continue to mature to late endosomes (Fig. 4.2E) and eventually fuse with the lysosome, where lysosomal enzymes degrade the cargo (Fig. 4.2F) [10]. Pathogen proteins have evolved mechanisms to exploit the

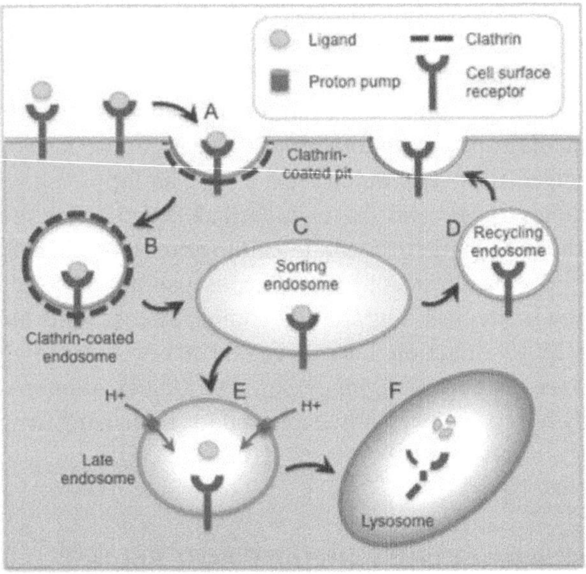

Figure 4.2 Clathrin-mediated internalization pathway. Receptor binding induces clathrin-coated pits (**A**) that pinch off into clathrin-coated vesicles (**B**), which shed their clathrin coats and deliver their cargo to early (or sorting) endosomes (**C**). Sorting endosomes direct cargo to either recycling endosomes, which deliver the cargo back to the cell surface (**D**), or to late endosomes, which undergo further acidification (**E**) and eventually fuse with the lysosome, where lysosomal enzymes degrade the cargo (**F**).

rapid internalization facilitated by clathrin-mediated endocytosis while escaping the degradative fate in the lysosme.

4.2.1 *Adenovirus*

The adenovirus (Ad) is one such pathogen whose highly efficient cell penetration and payload transport makes it one of the most widely used and studied viruses for gene delivery. Several capsid proteins that comprise the shell or protein coat of this nonenveloped virus facilitate this process [29]. Similar to many viruses, cell entry by Ad is a two-step process, initiated by binding of the Ad capsid fiber to a primary cell-surface protein (Fig. 4.3A) followed by secondary binding of the Ad capsid penton base to cell-surface integrins

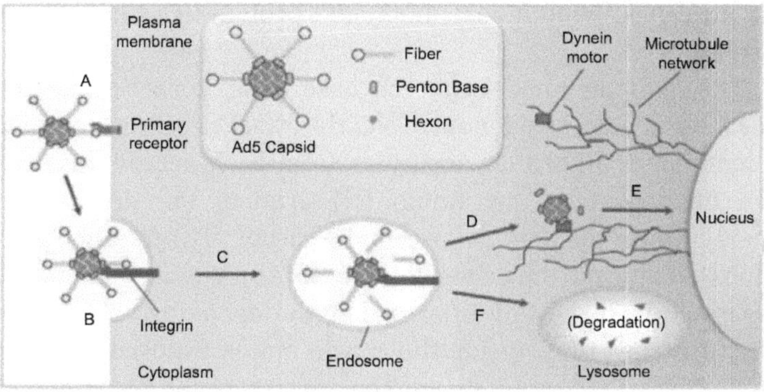

Figure 4.3 Adenovirus trafficking. Cell entry is initiated by binding of the Ad capsid fiber to a primary cell-surface protein (**A**) followed by secondary binding of the penton base to cell-surface integrins (**B**), triggering integrin-mediated endocytosis (**C**). Internalized virions escape the endosome (**D**) and hijack the dynein trafficking machinery to undergo microtubule-dependent transit to the nucleus (**E**), avoiding lysosome-mediated degradation (**F**). Reprinted with permission from Ref. 29, with permission from Future Science Ltd.

(Fig. 4.3B), triggering integrin-mediated endocytosis (Fig. 4.3C) [30]. The primary receptor differs depending on Ad serotype, and among the identified proteins are the coxsackievirus–adenovirus receptor (CAR) [31], heparan sulfate glycosaminoglycans (HS-GAG) [32–34], CD46 [35], and desmoglein [36]. Intracellular trafficking studies on Ad2 and Ad5 show that internalized virions escape the endosome (Fig. 4.3D) and hijack the dynein trafficking machinery to undergo microtubule-dependent transit to the nucleus (Fig. 4.3E) [37–39]. Early studies on Ad cell entry indicated that Ad-mediated endosomal penetration is induced by the low pH of the acidifying endosome [40]. The ability of replication-deficient Ad to facilitate endosomal escape of exogenously delivered DNA indicated that cytoplasmic entry was mediated by one or more of the capsid proteins [41, 42]. Studies comparing Ad with rhinovirus indicated that membrane penetration by Ad involved a generalized disruption of the endosomal membrane, allowing molecules of a wide range of sizes to escape into the cytoplasm, in contrast to the pore formation by the rhinovirus, which limited leakage of certain sized molecules out of the endosome [43]. More recent studies have identified the Ad capsid penton base, fiber, and protein VI as endosomolytic mediators.

The homopentameric penton base lying at each vertex of the icosahedral Ad capsid noncovalently attaches to an antenna-like homotrimeric fiber that protrudes from each penton base cluster. The fiber contains an amino (N)-terminal tail domain, which interacts with the penton base, and a carboxy (C)-terminal globular knob that binds to the primary receptor [44]. These two domains are separated by a shaft of differing length depending on Ad serotype [45]. Protein VI lies inside the capsid and is exposed upon capsid disassembly that occurs during the early stages of Ad infection [46].

A number of independent studies have indicated that the penton base possesses membrane penetration ability when studied as a soluble, isolated protein. The exposure and binding of hydrophobic domains in the protein to cell membranes at low pH suggested that the penton base may contribute to lysis of the acidifying endosome [47]. Electron microscopy of soluble recombinant Ad2 penton base showed penetration through the endosomal membrane and entry into the nucleus, which was corroborated by confocal microscopy, intracellular trafficking, and subcellular fractionation studies of recombinant Ad5 penton base protein [48, 49]. More recently, the ability of recombinant penton base to mediate gene and siRNA transfer and delivery of membrane impermeant toxic corrole molecules have suggested that this protein contains membrane-lytic activity [49, 50]. Gene transfer studies suggest that endosomal escape of soluble recombinant Ad5 penton base is not pH dependent, unlike the whole virus, as blocking endosome acidification by bafilomycin did not prevent gene expression [51].

The fiber protein and peptides derived from this protein can also undergo cell entry in certain circumstances. Soluble recombinant Ad5 fiber can enter HeLa cells and lacrimal gland cells of the eye via CAR and proteoglycan binding [33, 52]. HeLa cell entry, which was actin mediated and temperature independent, could traffic to the nucleus as well as facilitate gene transfer of protamine-condensed plasmid DNA [52]. Lacrimal cell entry could also be accomplished by the knob domain alone, in contrast to HeLa cell entry, and was associated with macropinocytosis [33]. The knob has been explored as a vehicle for CAR-dependent delivery and has been used to demonstrate delivery of elastin-like polypeptides in hepatocytes [53]. The trafficking of these complexes to lysosomes suggests that the knob does not bear membrane penetration activity. The tail domain of the Ad3 fiber also appears to undergo cell penetration, though its internalization route may differ from that observed of the Ad5 fiber [54]. In the context of the whole virus, the fiber may dictate the timing of endosome escape by acting as a probe for determining the pH at which escape occurs [55, 56]. This is serotype-dependent and could have implications in drug delivery: fiber proteins that signal escape at lower pH would release the endosomal cargo at later stages of intracellular trafficking and thus closer to the nucleus, enabling enhanced nuclear delivery of therapeutic payloads.

The dependence of virus-mediated membrane penetration on capsid uncoating suggests that an underlying factor becomes exposed for interaction with endosomal membranes [57, 58]. Protein VI, which lies inside the Ad capsid, is a factor that becomes revealed by uncoating and whose processing by the Ad protease exposes putative helical membrane-lytic domains [46]. Recent studies demonstrate that soluble recombinant protein VI facilitates entry of a nuclear dye into HeLa cells via the putative helical domain located at the N-terminus [59].

It is possible that the abovementioned three proteins may act in concert in the context of the whole virus to conduct endosomal escape. Nevertheless, independent studies show that these proteins bear the capacity to penetrate cell membranes on their own, independent of other viral proteins. Their utility for drug delivery has been demonstrated. Recombinant Ad5 penton base has been modified for integrin-targeted gene and siRNA delivery [49, 50, 60]. This protein could be re-directed to the human epidermal growth factor receptor (HER) by fusion to an exogenous targeting ligand, and deliver genes, siRNA, toxic corrole compounds, and the chemotherapeutic drug doxorubicin, all through noncovalent interactions enabling spontaneous self-assembly [51, 60–62]. Tumor-targeted therapeutic delivery could be accomplished *in vivo*, and importantly not just drug transport to the tumors was observed, but drug penetration into the tumor cells and subcellular delivery to the target organelle could also be observed [63]. The requirement for the membrane penetration capability of the viral protein to enhance drug transport has been evident: targeted protein constructs that lack the penton base domain exhibited negligible gene and drug delivery *in vitro* [51, 64]; and delivery by the protein constructs *in vivo* enabled >100-fold lower dose of therapeutic drug compared to the free untargeted drug [61, 62].

The Ad3 penton base has also been evaluated for gene and drug delivery. Unlike the Ad5-derived protein, the penton base from this serotype can noncovalently self-assemble into dodecahedrons (Dd) comprising 12 pentamers (or 60 penton base monomers) when overexpressed during viral infection and as a recombinant protein in insect cells [65, 66]. Dd can mimic viral cell entry by binding to integrins as well as heparan sulfate glycosaminoglycans (HS-GAG) [67, 68], undergoing receptor-mediated endocytosis and trafficking toward the nuclear periphery where Dds accumulate inside the cell. Dd can mediate the delivery of attached nucleic acids [65] and the drug bleomycin [69], *in vitro*. More recently, it has been explored as an antigen-delivery platform for tumor immunotherapy [70].

Finally, a peptide called JO-1, derived from the receptor-binding domain of the Ad3 fiber, has shown efficacy at augmenting drug penetration into tumor tissue [71]. This is due to the peptide's ability to bind the Ad3 receptor desmoglein, which facilitates tight junctions between epithelial cells and is upregulated in tumors [72, 73]. The high affinity (pM) binding of JO-1 to this receptor opens up tight junctions, thus enabling deeper penetration of therapeutic drugs into tumors [74].

4.2.2 *Listeria*

The discovery that *Listeria monocytogenes*, a bacterium that causes gastroenteritis, brain and maternofetal infections, is transported into cells by a clathrin-mediated mechanism was an unexpected finding, as estimates based on the physical constraints of the assembled clathrin triskelions had previously limited the diameter of the cargo that coated vesicles can accommodate to a maximum of 150 nm [75–77]. Studies on vesicular stomatitis virus (VSV), which is 200 nm long, show that actin-mediated entry occurs in endosomes that are partially coated with clathrin [78]. It is now appreciated that a number of large pathogens in addition to *Listeria*, including

Candida albicans and other types of bacteria, use clathrin-mediated endocytosis to gain cell entry [79, 80].

Internalization by *Listeria* engages a zipper mechanism in which the bacterium is tightly apposed to the plasma membrane, signaling actin enrichment at the contact site, forming actin "pedestals" whose nucleation is regulated by clathrin [81]. This is followed by actin-mediated endocytosis into a vacuole from which *Listeria* escapes into the cytosol by vacuole-lysis through the pore-forming toxin listeriolysin O (LLO) [82].

A novel mechanism of bacterial spread by *Listeria* and similar pathogens such as *Shigella* exploits the cytoskeleton to propel the bacterium inside the host cell and from cell to cell along an actin-rich structure resembling a comet tail [83]. A series of cellular proteins interacting with the ActA protein on the *Listeria* cell surface induce actin polymerization on one side of the bacterium, pushing the pathogen through the cytoplasm and sometimes out of one cell and into a neighboring cell [84]. Studies isolating the minimal proteins that contribute to this process have identified actin, the Arp2/3 complex, ADF/cofilin, an actin-capping protein, and the Listeria ActA protein, which activates the Arp2/3 complex [85, 86]. Studies probing this mechanism have shown that isolated ActA coated on polystyrene beads is sufficient to mediate the movement of the beads along actin comets in *Xenopus laevis* egg extracts [87]. Subsequently, Ng et al. [88] has shown that polymers and lipids commonly used in gene delivery complexes adopt cytoplasmic motility using actin comets when conjugated to ActA. These findings have implications on improving nonviral gene transfer, as the rapid propulsion mediated by these structures has the potential to reduce the possibility of DNA degradation by cytosolic nucleases and thus enhance nonviral gene therapy.

4.3 Retrograde Trafficking: Toxins Going Retro

Plant toxins such as ricin and abrin and bacterial toxins such as Shiga toxin (STx), cholera toxin (CTx), and *Pseudomonas* exotoxin A (PEx) require delivery to the cytoplasm where specific components of the protein synthesis machinery are the targets of toxic activity [89–

93]. To access the cytoplasm while avoiding lysosomal degradation, these toxins undergo retrograde trafficking [94], which describes the route of cell entry in reverse of the classical secretion pathway.

The classical route used by newly synthesized, secreted proteins involves budding from the endoplasmic reticulum (ER) and transit to the Golgi, followed by budding from the trans-Golgi network (TGN) and fusion with the plasma membrane [95]. To traffic this pathway in reverse, toxin entry is initiated by cell-surface binding (Fig. 4.4A) mediated by the B chain of the toxin, which is linked to the A-chain that comprises the catalytic or toxic domain [96]. The different types of toxins bind to different cell-surface molecules: STx and CTx bind host cell membrane glycolipids [97–101]; PEx binds α_2-macroglobulin receptor/low-density lipoprotein receptor-related protein [102]; and ricin binds β1-4-linked galactosides displayed on a wide range of cell-surface glycoproteins and glycolipids, thus enabling toxicity to a broad range of cells [103].

Endocytosis of the toxin after cell binding (Fig. 4.4B) is followed by vesicle delivery to the TGN from either early or late endosomes. PEx uses the late endosome pathway mediated by the small GTPase Rab9 (Fig. 4.4C) [104], whereas CTx, STx, and ricin use a Rab9-independent route (Fig. 4.4D) [105]. Each route supports vesicle fusion with the TGN using its own separate SNAREs (soluble *N*-ethylmaleimide-sensitive fusion attachment protein receptors) [106, 107].

The general route of toxin trafficking from the TGN to the ER is by traversing the Golgi cisternae and interacting with the KDEL receptor, which cycles between the TGN and ER for retrieval of escaped ER proteins from the Golgi (Fig. 4.4E) [108]. This trafficking relies on coatamer protein I (COPI), which coats vesicles budding from the Golgi apparatus and can direct transit based on interactions with the cytoplasmic domains of membrane proteins [109, 110]. PEx contains a KDEL-like sequence that is exposed after A-chain release by furin cleavage in early/recycling endosomes [111–115]. Ricin lacks a KDEL sequence, but it can bind the chaperone calreticulin, which has a KDEL motif, and undergoes COPI-dependent trafficking to the ER [116]. KDEL and/or Golgi stack-independent trafficking can occur as alternative and less well-defined routes [117].

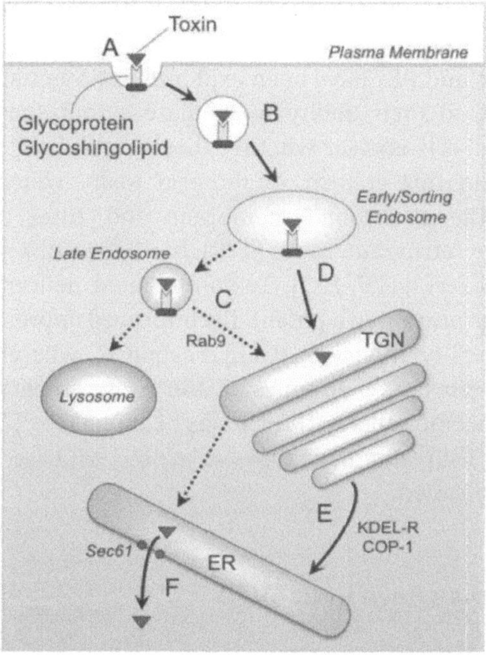

Figure 4.4 **Retrograde trafficking.** Plant and bacterial toxin cell entry is initiated by binding to a cell-surface receptor (**A**), triggering endocytosis (**B**) and vesicle delivery to the TGN from late or early endosomes, using either a Rab9-dependent (**C**) or Rab9-independent pathway (**D**). Trafficking from the TGN to the ER occurs by interaction with the KDEL receptor, which cycles between the TGN and ER for retrieval of escaped ER proteins from the Golgi (**E**) and is COPI dependent. Arrival at the ER is followed by transport into the cytoplasm via ERAD and retrotranslocation through the Sec61 dislocon (**F**). Reprinted from Ref. 118, Copyright 2007, with permission from Elsevier.

Arrival at the ER is followed by transport into the cytoplasm through a protein-auditing system known as ERAD (ER-associated protein degradation). This eliminates misfolded proteins from the ER by discarding them into the cytoplasm through a pore known as the Sec61 dislocon (Fig. 4.4F) [119, 120]. Translocated proteins are typically ubiquitinated and targeted to the proteasome for degradation; however, the low lysine content in toxins renders them poor substrates for ubiquitination [121], and thus degradation is avoided while exposure to the cytoplasm allows access to the protein synthesis machinery.

STx, CTx, and PEx have been explored as potential gene delivery vectors due to their ability to mediate cytoplasmic entry [122–127], though it is unclear whether these proteins retain retrograde trafficking in this context. Diphtheria toxin, which can directly penetrate the endosomal membrane and enter the cytoplasm avoiding the retrograde route [128], has been utilized for gene and peptide delivery [129–131]. Toxin-mediated delivery of antigenic peptides has presented a potentially improved approach for priming a cytotoxic T-lymphocyte (CTL) response, whereby retrograde trafficking can target epitopes to the proteasome for display on major histocompatibility (MHC) class I molecules [132, 133]. PEx [134], STx [135], and other types of toxins [96] have been explored for this application.

4.4 Caveolae Cave in to SV40

Simian virus 40 (SV40), a nonenveloped DNA virus of the papovavirus family, can bypass the endolysosomal pathway and utilize caveoli to enter the nucleus by way of the ER [136–139]. Caveolae are flask-shaped invaginations at the plasma membrane [140] coated on the cytoplasmic surface with caveolin-1 (Cav1) [141, 142], a palmitylated, cholesterol-binding protein [143, 144] that mediates a broad range of intracellular signaling pathways [145]. Polyomaviruses like SV40 and influenza, corona-, and echoviruses enter cells via caveolae [146–150].

SV40 binds MHC class I [151] and ganglioside GM1 [152] co-receptors (Fig. 4.5A), inducing tyrosine kinase phosphorylation, protein kinase C activation, and actin depolymerization to mediate caveolar endocytosis [136, 139, 153]. SV40 and GM1 internalize into caveolae (Fig. 4.5B), while MHC class I is shed from the cell surface via metallo-proteinase cleavage [154, 155]. Dynamin-2 pinches off caveolar vesicles, or cavicles (Fig. 4.5C), that traffic along microtubules [139, 156] and can fuse with sorting endosomes of the clathrin-mediated pathway [157]. The more likely route, however, is to caveosomes, Cav1-positive, pH neutral compartments rich in

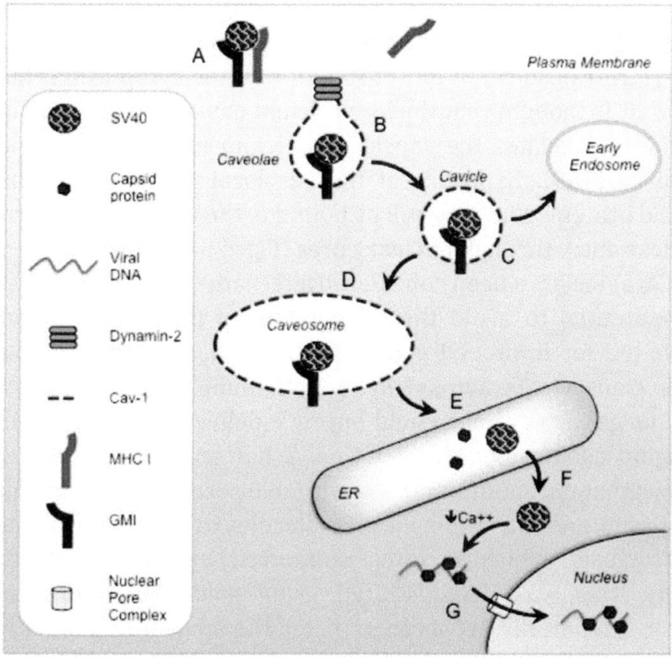

Figure 4.5 **Caveolar trafficking of SV40.** SV40 binds MHC class I and ganglioside GM1 co-receptors (**A**), inducing SV40 and GM1 internalization into caveolae (**B**), while MHC class I is shed from the cell surface. Dynamin-2 pinches off caveolar vesicles, or cavicles (**C**), that traffic along microtubules and can fuse with sorting endosomes of the clathrin-mediated pathway, or to caveosomes, which are Cav1-positive, pH neutral compartments rich in cholesterol and glycosphingolipids (**D**). Tubular structures pinching off from caveosomes traffic along microtubules to the ER (**E**), which may be mediated by COPI and COPII coat complexes. Protein-folding enzymes in the ER facilitate virion uncoating, triggering recognition by ERAD and retrotranslocation into the cytoplasm (**F**). The low-calcium environment of the cytosol further dissociates the capsid, while exposure of nuclear localization signals in viral capsid proteins still be bound to the viral genome facilitate nuclear entry through nuclear pores (**G**).

cholesterol and glycosphingolipids (Fig. 4.5D) [136]. SV40 can be retained in this compartment for several hours, and hence this pathway tends to be relatively slow [136].

Tubular structures pinching off from caveosomes traffic along microtubules to the ER (Fig. 4.5E), which may be mediated by COPI and COPII coat complexes [136, 158]. Protein-folding enzymes in the ER facilitate virion uncoating, triggering recognition by ERAD, and retrotranslocation out of the ER into the cytoplasm (Fig. 4.5F) [159]. It is thought that the low-calcium environment of the cytosol further dissociates the capsid, as calcium is required for particle stability [159]. Exposure of nuclear localization signals in viral capsid proteins that may still be bound to the viral genome facilitates nuclear entry through nuclear pores (Fig. 4.5G) [160–163].

Caveolae have been considered alternative cell-entry routes with the potential to avoid the endolysosomal pathway and could be exploited for improved gene and drug delivery. Their association with transcytosis across the endothelium raises the possibility that targeting caveolae could improve delivery of therapeutics and imaging agents to desired tissue. Schnitzer and colleagues have shown that the endothelium has variable permeability depending on the tissue, and thus transport of molecules to certain tissue requires transcytosis, which is often supported by caveolae trafficking [164]. This was demonstrated using gold-labeled probes that, when systemically delivered, targeted the lung endothelial caveolae and resulted in release at the lung interstitium within minutes after delivery [164]. More recently, endothelial receptors that bind albumin have exhibited endocytosis and transcytosis via caveolae [165] and have been exploited for ligand-directed nanoparticle delivery across the endothelium [166]. Gabrielson et al. [167] have taken advantage of the preferential uptake of the folate receptor via caveolae as a means to target nonviral gene delivery [167]. This study showed that uptake of these complexes avoided the lysosomal pathway, thus supporting the findings on SV40 trafficking that suggest caveolar routing may afford protection of therapeutic cargo from degradative compartments inside the cell and thus improve potential therapeutic efficacy.

4.5 Conclusions and Perspectives

The examples provided above highlight just a few of the varied pathogen strategies used to overcome cellular resistance to invasion and a number of useful molecules have been derived from these pathogens for the purpose of facilitating and improving molecular delivery. For example, studies of the rhinovirus and influenza virus membrane penetration mechanisms have given rise to the GALA and HA peptides, respectively, which form pores in endosomal membranes in response to low pH [43, 168–171] and have been incorporated into numerous gene delivery vectors to mediate cytosolic delivery [8, 9]. Studies of SV40 delivery have suggested that the use of pathways that avoid the endolysosomal route has the potential to reduce immunogenicity of delivery agents, as proteolytic processing of internalized proteins is essential for antigen display [172]. Hence, mimicking the cavaeolar or retrograde routes of polyomaviruses or bacterial toxins may augment therapeutic efficacy by reducing immune stimulation against the delivery agent. As nanomedicines continue to develop, an understanding of pathogen transport strategies has the potential to significantly improve the efficacy of drug delivery and thus highly impact disease treatment.

Acknowledgments

Work in Dr. Medina-Kauwe's lab has been supported by grants from the NIH/NCI (R01CA102126, R01CA140995, R01CA129822, and R21 CA116014), the DoD (W81XWH-06-1-0549), the Susan G. Komen Breast Cancer Foundation (BCTR02-1194), and the Donna and Jesse Garber award. The author thanks C. Rey, A. Kauwe, M. M.-Kauwe, and D. Revetto.

References

1. Shen, F., Chu, S., Bence, A.K., Bailey, B., Xue, X., Erickson, P.A., Montrose, M.H., Beck, W.T., and Erickson, L.C. (2008). Quantitation of doxorubicin uptake, efflux, and modulation of multidrug resistance (MDR) in MDR human cancer cells, *J. Pharm. Exp. Ther.*, **324**, pp. 95–102.
2. Goren, D., Horowitz, A.T., Tzemach, D., Tarshish, M., Zalipsky, S., and Gabizon, A. (2000). Nuclear delivery of doxorubicin via folate-targeted liposomes with bypass of multidrug-resistance efflux pump, *Clin. Cancer Res.*, **6**, pp. 1949–1957.

3. Mazel, M., Clair, P., Rousselle, C., Vidal, P., Scherrmann, J.M., Mathieu, D., and Temsamani, J. (2001). Doxorubicin-peptide conjugates overcome multidrug resistance, *Anticancer Drugs*, **12**, pp. 107–116.
4. Hu, C.M., and Zhang, L. (2009). Therapeutic nanoparticles to combat cancer drug resistance, *Curr. Drug Metab.*, **10**, pp. 836–841.
5. Grant, B.D., and Donaldson, J.G. (2009). Pathways and mechanisms of endocytic recycling, *Nat. Rev. Mol. Cell Biol.*, **10**, pp. 597–608.
6. Haag, R., and Kratz, F. (2006). Polymer therapeutics: Concepts and applications, *Angew. Chem. Int. Ed.*, **45**, pp. 1198–1215.
7. Cardoso, M.M., Peca, I.N., and Roque, A.C. (2012). Antibody-conjugated nanoparticles for therapeutic applications, *Curr. Med. Chem.*, **19**, pp. 3103–3127.
8. Medina-Kauwe, L.K. (2006). Gene therapy for neurological disorders. In *Non-Viral Mediated Gene Deliveryfor Therapeutic Applications*, Lowenstein, P., and Castro, M. (Eds.), CRC Press, Boca Rat on, pp. 115–140.
9. Medina-Kauwe, L.K., Xie, J., and Hamm-Alvarez, S. (2005). Intracellular trafficking of nonviral vectors, *Gene Ther.*, **12**, pp. 1734–1751.
10. Lechardeur, D., and Lukacs, G.L. (2002). Intracellular barriers to non-viral gene transfer, *Curr. Gene Ther.*, **2**, pp. 183–194.
11. Lukacs, G.L., Haggie, P., Seksek, O., Lechardeur, D., Freedman, N., and Verkman, A.S. (2000). Size-dependent DNA mobility in cytoplasm and nucleus, *J. Biol. Chem.*, **275**, pp. 1625–1629.
12. Pollard, H., Toumaniantz, G., Amos, J.L., Avet-Loiseau, H., Guihard, G., Behr, J.P., and Escande, D. (2001). Ca^{2+}-sensitive cytosolic nucleases prevent efficient delivery to the nucleus of injected plasmids, *J. Gene Med.*, **3**, pp. 153–164.
13. McMahon, H.T., and Boucrot, E. (2011). Molecular mechanism and physiological functions of clathrin-mediated endocytosis, *Nat. Rev. Mol. Cell Biol.*, **12**, pp. 517–533.
14. Ohno, H. (2006). Clathrin-associated adaptor protein complexes, *J. Cell Sci.*, **119**, pp. 3719–3721.
15. Kiss, A.L. (2012). Caveolae and the regulation of endocytosis, *Adv. Exp. Med. Biol.*, **729**, pp. 14–28.
16. Quest, A.F., Lobos-Gonzalez, L., Nunez, S., Sanhueza, C., Fernandez, J.G., Aguirre, A., Rodriguez, D., Leyton, L., and Torres, V. (2013). The caveolin-1 connection to cell death and survival, *Curr. Mol. Med.*, **13**, pp. 266–281.

17. Parton, R.G., and del Pozo, M.A. (2013). Caveolae as plasma membrane sensors, protectors and organizers, *Nat. Rev. Mol. Cell Biol.*, **14**, pp. 98–112.
18. Lim, J.P., and Gleeson, P.A. (2011). Macropinocytosis: An endocytic pathway for internalising large gulps, *Immunol. Cell Biol.*, **89**, pp. 836–843.
19. Groves, E., Dart, A.E., Covarelli, V., and Caron, E. (2008). Molecular mechanisms of phagocytic uptake in mammalian cells, *Cell. Mol. Life Sci.*, **65**, pp. 1957–1976.
20. Doherty, G.J., and McMahon, H.T. (2009). Mechanisms of endocytosis, *Annu, Rev. Biochem.*, **78**, pp. 857–902.
21. Glebov, O.O., Bright, N.A., and Nichols, B.J. (2006). Flotillin-1 defines a clathrin-independent endocytic pathway in mammalian cells, *Nat. Cell Biol.*, **8**, pp. 46–54.
22. Khan, A.G., Pickl-Herk, A., Gajdzik, L., Marlovits, T.C., Fuchs, R., and Blaas, D. (2010). Human rhinovirus 14 enters rhabdomyosarcoma cells expressing icam-1 by a clathrin-, caveolin-, and flotillin-independent pathway, *J. Virol.*, **84**, pp. 3984–3992.
23. Muro, S., Wiewrodt, R., Thomas, A., Koniaris, L., Albelda, S.M., Muzykantov, V.R., and Koval, M. (2003). A novel endocytic pathway induced by clustering endothelial ICAM-1 or PECAM-1, *J. Cell Sci.*, **116**, pp. 1599–1609.
24. Mousavi, S.A., Malerod, L., Berg, T., and Kjeken, R. (2004). Clathrin-dependent endocytosis, *Biochem. J.*, **377**, pp. 1–16.
25. Hinshaw, J.E., and Schmid, S.L. (1995). Dynamin self-assembles into rings suggesting a mechanism for coated vesicle budding, *Nature,* **374**, pp. 190–192.
26. Schmid, S.L., McNiven, M.A., and De Camilli, P. (1998). Dynamin and its partners: A progress report, *Curr. Opin. Cell Biol.*, **10**, pp. 504–512.
27. Rothman, J.E., and Schmid, S.L. (1986). Enzymatic recycling of clathrin from coated vesicles, *Cell,* **46**, pp. 5–9.
28. Van Dyke, R.W. (1996). Acidification of lysosomes and endosomes, *Subcell. Biochem.*, **27**, pp. 331–360.
29. Medina-Kauwe, L.K. (2013). Development of adenovirus capsid proteins for targeted therapeutic delivery, *Ther. Deliv.*, **4**, pp. 267–277.
30. Medina-Kauwe, L.K. (2003). Endocytosis of adenovirus and adenovirus capsid proteins, *Adv. Drug Deliv. Rev.*, **55**, pp. 1485–1496.
31. Bergelson, J.M., Cunningham, J.A., Droguett, G., Kurt-Jones, E.A., Krithivas, A., Hong, J.S., Horwitz, M.S., Crowell, R.L., and Finberg, R.W.

(1997). Isolation of a common receptor for Coxsackie B viruses and adenoviruses 2 and 5, *Science*, **275**, pp. 1320–1323.

32. Lenaerts, L., van Dam, W., Persoons, L., and Naesens, L. (2012). Interaction between mouse adenovirus type 1 and cell surface heparan sulfate proteoglycans, *PLoS One*, **7**, pp. e31454.
33. Xie, J., Chiang, L., Contreras, J., Wu, K., Garner, J.A., Medina-Kauwe, L., and Hamm-Alvarez, S.F. (2006). Novel fiber-dependent entry mechanism for adenovirus serotype 5 in lacrimal acini, *J. Virol.*, **80**, pp. 11833–11851.
34. Bayo-Puxan, N., Cascallo, M., Gros, A., Huch, M., Fillat, C., and Alemany, R. (2006). Role of the putative heparan sulfate glycosaminoglycan-binding site of the adenovirus type 5 fiber shaft on liver detargeting and knob-mediated retargeting, *J. Gen. Virol.*, **87**, pp. 2487–2495.
35. Gaggar, A., Shayakhmetov, D.M., and Lieber, A. (2003). CD46 is a cellular receptor for group B adenoviruses, *Nat. Med.*, **9**, pp. 1408–1412.
36. Wang, H., Li, Z.Y., Liu, Y., Persson, J., Beyer, I., Moller, T., Koyuncu, D., Drescher, M.R., Strauss, R., Zhang, X.B., Wahl, J.K., 3rd, Urban, N., Drescher, C., Hemminki, A., Fender, P., and Lieber, A. (2011). Desmoglein 2 is a receptor for adenovirus serotypes 3, 7, 11 and 14, *Nat. Med.*, **17**, pp. 96–104.
37. Kelkar, S.A., Pfister, K.K., Crystal, R.G., Leopold, P.L., Kelkar, S., De, B.P., Gao, G., and Wilson, J.M. (2004). Cytoplasmic dynein mediates adenovirus binding to microtubules, *J. Virol.*, **78**, pp. 10122–10132.
38. Leopold, P.L., Kreitzer, G., Miyazawa, N., Rempel, S., Pfister, K.K., Rodriguez-Boulan, E., and Crystal, R.G. (2000). Dynein- and microtubule-mediated translocation of adenovirus serotype 5 occurs after endosomal lysis, *Hum. Gene Ther.*, **11**, pp. 151–165.
39. Suomalainen, M., Nakano, M.Y., Keller, S., Boucke, K., Stidwill, R.P., and Greber, U.F. (1999). Microtubule-dependent plus- and minus end-directed motilities are competing processes for nuclear targeting of adenovirus, *J. Cell Biol.*, **144**, pp. 657–672.
40. Greber, U.F., Willetts, M., Webster, P., and Helenius, A. (1993). Stepwise dismantling of adenovirus 2 during entry into cells, *Cell*, **75**, pp. 477–486.
41. Curiel, D.T., Agarwal, S., Wagner, E., and Cotten, M. (1991). Adenovirus enhancement of transferrin-polylysine-mediated gene delivery, *Proc. Natl. Acad. Sci. U.S.A.*, **88**, pp. 8850–8854.
42. Wagner, E., Zatloukal, K., Cotten, M., Kirlappos, H., Mechtler, K., Curiel, D.T., and Birnstiel, M.L. (1992). Coupling of adenovirus to transferrin-

polylysine/DNA complexes greatly enhances receptor-mediated gene delivery and expression of transfected genes, *Proc. Natl. Acad. Sci. U.S.A.*, **89**, pp. 6099–7103.

43. Prchla, E., Plank, C., Wagner, E., Blaas, D., and Fuchs, R. (1995). Virus-mediated release of endosomal content *in vitro*: Different behavior of adenovirus and rhinovirus serotype 2, *J. Cell Biol.*, **131**, pp. 111–123.
44. Boudin, M.L., and Boulanger, P. (1982). Assembly of adenovirus penton base and fiber, *Virology*, **116**, pp. 589–604.
45. Shayakhmetov, D.M., and Lieber, A. (2000). Dependence of adenovirus infectivity on length of the fiber shaft domain, *J. Virol.*, **74**, pp. 10274–10286.
46. Wiethoff, C.M., Wodrich, H., Gerace, L., and Nemerow, G.R. (2005). Adenovirus protein VI mediates membrane disruption following capsid disassembly, *J. Virol.*, **79**, pp. 1992–2000.
47. Seth, P., Fitzgerald, D., Ginsberg, H., Willingham, M., and Pastan, I. (1984). Evidence that the penton base of adenovirus is involved in potentiation of toxicity of Pseudomonas exotoxin conjugated to epidermal growth factor, *Mol. Cell. Biol.*, **4**, pp. 1528–1533.
48. Hong, S.S., Gay, B., Karayan, L., Dabauvalle, M.C., and Boulanger, P. (1999). Cellular uptake and nuclear delivery of recombinant adenovirus penton base, *Virology*, **262**, pp. 163–177.
49. Rentsendorj, A., Xie, J., MacVeigh, M., Agadjanian, H., Bass, S., Kim, D.H., Rossi, J., Hamm-Alvarez, S.F., and Medina-Kauwe, L.K. (2006). Typical and atypical trafficking pathways of Ad5 penton base recombinant protein: Implications for gene transfer, *Gene Ther.*, **13**, pp. 821–836.
50. Medina-Kauwe, L.K., Kasahara, N., and Kedes, L. (2001). 3PO, a novel non-viral gene delivery system using engineered Ad5 penton proteins, *Gene Ther.*, **8**, pp. 795–803.
51. Medina-Kauwe, L.K., Maguire, M., Kasahara, N., and Kedes, L. (2001). Non-viral gene delivery to human breast cancer cells by targeted Ad5 penton proteins, *Gene Ther.*, **8**, pp. 1753–1761.
52. Rentsendorj, A., Agadjanian, H., Chen, X., Cirivello, M., Macveigh, M., Kedes, L., Hamm-Alvarez, S., and Medina-Kauwe, L.K. (2005). The Ad5 fiber mediates nonviral gene transfer in the absence of the whole virus, utilizing a novel cell entry pathway, *Gene Ther.*, **12**, pp. 225–237.
53. Sun, G., Hsueh, P.Y., Janib, S.M., Hamm-Alvarez, S., and Andrew MacKay, J. (2011). Design and cellular internalization of genetically engineered polypeptide nanoparticles displaying adenovirus knob domain, *J. Control. Release*, **155**, pp. 218–226.

54. Zhang, F., Andreassen, P., Fender, P., Geissler, E., Hernandez, J.F., and Chroboczek, J. (1999). A transfecting peptide derived from adenovirus fiber protein, *Gene Ther.*, **6**, pp. 171–181.
55. Miyazawa, N., Crystal, R.G., and Leopold, P.L. (2001). Adenovirus serotype 7 retention in a late endosomal compartment prior to cytosol escape is modulated by fiber protein, *J. Virol.*, **75**, pp. 1387–1400.
56. Miyazawa, N., Leopold, P.L., Hackett, N.R., Ferris, B., Worgall, S., Falck-Pedersen, E., and Crystal, R.G. (1999). Fiber swap between adenovirus subgroups B and C alters intracellular trafficking of adenovirus gene transfer vectors, *J. Virol.*, **73**, pp. 6056–6065.
57. Cotten, M., and Weber, J.M. (1995). The adenovirus protease is required for virus entry into host cells, *Virology*, **213**, pp. 494–502.
58. Greber, U.F., Webster, P., Weber, J., and Helenius, A. (1996). The role of the adenovirus protease on virus entry into cells, *EMBO J.*, **15**, pp. 1766–1777.
59. Maier, O., Galan, D.L., Wodrich, H., and Wiethoff, C.M. (2010). An N-terminal domain of adenovirus protein VI fragments membranes by inducing positive membrane curvature, *Virology*, **402**, pp. 11–19.
60. Rentsendorj, A., Agadjanian, H., Ma, J., Valluripalli, V., Kim, D., Rossi, J., and Medina-Kauwe, L.K. (2008). New approach for targeting rnai-mediated toxicity to cancer cells, *Mol. Ther.*, **16**, pp. S217–S218.
61. Agadjanian, H., Chu, D., Hwang, J.Y., Wachsmann-Hogiu, S., Rentsendorj, A., Song, L., Valluripalli, V., Lubow, J., Ma, J., Sharifi, B., Farkas, D.L., and Medina-Kauwe, L.K. (2012). Chemotherapy targeting by DNA capture in viral protein particles, *Nanomedicine*, **7**, pp. 335–352.
62. Agadjanian, H., Ma, J., Rentsendorj, A., Valluripalli, V., Hwang, J.Y., Mahammed, A., Farkas, D.L., Gray, H.B., Gross, Z., and Medina-Kauwe, L.K. (2009). Tumor detection and elimination by a targeted gallium corrole, *Proc. Natl. Acad. Sci. U.S.A.*, **106**, pp. 6105–6110. Epub 2009 Apr 2.
63. Hwang, J.Y., Park, J., Kang, B.J., Lubow, D.J., Chu, D., Farkas, D.L., Shung, K.K., and Medina-Kauwe, L.K. (2012). Multimodality imaging *in vivo* for preclinical assessment of tumor-targeted doxorubicin nanoparticles, *PLoS One*, **7**, pp. e34463.
64. Hwang, J.Y., Lubow, J., Chu, D., Ma, J., Agadjanian, H., Sims, J., Gray, H.B., Gross, Z., Farkas, D.L., and Medina-Kauwe, L.K. (2011). A mechanistic study of tumor-targeted corrole toxicity, *Mol. Pharm.*, **8**, pp. 2233–2243.

65. Fender, P., Ruigrok, R.W., Gout, E., Buffet, S., and Chroboczek, J. (1997). Adenovirus dodecahedron, a new vector for human gene transfer, *Nat. Biotechnol.*, **15**, pp. 52–56.

66. Szolajska, E., Burmeister, W.P., Zochowska, M., Nerlo, B., Andreev, I., Schoehn, G., Andrieu, J.-P., Fender, P., Naskalska, A., Zubieta, C., Cusack, S., and Chroboczek, J. (2012). The structural basis for the integrity of adenovirus Ad3 dodecahedron, *PLoS One*, **7**, pp. e46075.

67. Vives, R.R., Lortat-Jacob, H., Chroboczek, J., and Fender, P. (2004). Heparan sulfate proteoglycan mediates the selective attachment and internalization of serotype 3 human adenovirus dodecahedron, *Virology*, 3**21**, pp. 332–340.

68. Fender, P., Schoehn, G., Perron-Sierra, F., Tucker, G.C., and Lortat-Jacob, H. (2008). Adenovirus dodecahedron cell attachment and entry are mediated by heparan sulfate and integrins and vary along the cell cycle, *Virology*, **371**, pp. 155–164.

69. Zochowska, M., Paca, A., Schoehn, G., Andrieu, J.P., Chroboczek, J., Dublet, B., and Szolajska, E. (2009). Adenovirus dodecahedron, as a drug delivery vector, *PLoS One*, **4**, pp. e5569.

70. Villegas-Mendez, A., Garin, M.I., Pineda-Molina, E., Veratti, E., Bueren, J.A., Fender, P., and Lenormand, J.-L. (2010). *In vivo* delivery of antigens by adenovirus dodecahedron induces cellular and humoral immune responses to elicit antitumor immunity, *Mol. Ther.*, **18**, pp. 1046–1053.

71. Beyer, I., Cao, H., Persson, J., Song, H., Richter, M., Feng, Q., Yumul, R., van Rensburg, R., Li, Z., Berenson, R., Carter, D., Roffler, S., Drescher, C., and Lieber, A. (2012). Coadministration of epithelial junction opener JO-1 improves the efficacy and safety of chemotherapeutic drugs, *Clin. Cancer Res.*, **18**, pp. 3340–3351.

72. Biedermann, K., Vogelsang, H., Becker, I., Plaschke, S., Siewert, J.R., Hofler, H., and Keller, G. (2005). Desmoglein 2 is expressed abnormally rather than mutated in familial and sporadic gastric cancer, *J. Pathol.*, **207**, pp. 199–206.

73. Harada, H., Iwatsuki, K., Ohtsuka, M., Han, G.W., and Kaneko, F. (1996). Abnormal desmoglein expression by squamous cell carcinoma cells, *Acta. Derm. Venereol.*, **76**, pp. 417–420.

74. Wang, H., Li, Z., Yumul, R., Lara, S., Hemminki, A., Fender, P., and Lieber, A. (2011). Multimerization of adenovirus serotype 3 fiber knob domains is required for efficient binding of virus to desmoglein 2 and subsequent opening of epithelial junctions, *J. Virol.*, **85**, pp. 6390–6402.

75. Ehrlich, M., Boll, W., Van Oijen, A., Hariharan, R., Chandran, K., Nibert, M.L., and Kirchhausen, T. (2004). Endocytosis by random initiation and stabilization of clathrin-coated pits, *Cell*, **118**, pp. 591–605.
76. Fotin, A., Kirchhausen, T., Grigorieff, N., Harrison, S.C., Walz, T., and Cheng, Y. (2006). Structure determination of clathrin coats to subnanometer resolution by single particle cryo-electron microscopy, *J. Struct. Biol.*, **156**, pp. 453–460.
77. Pizarro-Cerdá, J., Bonazzi, M., and Cossart, P. (2010). Clathrin-mediated endocytosis: What works for small, also works for big, *BioEssays*, **32**, pp. 496–504.
78. Cureton, D.K., Massol, R.H., Saffarian, S., Kirchhausen, T.L., and Whelan, S.P. (2009). Vesicular stomatitis virus enters cells through vesicles incompletely coated with clathrin that depend upon actin for internalization, *PLoS Pathog.*, **5**, pp. e1000394.
79. Moreno-Ruiz, E., Galan-Diez, M., Zhu, W., Fernandez-Ruiz, E., d'Enfert, C., Filler, S.G., Cossart, P., and Veiga, E. (2009). Candida albicans internalization by host cells is mediated by a clathrin-dependent mechanism, *Cell Microbiol.*, **11**, pp. 1179–1189.
80. Veiga, E., Guttman, J.A., Bonazzi, M., Boucrot, E., Toledo-Arana, A., Lin, A.E., Enninga, J., Pizarro-Cerdá, J., Finlay, B.B., Kirchhausen, T., and Cossart, P. (2007). Invasive and adherent bacterial pathogens co-opt host clathrin for infection, *Cell Host Microbe*, **2**, pp. 340–351.
81. Cossart, P., and Veiga, E. (2008). Non-classical use of clathrin during bacterial infections, *J. Microsc.*, **231**, pp. 524–528.
82. Schnupf, P., and Portnoy, D.A. (2007). Listeriolysin O: A phagosome-specific lysin, *Microbes Infect.*, **9**, pp. 1176–1187.
83. Tilney, L.G., and Portnoy, D.A. (1989). Actin filaments and the growth, movement, and spread of the intracellular bacterial parasite, Listeria monocytogenes, *J. Cell Biol.*, **109**, pp. 1597–1608.
84. Cossart, P., and Sansonetti, P.J. (2004). Bacterial invasion: The paradigms of enteroinvasive pathogens, *Science*, **304**, pp. 242–248.
85. Loisel, T.P., Boujemaa, R., Pantaloni, D., and Carlier, M.F. (1999). Reconstitution of actin-based motility of Listeria and Shigella using pure proteins, *Nature*, **401**, pp. 613–616.
86. Wiesner, S., Helfer, E., Didry, D., Ducouret, G., Lafuma, F., Carlier, M.F., and Pantaloni, D. (2003). A biomimetic motility assay provides insight into the mechanism of actin-based motility, *J. Cell Biol.*, **160**, pp. 387–398.

87. Cameron, L.A., Footer, M.J., Van Oudenaarden, A., and Theriot, J.A. (1999). Motility of ActA protein-coated microspheres driven by actin polymerization, *Proc. Natl. Acad. Sci. U.S.A.*, **96**, pp. 4908–4913.

88. Ng, C.P., Goodman, T.T., Park, I.-K., and Pun, S.H. (2009). Biomimetic surface engineering of plasmid-loaded nanoparticles for active intracellular trafficking by actin comet-tail motility, *Biomaterials*, **30**, pp. 951–958.

89. Endo, Y., and Tsurugi, K. (1987). RNA N-glycosidase activity of ricin A-chain. Mechanism of action of the toxic lectin ricin on eukaryotic ribosomes, *J. Biol. Chem.*, **262**, pp. 8128–8130.

90. Endo, Y., Tsurugi, K., Yutsudo, T., Takeda, Y., Ogasawara, T., and Igarashi, K. (1988). Site of action of a Vero toxin (VT2) from *Escherichia coli* O157:H7 and of Shiga toxin on eukaryotic ribosomes. RNA N-glycosidase activity of the toxins, *Eur. J. Biochem.*, **171**, pp. 45–50.

91. Moss, J., and Vaughan, M. (1977). Mechanism of action of choleragen. Evidence for ADP-ribosyltransferase activity with arginine as an acceptor, *J. Biol. Chem.*, **252**, pp. 2455–2457.

92. Kaper, J.B., Morris, J.G., Jr., and Levine, M.M. (1995). Cholera, *Clin. Microbiol. Rev.*, **8**, pp. 48–86.

93. Iglewski, B.H., and Kabat, D. (1975). NAD-dependent inhibition of protein synthesis by *Pseudomonas aeruginosa* toxin, *Proc. Natl. Acad. Sci. U.S.A.*, **72**, pp. 2284–2288.

94. Lord, J.M., and Roberts, L.M. (1998). Toxin entry: Retrograde transport through the secretory pathway, *J. Cell Biol.*, **140**, pp. 733–736.

95. Zhang, M., and Schekman, R. (2013). Unconventional secretion, unconventional solutions, *Science*, **340**, pp. 559–561.

96. Sandvig, K., and van Deurs, B. (2000). Entry of ricin and Shiga toxin into cells: Molecular mechanisms and medical perspectives, *EMBO J.*, **19**, pp. 5943–5950.

97. Waddell, T., Cohen, A., and Lingwood, C.A. (1990). Induction of verotoxin sensitivity in receptor-deficient cell lines using the receptor glycolipid globotriosylceramide, *Proc. Natl. Acad. Sci. U.S.A.*, **87**, pp. 7898–7901.

98. Jacewicz, M., Clausen, H., Nudelman, E., Donohue-Rolfe, A., and Keusch, G.T. (1986). Pathogenesis of shigella diarrhea. XI. Isolation of a shigella toxin-binding glycolipid from rabbit jejunum and HeLa cells and its identification as globotriaosylceramide, *J. Exp. Med.*, **163**, pp. 1391–1404.

99. Lindberg, A.A., Brown, J.E., Stromberg, N., Westling-Ryd, M., Schultz, J.E., and Karlsson, K.A. (1987). Identification of the carbohydrate receptor for Shiga toxin produced by Shigella dysenteriae type 1, *J. Biol. Chem.*, **262**, pp. 1779–1785.

100. Holmgren, J., Lonnroth, I., and Svennerholm, L. (1973). Tissue receptor for cholera exotoxin: Postulated structure from studies with GM1 ganglioside and related glycolipids, *Infect. Immun.*, **8**, pp. 208–214.

101. Holmgren, J., Lonnroth, I., and Svennerholm, L. (1973). Fixation and inactivation of cholera toxin by GM1 ganglioside, *Scand. J. Infect. Dis.*, **5**, pp. 77–78.

102. Kounnas, M.Z., Morris, R.E., Thompson, M.R., FitzGerald, D.J., Strickland, D.K., and Saelinger, C.B. (1992). The alpha 2-macroglobulin receptor/low density lipoprotein receptor-related protein binds and internalizes Pseudomonas exotoxin A, *J. Biol. Chem.*, **267**, pp. 12420–12423.

103. Olsnes, S., and Pihl, A. (1982). Molecular action of toxins and viruses. In *Toxic Lectins and Related Proteins*, Cohen, P., and van Heyningen, S. (Eds.), Elsevier, Amsterdam, pp. 51–105.

104. Lombardi, D., Soldati, T., Riederer, M.A., Goda, Y., Zerial, M., and Pfeffer, S.R. (1993). Rab9 functions in transport between late endosomes and the trans-Golgi network, *EMBO J.*, **12**, pp. 677–682.

105. Sandvig, K., Grimmer, S., Lauvrak, S.U., Torgersen, M.L., Skretting, G., van Deurs, B., and Iversen, T.G. (2002). Pathways followed by ricin and Shiga toxin into cells, *Histochem. Cell Biol.*, **117**, pp. 131–141. Epub 2001 Nov 20.

106. Bonifacino, J.S., and Rojas, R. (2006). Retrograde transport from endosomes to the trans-Golgi network, *Nat. Rev. Mol. Cell Biol.*, **7**, pp. 568–579.

107. Lewis, M.J., Nichols, B.J., Prescianotto-Baschong, C., Riezman, H., and Pelham, H.R. (2000). Specific retrieval of the exocytic SNARE Snc1p from early yeast endosomes, *Mol. Biol. Cell*, **11**, pp. 23–38.

108. Miesenbock, G., and Rothman, J.E. (1995). The capacity to retrieve escaped ER proteins extends to the trans-most cisterna of the Golgi stack, *J. Cell Biol.*, **129**, pp. 309–319.

109. Cosson, P., and Letourneur, F. (1997). Coatomer (COPI)-coated vesicles: Role in intracellular transport and protein sorting, *Curr. Opin. Cell Biol.*, **9**, pp. 484–487.

110. Letourneur, F., Gaynor, E.C., Hennecke, S., Demolliere, C., Duden, R., Emr, S.D., Riezman, H., and Cosson, P. (1994). Coatomer is essential for

retrieval of dilysine-tagged proteins to the endoplasmic reticulum, *Cell,* **79**, pp. 1199–1207.

111. Smith, D.C., Spooner, R.A., Watson, P.D., Murray, J.L., Hodge, T.W., Amessou, M., Johannes, L., Lord, J.M., and Roberts, L.M. (2006). Internalized Pseudomonas exotoxin A can exploit multiple pathways to reach the endoplasmic reticulum, *Traffic,* **7**, pp. 379–393.
112. Chaudhary, V.K., Jinno, Y., FitzGerald, D., and Pastan, I. (1990). Pseudomonas exotoxin contains a specific sequence at the carboxyl terminus that is required for cytotoxicity, *Proc. Natl. Acad. Sci. U.S.A.,* **87**, pp. 308–312.
113. Jackson, M.E., Simpson, J.C., Girod, A., Pepperkok, R., Roberts, L.M., and Lord, J.M. (1999). The KDEL retrieval system is exploited by Pseudomonas exotoxin A, but not by Shiga-like toxin-1, during retrograde transport from the Golgi complex to the endoplasmic reticulum, *J. Cell Sci.,* **112**, pp. 467–475.
114. Kreitman, R.J., and Pastan, I. (1995). Importance of the glutamate residue of KDEL in increasing the cytotoxicity of Pseudomonas exotoxin derivatives and for increased binding to the KDEL receptor, *Biochem. J.,* **307**, pp. 29–37.
115. Seetharam, S., Chaudhary, V.K., FitzGerald, D., and Pastan, I. (1991). Increased cytotoxic activity of Pseudomonas exotoxin and two chimeric toxins ending in KDEL, *J. Biol. Chem.,* **266**, pp. 17376–17381.
116. Day, P.J., Owens, S.R., Wesche, J., Olsnes, S., Roberts, L.M., and Lord, J.M. (2001). An interaction between ricin and calreticulin that may have implications for toxin trafficking, *J. Biol. Chem.,* **276**, pp. 7202–7208. Epub 2000 Dec 11.
117. Spooner, R.A., Smith, D.C., Easton, A.J., Roberts, L.M., and Lord, J.M. (2006). Retrograde transport pathways utilised by viruses and protein toxins, *Virol. J.,* **3**, pp. 26.
118. Medina-Kauwe, L.K. (2007). "Alternative" endocytic mechanisms exploited by pathogens: New avenues for therapeutic delivery?, *Advanced Drug Delivery Reviews,* **59**(8), 798–809. PMID: 17707545.
119. Clemons, W.M., Jr., Menetret, J.F., Akey, C.W., and Rapoport, T.A. (2004). Structural insight into the protein translocation channel, *Curr. Opin. Struct. Biol.,* **14**, pp. 390–396.
120. Van den Berg, B., Clemons, W.M., Jr., Collinson, I., Modis, Y., Hartmann, E., Harrison, S.C., and Rapoport, T.A. (2004). X-ray structure of a protein-conducting channel, *Nature,* **427**, pp. 36–44. Epub 2003 Dec 3.

121. Hazes, B., and Read, R.J. (1997). Accumulating evidence suggests that several AB-toxins subvert the endoplasmic reticulum-associated protein degradation pathway to enter target cells, *Biochemistry*, **36**, pp. 11051–11054.

122. Fominaya, J., Uherek, C., and Wels, W. (1998). A chimeric fusion protein containing transforming growth factor-α mediates gene transfer via binding to the EGF receptor, *Gene Ther.*, **5**, pp. 521–530.

123. Fominaya, J., and Wels, W. (1996). Target cell-specific DNA transfer mediated by a chimeric multidomain protein. Novel non-viral gene delivery system, *J. Biol. Chem.*, **271**, pp. 10560–10568.

124. Barrett, L.B., Berry, M., Ying, W.B., Hodgkin, M.N., Seymour, L.W., Gonzalez, A.M., Read, M.L., Baird, A., and Logan, A. (2004). CTb targeted non-viral cDNA delivery enhances transgene expression in neurons, *J. Gene Med.*, **6**, pp. 429–438.

125. Barrett, L.B., Logan, A., Berry, M., Ying, W., Gonzalez, A.M., Baird, A., and Seymour, L.W. (1999). Targeted transfection of neuronal cells using a poly(D-lysine)-cholera-toxin b chain conjugate, *Biochem. Soc. Trans.*, **27**, pp. 851–857.

126. Gaur, R., Gupta, P.K., Goyal, A., Wels, W., and Singh, Y. (2002). Delivery of nucleic acid into mammalian cells by anthrax toxin, *Biochem. Biophys. Res. Commun.*, **297**, pp. 1121–1127.

127. Facchini, L.M., and Lingwood, C.A. (2001). A verotoxin 1 B subunit-lambda CRO chimeric protein specifically binds both DNA and globotriaosylceramide (Gb(3)) to effect nuclear targeting of exogenous DNA in Gb(3) positive cells, *Exp. Cell. Res.*, **269**, pp. 117–129.

128. Sandvig, K., and van Deurs, B. (2005). Delivery into cells: Lessons learned from plant and bacterial toxins, *Gene Ther.*, **12**, pp. 865–872.

129. Fisher, K.J., and Wilson, J.M. (1997). The transmembrane domain of diphtheria toxin improves molecular conjugate gene transfer, *Biochem. J.*, **321**, pp. 49–58.

130. Madshus, I.H., Olsnes, S., and Stenmark, H. (1992). Membrane translocation of diphtheria toxin carrying passenger protein domains, *Infect. Immun.*, **60**, pp. 3296–3302.

131. Stenmark, H., Moskaug, J.O., Madshus, I.H., Sandvig, K., and Olsnes, S. (1991). Peptides fused to the amino-terminal end of diphtheria toxin are translocated to the cytosol, *J. Cell Biol.*, **113**, pp. 1025–1032.

132. Smith, D.C., Lord, J.M., Roberts, L.M., Tartour, E., and Johannes, L. (2002). 1st class ticket to class I: Protein toxins as pathfinders for antigen presentation, *Traffic*, **3**, pp. 697–704.

133. Cabiaux, V. (2004). pH-sensitive toxins: Interactions with membrane bilayers and application to drug delivery, *Adv. Drug Deliv. Rev.,* **56**, pp. 987–997.

134. Donnelly, J.J., Ulmer, J.B., Hawe, L.A., Friedman, A., Shi, X.P., Leander, K.R., Shiver, J.W., Oliff, A.I., Martinez, D., Montgomery, D., and Liu, M.A. (1993). Targeted delivery of peptide epitopes to class I major histocompatibility molecules by a modified Pseudomonas exotoxin, *Proc. Natl. Acad. Sci. U.S.A.,* **90**, pp. 3530–3534.

135. Haicheur, N., Bismuth, E., Bosset, S., Adotevi, O., Warnier, G., Lacabanne, V., Regnault, A., Desaymard, C., Amigorena, S., Ricciardi-Castagnoli, P., Goud, B., Fridman, W.H., Johannes, L., and Tartour, E. (2000). The B subunit of Shiga toxin fused to a tumor antigen elicits CTL and targets dendritic cells to allow MHC class I-restricted presentation of peptides derived from exogenous antigens, *J. Immunol.,* **165**, pp. 3301–3308.

136. Pelkmans, L., Kartenbeck, J., and Helenius, A. (2001). Caveolar endocytosis of simian virus 40 reveals a new two-step vesicular-transport pathway to the ER, *Nat. Cell Biol.*, **3**, pp. 473–483.

137. Damm, E.M., Pelkmans, L., Kartenbeck, J., Mezzacasa, A., Kurzchalia, T., and Helenius, A. (2005). Clathrin- and caveolin-1-independent endocytosis: Entry of simian virus 40 into cells devoid of caveolae, *J. Cell Biol.,* **168**, pp. 477–488. Epub 2005 Jan 24.

138. Anderson, H.A., Chen, Y., and Norkin, L.C. (1996). Bound simian virus 40 translocates to caveolin-enriched membrane domains, and its entry is inhibited by drugs that selectively disrupt caveolae, *Mol. Biol. Cell,* **7**, pp. 1825–1834.

139. Pelkmans, L., Puntener, D., and Helenius, A. (2002). Local actin polymerization and dynamin recruitment in SV40-induced internalization of caveolae, *Science,* **296**, pp. 535–539.

140. Yamada, E. (1955). The fine structure of the gall bladder epithelium of the mouse, *J. Biophys. Biochem. Cytol.,* **1**, pp. 445–458.

141. Rothberg, K.G., Heuser, J.E., Donzell, W.C., Ying, Y.S., Glenney, J.R., and Anderson, R.G. (1992). Caveolin, a protein component of caveolae membrane coats, *Cell,* **68**, pp. 673–682.

142. Kurzchalia, T.V., Dupree, P., Parton, R.G., Kellner, R., Virta, H., Lehnert, M., and Simons, K. (1992). VIP21, a 21-kD membrane protein is an integral component of trans-Golgi-network-derived transport vesicles, *J. Cell Biol.,* **118**, pp. 1003–1014.

143. Dietzen, D.J., Hastings, W.R., and Lublin, D.M. (1995). Caveolin is palmitoylated on multiple cysteine residues. Palmitoylation is not

necessary for localization of caveolin to caveolae, *J. Biol. Chem.*, **270**, pp. 6838–6842.

144. Murata, M., Peranen, J., Schreiner, R., Wieland, F., Kurzchalia, T.V., and Simons, K. (1995). VIP21/caveolin is a cholesterol-binding protein, *Proc. Natl. Acad. Sci. U.S.A.*, **92**, pp. 10339–10343.
145. Cohen, A.W., Hnasko, R., Schubert, W., and Lisanti, M.P. (2004). Role of caveolae and caveolins in health and disease, *Physiol. Rev.*, **84**, pp. 1341–1379.
146. Richterova, Z., Liebl, D., Horak, M., Palkova, Z., Stokrova, J., Hozak, P., Korb, J., and Forstova, J. (2001). Caveolae are involved in the trafficking of mouse polyomavirus virions and artificial VP1 pseudocapsids toward cell nuclei, *J. Virol.*, **75**, pp. 10880–10891.
147. Pietiainen, V., Marjomaki, V., Upla, P., Pelkmans, L., Helenius, A., and Hyypia, T. (2004). Echovirus 1 endocytosis into caveosomes requires lipid rafts, dynamin II, and signaling events, *Mol. Biol. Cell.*, **15**, pp. 4911–4925. Epub 2004 Sep 8.
148. Marjomaki, V., Pietiainen, V., Matilainen, H., Upla, P., Ivaska, J., Nissinen, L., Reunanen, H., Huttunen, P., Hyypia, T., and Heino, J. (2002). Internalization of echovirus 1 in caveolae, *J. Virol.*, **76**, pp. 1856–1865.
149. Nunes-Correia, I., Eulalio, A., Nir, S., and Pedroso de Lima, M.C. (2004). Caveolae as an additional route for influenza virus endocytosis in MDCK cells, *Cell. Mol. Biol. Lett.*, **9**, pp. 47–60.
150. Nomura, R., Kiyota, A., Suzaki, E., Kataoka, K., Ohe, Y., Miyamoto, K., Senda, T., and Fujimoto, T. (2004). Human coronavirus 229E binds to CD13 in rafts and enters the cell through caveolae, *J. Virol.*, **78**, pp. 8701–8708.
151. Bernacchi, S., Mueller, G., Langowski, J., and Waldeck, W. (2004). Characterization of simian virus 40 on its infectious entry pathway in cells using fluorescence correlation spectroscopy, *Biochem. Soc. Trans.*, **32**, pp. 746–749.
152. Tsai, B., Gilbert, J.M., Stehle, T., Lencer, W., Benjamin, T.L., and Rapoport, T.A. (2003). Gangliosides are receptors for murine polyoma virus and SV40, *EMBO J.*, **22**, pp. 4346–4355.
153. Dangoria, N.S., Breau, W.C., Anderson, H.A., Cishek, D.M., and Norkin, L.C. (1996). Extracellular simian virus 40 induces an ERK/MAP kinase-independent signalling pathway that activates primary response genes and promotes virus entry, *J. Gen. Virol.*, **77**, pp. 2173–2182.
154. Anderson, H.A., Chen, Y., and Norkin, L.C. (1998). MHC class I molecules are enriched in caveolae but do not enter with simian virus 40, *J. Gen. Virol.*, **79**, pp. 1469–1477.

155. Stang, E., Kartenbeck, J., and Parton, R.G. (1997). Major histocompatibility complex class I molecules mediate association of SV40 with caveolae, *Mol. Biol. Cell,* **8**, pp. 47–57.

156. Mundy, D.I., Machleidt, T., Ying, Y.S., Anderson, R.G., and Bloom, G.S. (2002). Dual control of caveolar membrane traffic by microtubules and the actin cytoskeleton, *J. Cell Sci.*, **115**, pp. 4327–4339.

157. Pelkmans, L., Burli, T., Zerial, M., and Helenius, A. (2004). Caveolin-stabilized membrane domains as multifunctional transport and sorting devices in endocytic membrane traffic, *Cell,* **118**, pp. 767–780.

158. Richards, A.A., Stang, E., Pepperkok, R., and Parton, R.G. (2002). Inhibitors of COP-mediated transport and cholera toxin action inhibit simian virus 40 infection, *Mol. Biol. Cell,* **13**, pp. 1750–1764.

159. Schelhaas, M., Malmstrom, J., Pelkmans, L., Haugstetter, J., Ellgaard, L., Grunewald, K., and Helenius, A. (2007). Simian Virus 40 depends on ER protein folding and quality control factors for entry into host cells, *Cell,* **131**, pp. 516–529.

160. Ishii, N., Minami, N., Chen, E.Y., Medina, A.L., Chico, M.M., and Kasamatsu, H. (1996). Analysis of a nuclear localization signal of simian virus 40 major capsid protein Vp1, *J. Virol.*, **70**, pp. 1317–1322.

161. Kasamatsu, H., and Nakanishi, A. (1998). How do animal DNA viruses get to the nucleus? *Annu. Rev. Microbiol.*, **52**, pp. 627–686.

162. Nakanishi, A., Clever, J., Yamada, M., Li, P.P., and Kasamatsu, H. (1996). Association with capsid proteins promotes nuclear targeting of simian virus 40 DNA, *Proc. Natl. Acad. Sci. U.S.A.*, **93**, pp. 96–100.

163. Yamada, M., and Kasamatsu, H. (1993). Role of nuclear pore complex in simian virus 40 nuclear targeting, *J. Virol.*, **67**, pp. 119–130.

164. Oh, P., Borgstrom, P., Witkiewicz, H., Li, Y., Borgstrom, B.J., Chrastina, A., Iwata, K., Zinn, K.R., Baldwin, R., Testa, J.E., and Schnitzer, J.E. (2007). Live dynamic imaging of caveolae pumping targeted antibody rapidly and specifically across endothelium in the lung, *Nat. Biotechnol.,* **25**, pp. 327–337. Epub 2007 Mar 4.

165. Wang, Z., and Malik, A.B. (2013). Nanoparticles squeezing across the blood-endothelial barrier via caveolae, *Ther. Deliv.*, **4**, pp. 131–133.

166. Wang, Z., Tiruppathi, C., Minshall, R.D., and Malik, A.B. (2009). Size and dynamics of caveolae studied using nanoparticles in living endothelial cells, *ACS Nano*, **3**, pp. 4110–4116.

167. Gabrielson, N.P., and Pack, D.W. (2009). Efficient polyethylenimine-mediated gene delivery proceeds via a caveolar pathway in HeLa cells, *J. Control. Release*, **136**, pp. 54–61.

168. Zauner, W., Blaas, D., Kuechler, E., and Wagner, E. (1995). Rhinovirus-mediated endosomal release of transfection complexes, *J. Virol.,* **69**, pp. 1085–1092.

169. Bullough, P.A., Hughson, F.M., Skehel, J.J., and Wiley, D.C. (1994). Structure of influenza haemagglutinin at the pH of membrane fusion, *Nature,* **371**, pp. 37–43.

170. White, J.M., and Wilson, I.A. (1987). Anti-peptide antibodies detect steps in a protein conformational change: Low-pH activation of the influenza virus hemagglutinin, *J. Cell Biol.,* **105**, pp. 2887–2896.

171. Wagner, E., Plank, C., Zatloukal, K., Cotten, M., and Birnstiel, M.L. (1992). Influenza virus hemagglutinin HA-2 N-terminal fusogenic peptides augment gene transfer by transferrin-polylysine-DNA complexes: Toward a synthetic virus-like gene-transfer vehicle, *Proc. Natl. Acad. Sci. U.S.A.,* **89**, pp. 7934–7938.

172. Norkin, L.C. (2001). Caveolae in the uptake and targeting of infectious agents and secreted toxins, *Adv. Drug Deliv. Rev.*, **49**, pp. 301–315.

Section B

Strategies for Drug Penetration Across Tissue Compartments

Chapter 5

Drug Transport across Skin

Renata Vidor Contri,[a] Tatiele Katzer,[b] Silvia S. Guterres,[a] and Adriana R. Pohlmann[a,c]

[a]*Programa de Pós-Graduação em Ciências Farmacêuticas, Universidade Federal do Rio Grande do Sul, Av Ipiranga, 2752 Porto Alegre, RS 90610-000, Brazil*
[b]*Departamento de Biologia e Farmácia, Universidade de Santa Cruz do Sul, Av Independência, 2293 Santa Cruz do Sul, RS 96815-900, Brazil*
[c]*Departamento de Química Orgânica, Instituto de Química, Universidade Federal do Rio Grande do Sul, Av Bento Gonçalves, 9400 Porto Alegre, RS 91501-970, Brazil*
renata.contri@gmail.com

The presence of an efficient physical barrier—*stratum corneum*, the outermost layer of the epidermis—impairs the optimal permeation and penetration of a substance across the skin. Many strategies are being researched to improve the delivery of substances to the viable layers of the epidermis as well as the dermis and blood circulation, after topical administration. This chapter reviews the skin structure and its possible pathways for penetration and permeation, as well as the most relevant alternatives regarding chemical and physical methods to modify drug transport across the skin, with a focus on the role of composite nanoparticles based on lipids and/or polymers in this process. To comprehend and validate experimental research, a brief review about the available *in vitro* and *in vivo* methods to determine drug transport across skin is also included.

Drug Delivery Across Physiological Barriers
Edited by Silvia Muro

ISBN 978-981-4669-40-5 (Hardcover), 978-981-4669-41-2 (eBook)
www.panstanford.com

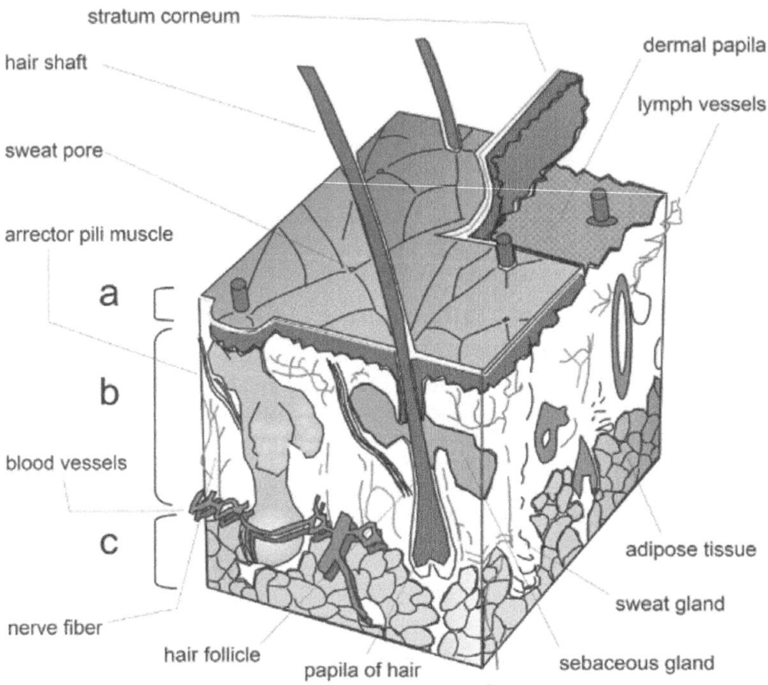

Figure 5.1 Skin structure (a: epidermis; b: dermis; c: hypodermis).

5.1 Introduction

The skin is the largest organ of the human body, acting as a barrier between the organism and the external environment. It is histologically divided into epidermis and dermis, besides the subcutaneous tissue (Fig. 5.1). The epidermis is subdivided into several strata, distinguished by changes that occur in keratinocytes as they move up from the dermis–epidermis junction to the outer surface of the epidermis, the *stratum corneum* (SC). The SC is almost entirely responsible for the skin barrier function. A determinant key of this barrier property is the dense lipid matrix that fills its extracellular space, which is composed of a heterogeneous mixture of ceramides, free fatty acids, and cholesterol. Lipids are disposed in lamellae located in the intercellular regions in the SC, contouring the surfaces of the cells and arranged in a crystalline sublattice, so that only a small proportion of lipids form a liquid phase.

Transport across the SC barrier occurs essentially by passive diffusion and can take place by any or a combination of three pathways: intercellular (substances permeate via contouring the cells, through the lipid matrix), transcellular (through the cells and lipid matrix), and through appendages (via the sweat and sebaceous glands or hair follicles). Through the transcellular pathway, a drug has to diffuse from an intercellular hydrophobic medium to an intracellular hydrophilic medium, several times. Diffusional resistance is found either at the cell wall or through intracellular keratin. Despite the small surface area, the intercellular pathway is the main way by which lipophilic substances and small molecules diffuse. Compared to the transcellular path, the diffusional resistance of this pathway is low. The skin appendages represent a route of penetration for ions, polar compounds, and large molecules, besides particles [1]. It is important to keep in mind that the presence of a substance into a skin appendage is not yet an absorption process itself, but the accumulation in such structures may lead to a faster and more efficient uptake due to altered barrier morphology in the appendages [2].

Percutaneous transport across the skin is often divided into penetration (the entry of a substance into a particular layer or structure), permeation (the penetration through one layer into a second layer that is functionally and structurally different from the first layer), and resorption (the uptake of a substance into the skin lymph and local vascular system leading to its systemic absorption) [3]. After reaching the viable epidermis, this layer will not provide much of a barrier, unless the drug is very lipophilic. Once the permeating species have crossed this region, the epidermal–dermal interface is reached, where the small blood vessels are situated and can remove the drug very efficiently. Recent models suggest that transport to lower dermis layers occurs mostly by blood/lymphatic flow rather than molecular diffusion alone [4]. When reaching the subcutaneous tissue, by diffusion or injection, this layer can serve as a drug reservoir, especially for lipophilic substances.

Transdermal drug delivery is an important route for drugs with local action in the skin, such as antimicrobial and anti-inflammatory substances and drugs meant to be taken up by the systemic circulation, such as hormones. The advantages of transdermal delivery include better patient compliance, avoidance of the pain and infection related to injections, and no gastrointestinal degradation and first-pass metabolism via the oral route, which can lead to considerable loss of drug. Also, the skin has emerged as a potential alternative route for noninvasive delivery of vaccines. It has been recognized as a highly immune-reactive tissue containing an abundance of antigen-presenting cells, especially within the epidermis. Topical immunization combines the advantages of needle-free delivery, while targeting the immunologically rich milieu of the skin. Vaccination through the skin may be particularly advantageous as immunocompetent Langerhans cells are found in abundance along the transdermal penetration pathways and these cells are aligned along the pores through which pathogens are likely to invade the body [5].

This chapter discusses the main factors influencing and modifying transport across the skin, since the difficulty in crossing the barrier is the highest limitation for the cutaneous route of drug administration. Strategies such as the use of permeation enhancers, physical methods, and the role of nanotechnology-based topical formulations to increase and modulate the penetration into the skin are presented.

5.2 Factors Influencing Transport Across Skin

Drugs usually present difficulties of penetrating the skin, especially due to low diffusion rate across the SC. Factors that influence this transport can be related to formulation, application, as well as skin composition and condition. All the main factors that can interfere in the drug penetration/permeation across the skin, as well as the main pathways for drug penetration into the tissue, are shown in Fig. 5.2.

In formulation, both lipophilicity and molecular size of the permeant are important factors [6]. The higher the lipophilicity and the lower the molecular size, the more penetrating capability the substance has. Lipophilicity is represented by logP (octanol/water partition coefficient) and influences the penetration due to increase

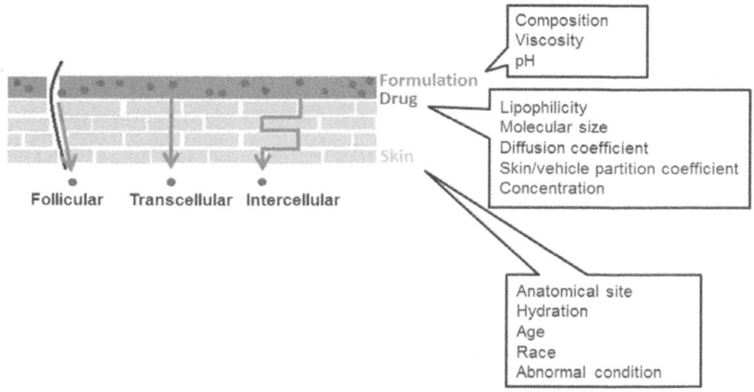

Figure 5.2 Main pathways for skin penetration and factors that interfere in the process.

in the interaction with the SC lipids. However, the skin/vehicle partition coefficient and the diffusion coefficient (mainly related with the concentration of the drug in the vehicle) are also influencing factors [7]. The supersaturation of a drug in a formulation, which increases the thermodynamic activity of the substance in its vehicle, can be used as a strategy to increase drug penetration/permeation. However, supersaturated formulations are unstable and drug crystallization can take place [8]. The pH, viscosity, and composition of the vehicle are of extreme importance, since the ionization of the drug, occlusion of the skin surface, modulation of drug release, and disruption of skin surface can occur. The use of permeation enhancers as well as the incorporation of drug in carrier systems is further discussed in Sections 5.4.1 and 5.4.3.

The skin composition and condition are parameters very closely related to drug transport across the skin. Changes in the SC hydration and lipid content are well correlated with skin permeability. Essential fatty acid deficiency can lead to abnormal skin barrier function. The level of cholesterol is more closely related with skin permeability than the levels of ceramide or free fatty acid, where increased cholesterol levels are responsible for a reduction in permeability [9]. However, ceramides also play an important role in skin barrier. It has been described that shorter-chain ceramides result in a significantly higher flux of model drugs compared to medium- or large-chain ceramides [10]. The protein composition of a tissue can also alter skin transport. The proteins of tight junctions, localized beneath the SC, represent another barrier when the surface is impaired [11]. Race and age should also be considered to predict skin permeability. Comparing East Asian (first-generation immigrants from China, Japan, and Korea), African American (Americans with origins in any of the black racial groups of Africa), and Caucasians, the first group presents the weakest barrier properties, while African Americans display a stronger barrier function [12], indicating higher permeability expected for light-colored skin (specially Asian origins) and lower permeability expected for dark-colored skin. Dark or black skin shows increased intracellular cohesion and higher lipid content compared to light or white skin [13]. Although the influence of age on the skin penetration represents a currently conflicting area, age seems to influence penetration due to differences in blood flow perfusion, pH, skin thickness, lipids, proteins and water content, and hair and pore densities between the elderly and the young [14]. Aged people may exhibit an increased barrier function due to increased amount of corneocytes on the skin surface. On the other side, infants of age less than 38 weeks show a very poor barrier [13].

Under some unusual conditions, the skin does not present its typical effective barrier function and becomes more permeable. Some diseases in which the skin is dry and the barrier function is defective include atopic dermatitis and ictiosis vulgaris. Psoriasis, which is characterized by an increase in keratinocyte proliferation and changes in the thickness of some epidermis layers, is also related to the increased skin permeability [13]. Some skin infections can also cause a temporary loss of skin barrier function [13].

The anatomical site of application of a drug can also influence its penetration across the skin [15], due to differences in the morphology of the tissue, skin hydration level, and presence of hair follicles. It has been shown that when substances are applied on the abdomen or on the dorsum of hand, their permeability is increased twofold, and this is fourfold increased when applied on the forehead and scalp, compared to the application on the forearm and on the hand palm [16]. Massaging the application area also increases the skin permeation of a substance. Other physical methods to modify drug permeation are described in Section 5.4.2.

5.3 Determining Drug Transport Across Skin

Penetration of chemicals and their subsequent permeation across the skin into a fluid reservoir can be determined through *in vitro* systems [3]. Generally, *in vitro* determination of drug transport across the skin is carried out using a Franz-cell dispositive, containing two compartments, a donor and a receptor, divided by a membrane. The membrane is usually human or pig skin, due to their anatomical similarities. The use of human skin grants a higher *in vitro–in vivo* correlation and, hence, it is considered the golden standard for *in vitro* penetration/permeation studies [3]. However, the use of pig skin or, more recently, reconstructed skin models presents the advantage of rendering lower variability. Full-thickness (epidermis and dermis), split-thickness (dermatomized), or isolated epidermis can be used [3]. A formulation is applied on the epidermis and, after a predetermined time and also after cleaning the skin surface, the amount of substance retained in the skin layers or permeated through the skin is analyzed by suitable methods such as tape stripping (penetration into SC), skin extraction (penetration into skin layers), or evaluation of the receptor fluid (permeation). The receptor fluid is usually composed of an isotonic saline solution or phosphate buffer with the addition of solubilizers [8, 17], maintained under stirring and at a fixed temperature. Different approaches can be taken regarding the amount of formulation to be applied. In the case of infinite dosing, the dose is considered constant, since the depletion of the drug in the donor chamber caused, for example, by its penetration across the barrier is negligible. On the other hand, the finite dosing, where a limited amount of the formulation is applied on the skin surface, is more reflective of a real *in vivo* situation, since evaporation of excipients and penetration of the drug affect the penetration/permeation profile. However, the use of finite dose regimen *in vitro* also associates with difficulties such as nonhomogeneous application and drug-detection problems [17].

The main advantage of performing an *in vivo* study over an *in vitro* one is that the former study uses a physiologically and metabolically intact system [18]. *In vivo* studies are carried out in laboratory animals, usually rodents (rabbit, mice, and hairless rats) and in human volunteers (considered the gold standard) [3]. Laboratory animals present higher skin permeability and systemic disposition compared to humans [3]. So an *in vivo* study with animals is likely to overestimate dermal absorption in humans [18]. The human studies, for technical and ethical reasons, are less often conducted [3]. The formulation is applied to a designated area of the skin, commonly the forearm or the back in humans and the back in laboratory animals [18]. Samples of body fluids, tissue, or excreta (urine, feces, exhaled air) are collected at predefined intervals. The chemical agent and/or its metabolites are measured by a suitable analytical procedure. Cutaneous microdialysis is an *in vivo* sampling technique used to measure endogenous and exogenous substances in the extracellular space beneath the skin (dermis, subcutaneous tissue, or blood vessels) using perfused dialysis. It allows monitoring the percutaneous absorption of various substances and can be used in human volunteers and in laboratory animal experiments [3].

The penetration of drugs across the skin and their percutaneous delivery are limited by the barrier function of the enormously organized structure of the SC, besides the multiple lipid bilayers of ceramides, fatty acids, cholesterol, and cholesterol esters that compose the "mortar" around keratinocytes. The following sections focus on the recent advancements in the transdermal drug delivery, which include chemical and physical methods, besides composite nanoparticles.

5.4 Modifying Drug Transport Across Skin

Transport of a substance across the skin can be modified by several approaches. The use of penetration enhancers, physical methods during application, or carrier systems are discussed in this section. Examples of these approaches are given in Table 5.1.

Table 5.1 Modifying drug transport across the skin through the use of chemical substances, physical methods, and composite nanoparticles

Approaches to modify the skin penetration of substances	Examples
Chemical enhancers [19–29]	Alcohols (ethanol, isopropyl alcohol), amides (Azone®, pyrrolidones), esters (ethyl acetate, octyl salicylate, isopropyl myristate), glycols (propylene glycol), glycol ethers (Transcutol®), fatty acids (oleic acid), sulfoxides (dimethylsulfoxide), surfactants (sodium lauryl sulfate, cetyltrimethyl ammonium bromide), terpenes (eugenol, limonene, menthol)
Physical methods [22, 26, 28–36]	Transdermal patches, microneedles, iontophoresis, electroporation, phonophoresis, microdermabrasion, laser, magnetophoresis, microwaves
Composite nanoparticles [37–63]	Liposomes, ethosomes, vesicles, nanoemulsions, solid lipid nanoparticles, nanostructured carrier systems, polymeric nanocapsules, polymeric nanospheres, hybrid nanoparticles

5.4.1 *Penetration Enhancers*

Penetration enhancers are chemicals added to topical formulations, which interact somehow with skin components, such as intercellular lipids, keratin domains, and/or desmosomes, increasing transport of drugs across the skin. Therefore, they are capable of increasing the permeability of the skin, influencing the diffusivity and/or solubility of a drug in the skin. Chemical enhancers are normally cheap and easy to formulate [19]. There is, so far, no ideal chemical enhancer, which satisfies all of the desirable properties such as no pharmacological effect, no toxicity, good cosmetic properties, chemical and physical compatibility, and fast and temporary effect on the barrier function [20]. Skin irritation is commonly observed after the application of penetration enhancers, due to their penetration up to the viable skin layers.

Penetration enhancers are substances from different chemical classes (alcohols, amides, esters, glycols, glycol ethers, fatty acids, sulfoxides, and surfactants), acting by different mechanisms to increase drug penetration [21]. A crucial point influencing the effectiveness of the majority of chemical enhancers is the interaction with the SC intercellular lipids by means of lipid extraction, lipid fluidization, phase separation, or polarity alteration [20]. Effects on intracellular keratin domains, on desmosome connections, on the metabolic activity within the skin, on the drug thermodynamic activity in the vehicle, and on the amount of drug in the skin are also possible ways to enhance penetration [22].

Alcohols, glycols, and surfactants are commonly added in dosage forms and patches [20]. The main long- or short-chain alcohols used as permeation enhancers are ethanol, isopropyl alcohol, hexanol, and octanol, among others [21]. Propylene glycol is commonly used in combination with other chemical enhancers. Surfactants solubilize lipids in the SC and interact with intracellular keratin. They present anionic (sodium lauryl sulfate), cationic (cetyltrimethyl ammonium bromide), or nonionic properties (polyoxyethylene sorbitan fatty acid ester), the latter presenting the lowest penetration enhancement effect [20, 21].

Dimethylsulfoxide, a powerful aprotic solvent, changes the conformation of intercellular keratin, causing increase in skin permeability, also interacting with the intercellular lipids [16]. However, it presents limited commercial utility due to its toxicity [21]. Hence, chemically related solvents such as dimethylformamide have been studied as substitutive agents [20]. Among the amides' group, Azone® is the first synthetic permeation enhancer, presenting effect due to its direct interaction with ceramides on the skin surface [21]. Compared to ethanol and polyethylene glycol, Azone® presents the greatest effect on the skin penetration of salbutamol [23].

Among the esters, ethyl acetate, octyl salicylate, and isopropyl myristate have been particularly studied [21]. The mechanism of action of ethyl acetate has not been elucidated yet, while octyl salicylate seems to reduce the conformational order of the SC lipid bilayer [21]. Isopropyl myristate and other esters of fatty acid have been widely studied and used in transdermal patches. They act by partitioning themselves in the lipid phase of the skin surface [19, 21]. Transcutol® is a glycol ether, which can be found in several cosmetic and dermatological products. However, the determination of its exact mechanism of action needs further investigation [21].

Fatty acids, mainly represented by oleic acid, have been proven to enhance the skin permeation of several substances, by means of interaction and modification of the SC lipids [20].

Water is also considered a natural permeation enhancer, and it is indeed the least toxic one [19]. The increase in permeability usually follows the process of occluding the skin through the use of formulations such as ointments [20]. The use of urea in cutaneous formulations presents some effect on skin penetration enhancement, due to hydration of the SC. In the same way, the use of phospholipids in nonvesicular form can occlude the skin causing increase in skin hydration and skin penetration [20].

The association of different chemical enhancers can lead to increase in efficacy, in a synergic way, especially with ternary mixtures [24]. Solvent mixtures, eutectic mixtures, inclusion complexes (cyclodextrins), and microemulsions are examples of the association of different permeation enhancers [19]. Cyclic oligosaccharides, or cyclodextrins, are permeation enhancer systems that increase skin penetration of lipophilic drugs in the presence of water by increasing the amount of drug at the skin surface [25] and also by interacting with free lipids on the skin surface [26]. Microemulsions are isotropic mixtures of oil, water, surfactant, and co-surfactant that have been used as vehicles for the transdermal delivery of hydrophilic and lipophilic drugs [27].

Several studies have analyzed the effect of the combination between chemical penetration enhancers and physical methods to increase skin permeability. Besides the already mentioned combination between penetration enhancers and patches, synergic effects have been described due to the combination of chemical substances and iontophoresis [23, 28, 29] for increasing skin penetration of drugs.

5.4.2 *Physical Methods*

To overcome the main limitation of topical and transdermal drug delivery (low permeability of the skin), physical means have been explored to improve drug (especially hydrophilic, charged, and/or large ones) penetration into the skin. The use of transdermal patches, microneedles, iontophoresis, electroporation, phonophoresis, microdermabrasion, and laser are the main approaches. They have been used in the clinics for diverse applications, including pain relief, inflammation treatment, and aesthetical purposes.

Transdermal patches are alternatives for the delivery of drugs to the body across the patient's skin. The occlusive effect of such patches when in contact with the skin is thought to contribute to drug penetration, since this occurs more rapidly through occluded hydrated skin [22]. The most common therapeutic indications for transdermal patches include hormone replacement, analgesia, pain relief, smoking cessation, neurological disorders, and, more recently, contraception and urinary incontinence [22, 26]. The disadvantages associated with the use of patches are local skin irritation, poor adhesion of the patches to the skin [22], limitations related to drug dose, and variability of the drug absorption rate based on the site of application, skin condition, heat, type of vehicles, and the properties of the drugs (molecular mass, pK_a, partition coefficient) [22, 30].

Another interesting approach is the use of microneedles. Hundreds of micron-sized needles pierce the SC, allowing the drug to pass this major barrier through microchannels [22, 26]. This strategy is considered minimally invasive, since the microneedles penetrate around 15 μm in depth keeping a safe and long distance from the papillary dermis, where the nerve fibers are predominately located, explaining the no-pain application. Besides, no bleeding is observed after microneedle insertion [22, 26, 30]. Drugs can be delivered in various ways by this technique, such as: (1) coating solid silicon microneedles with drug followed by their insertion into the skin for subsequent dissolution of the drug within the skin; (2) filling hollow metal microneedles with drug to obtain a drug deposit within the skin; (3) piercing an array of solid microneedles into the skin followed by the application of a drug patch; and (4) entrapping drug within biodegradable polymeric microneedles followed by insertion into the skin for controlled drug release [22, 26]. The poor safety and inconvenience regarding self-administration are probably the main reasons why microneedles-based transdermal drug delivery systems are not commercialized yet [26].

Iontophoresis is another technique currently used to facilitate transport of permeates across the skin. This strategy involves administration of ionizable drugs on the skin surface under an electrode bearing the same charge as the penetrant. An indifferent counter electrode is positioned elsewhere in the body, and a voltage source supplies the low-level electric current (0.5 mA/cm^2 or less) resulting in the repulsion of the active substance from the active electrode, forcing the drug into the skin [22, 30, 31]. The mechanisms of transdermal iontophoresis include electrophoresis (the electrically induced movement of charged particles), electroosmosis (electric field induced fluid flow), and electroporation (increasing the porosity of skin due to electric field) [26].

Electroporation may provide advantages over iontophoresis, such as faster transport and higher efficiency. Electroporation (or electropermeabilization) is a process in which brief intense electric pulses create small hydrophilic pores in the phospholipid bilayer of skin cell membranes. It has been suggested that electroporation affects the organization of the intercellular lipid bilayers of the SC as well [32]. This momentary structural perturbation provides the transdermal delivery of drugs. The efficacy of the transport depends on the electrical parameters (waveform, voltage, duration, and interval between pulses) and on the physicochemical properties of drugs [26]. Longer duration pulses ($>$10 ms) make possible two secondary effects: electrophoresis and Joule heating. Electrophoresis is the primary contributor in transdermal delivery of large charged molecules, while in Joule heating, the temperature rises above lipid phase transition temperatures, contributing to the changes in the SC lipid structure and permeability [32]. After the application of electric pulses, there is a transiently transepidermal water loss and an increased blood flow (noticed as an erythema), as well as a decrease in skin resistance [22]. The changes tend to decrease within a few hours after treatment. A possible explanation for the higher efficacy of electroporation over iontophoresis is that the pathways created at high voltage (electroporation) present less steric and/or electrical hindrance to transport than those undertaken during iontophoresis. The combination of these techniques (electroporation followed by iontophoresis) usually shows synergic effects.

Among the noninvasive methods, low-frequency ultrasound, widely known as sonophoresis or phonophoresis, has shown an enhancing effect on the transdermal delivery of various molecules, both *in vivo* and *in vitro*. This is an active form of transdermal delivery, which enhances the transport of drugs through cell membranes as a result of ultrasonic energy [30]. Cavitation is the well-accepted theory to explain the biological effect of ultrasound [33]. The formation of gaseous cavities/bubbles leads to the disruption of the lipid bilayers of the SC, thereby influencing the influx of drugs [22]. The disorder caused in the SC lipids might cause the formation of aqueous channels through the intercellular lipids, resulting in significant water penetration [33].

Microdermabrasion, another technique typically used with aesthetic purposes (smoothing the appearance of fine lines, wrinkles, and scars), has been experimentally tested to increase skin permeability. Microdermabrasion is a superficial cosmetic resurfacing technique that removes the skin layers by bombarding them with high-velocity abrasive particles [34].

Other physical methods certainly deserve further investigations, such as magnetophoresis (application of magnetic field) [35] and microwaves [36].

5.4.3 *Composite Nanoparticles Based on Lipids and/or Polymers*

Drug nanoencapsulation is a useful pharmaceutical strategy to modify the physicochemical properties of an encapsulated molecule and offers means to facilitate the percutaneous delivery of difficult-to-uptake substances. Composite nanoparticles usually present mean diameters in the range from 100 to 300 nm and are composed by organic substances (lipids and/or polymers) [37, 38]. They are separately presented in the following sections.

5.4.3.1 Lipid nanoparticles and nanoemulsions

The majority of drug delivery systems studied for skin use are composed of lipid nanoparticles and nanoemulsions [38]. The main kinds of lipid nanoparticles are solid lipid nanoparticles (SLNs), nanostructured lipid carriers, and vesicular systems, such as liposomes.

Liposomes were the first carrier systems introduced in the dermal area, starting in the cosmetic market in 1986 [39].

Liposomes are vesicular systems composed of lipid bilayers capable of encapsulating either hydrophobic (in the lipid bilayer) or hydrophilic drugs (in the aqueous core). They can enhance the amount of drug able to penetrate deeper skin layers, as well as sustain the drug delivery. The increased skin penetration property of vesicular systems seems to be more pronounced for hydrophilic drugs, compared to lipophilic ones, due to an increase in the partitioning of hydrophilic drugs into the SC [40]. The penetration of individual lipid components and their interaction with the skin lipid bilayer render a penetration enhancer effect for liposomes [41]. Moreover, liposomes may create a lipid-enriched environment in the SC, facilitating penetration of hydrophobic drugs [41].

Nanoemulsions with oil droplets lower than 200 nm present higher stability than conventional emulsions [42]. They have been proposed for cosmetic and pharmaceutical areas and might modulate (increase or control) drug penetration. However, compared to other lipid nanoparticles, the potential of nanoemulsions for controlling drug release is lower [43]. Skin uptake of a nanoencapsulated substance seems to be related to the charge of nanoemulsions. Cationic-charged nanodroplets present improved properties when compared to the anionic ones. [44]. Drug loading is usually limited to the solubility of the drug in the oily phase, but the surfactant can also play a role in increasing drug-loading capability.

At the beginning of the 1990s, SLNs were developed by simple exchange of liquid lipid in the colloidal phase of the formulation for a solid lipid at body and room temperatures [39]. These particles have broad applications in the cosmetic area and in the treatment of skin diseases. Compared to other types of lipid nanoparticles, such as nanoemulsions, the release of substances incorporated in these systems can be retarded due to its solid components [39, 43]. Additionally, when SLNs are applied on the skin, they can lose their shape, due to interactions with lipid components of the skin [45], occluding the skin surface or/and increasing skin permeability [39]. Greater skin uptake of substances was observed through the use of SLNs, indicating skin penetration enhancing properties. Surface modification and lipid composition of SLNs can change the skin penetration profile of the encapsulated drug [46, 47]. The incorporation of drugs in SLNs is affected by many factors, but the prerequisite to obtain a sufficient loading capacity is a sufficiently high solubility of the drug in the lipid melt. Besides, the chemical nature of the lipid is also important because lipids that form highly crystalline particles with a perfect lattice lead to drug expulsion [48], which is contoured by the addition of a liquid lipid.

Nanostructured lipid carriers (NLCs) are considered an evolution in the development of lipid carriers because they have high drug-loading capacity and stability during storage compared to SLNs [49, 50]. NLCs are formed by a blend of solid and liquid lipids [50]. Regarding the property of skin penetration, NLCs were described as less efficient than SLNs [51], but more efficient than nanoemulsions [52] when encapsulating Nile red dye.

5.4.3.2 Polymeric and hybrid nanoparticles

For drug delivery, the most promising and applied carrier technologies are based on polymeric materials [2]. Regarding skin applications, controlled release of drugs plays a major role, since encapsulated active ingredients usually have to diffuse through the polymeric matrix to permeate the skin [37]. Nanocapsules and nanospheres are the most common types of polymeric nanoparticles [53], which are carrier systems presenting diameters lower than 1 μm (usually around 200 to 300 nm). They can be differentiated according to their composition and structural organizations. Polymeric nanocapsules are reservoir or vesicular systems, composed of oily nanodroplets surrounded by a polymeric wall, whereas nanospheres do not have oil in their composition being represented as a polymeric matrix [53]. One of the advantages of nanocapsules over nanospheres is that drug loading can be greatly increased if the core material is a good solvent for the drug in question [53]. A modification in the core composition of polymeric nanocapsules led to a hybrid particle called lipid-core nanocapsule (LNC). Such hybrid nanocapsules have an organogel as core, composed of sorbitan monostearate and caprylic/capric trygliceride, a polymer wall, and polysorbate 80 micelles as stabilizers. LNCs are able to modulate and control the release profile of drugs [54].

Some of the most widely used polymers to produce polymeric nanoparticles are poly(ε-caprolactone), poly(lactic acid), poly(glycolic acid), and copolymers such as poly(lactide-co-glycolide) or poly(methyl methacrylate-co-ethyl acrylate). The chemical composition, size, surface charge, and the high specific surface area of the nanoparticles may play a role in determining their skin penetration and cellular effects [55]. For instance, poly(ε-caprolactone) nanocapsules, when coated with chitosan, presented differentiated drug penetration/permeation profile, probably due to the cationic charge [56].

Couvreur and coworkers (2002) [53] differentiate two important concepts: drug loading and encapsulation efficiency. The first expresses the percentage of drug by weight in the final formulation, and the last refers to the amount of drug encapsulated in the final nanoparticle product as the percentage of the total amount of drug added in the process. The most appropriate technique to evaluate encapsulation efficiency is the ultrafiltration–centrifugation method; however, this method cannot distinguish the mechanism of drug encapsulation. Considering encapsulation mechanisms associated with polymeric nanoparticles, a drug can be entrapped, dispersed, dissolved within, or adsorbed on the nanoparticles [57].

The ability of polymeric nanoparticles in controlling the skin penetration of encapsulated substances is well described. Therefore, polymeric nanocapsules are useful to retain active substances in the SC region [37, 58]. This characteristic is highly appropriated for sunscreen formulations, especially considering that the nanoparticles can act as physical UV sunscreens [57]. Incorporation in nanoparticles has also been described to increase penetration/permeation through the skin [59]. Moreover, combination of nanoparticles with iontophoresis leads to an increase in drug diffusion through the skin. Hence, physical penetration enhancers can be used along with nanoparticles to improve drug penetration [60].

Considering the skin penetration of the polymeric nanoparticles themselves (Fig. 5.3), there are no indications that particles bigger than 100 nm in size can pass the healthy skin barrier [2], which can be explained by the small intercellular spaces and lipids of the SC. However, there are reports of penetration of polymeric nanoparticles through the epidermis and dermis of rat skin (188 nm) [61], or accumulation into the epidermis of human skin (122 nm) [62]. It is important to mention that the sizes informed are

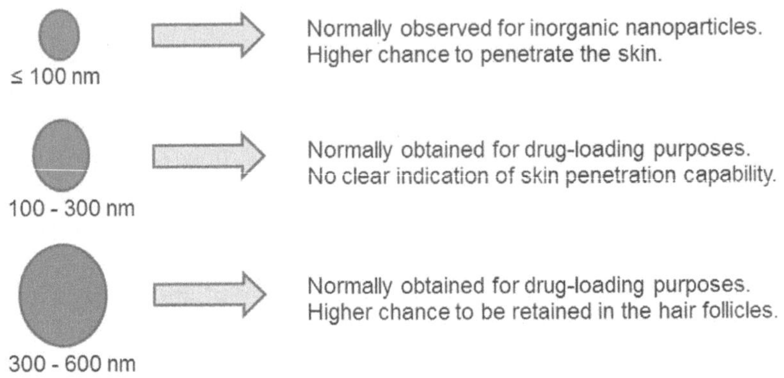

Figure 5.3 Expected penetration of nanoparticles into the skin.

mean values and the presence of smaller particles in a sample should be considered. Nanoparticle penetration is most likely along two possible routes: the intercellular (following the lipid channels between corneocytes) or the appendage route (hair follicles and sweat glands) [2, 63]. Hair follicles represent an important long-term reservoir for nanoparticles, possibly leading to a controlled and prolonged release of the drug into the viable layers. It has been shown that the penetration of particles into the hair follicles is size dependent, being the optimum size estimated to be between 300 nm and 600 nm. Besides, nanoparticles were stored significantly longer into the hair follicles than in the SC [63].

5.5 Conclusions and Perspectives

Optimal drug penetration/permeation across the skin is a difficult task to achieve due to the SC barrier. The process can occur transcellularly, intercellularly, or via appendages such as hair follicles. Several factors related to the formulation, drug, and skin, can interfere in drug penetration/permeation through this tissue. The most usual approaches to modify drug transport across the skin include the use of chemical enhancers and physical methods. Chemical enhancers such as surfactants and organic solvents physically or chemically change the SC inducing irritation, causing damage, and reducing the skin barrier function. Physical methods have the drawbacks of high cost and difficulty in accessibility. Thus, composite nanoparticles, which are nanoscale multicomponent particles, are promising strategies to modify skin penetration and permeation profile of drugs. Future goals to be achieved in this area include nanotoxicology studies to prove the absence of dermal toxicity due to the application of nanostructured formulations, including the evaluation of nanoparticles penetration into the skin, especially into impaired skin, and the effects related to the chronic use of topically applied nanoparticles.

References

1. Cullander, C., and Guy, R.H. (1992). Routes of delivery: Case studies. (6) Transdermal delivery of peptides and proteins, *Adv. Drug Deliv. Rev.*, **8**, pp. 291–329.
2. Schneider, M., Stracke, F., Hansen, S., and Schaefer, U.F. (2009). Nanoparticles and their interactions with the dermal barrier, *Dermato-Endocrinology*, **1**, pp. 197–206.
3. World Health Organization (WHO) (2006). *Environmental Health Criteria 235: Dermal Absorption*, International Labour Organization and the World Health Organization, pp. 1–197.
4. Jepps, O.G., Dancik, Y., Anissimov, Y.G., and Roberts, M.S. (2013). Modeling the human skin barrier: Towards a better understanding of dermal absorption, *Adv. Drug Deliv. Rev.*, **65**, pp. 152–168.
5. Paul, A., and Cevc, G. (1995). Noninvasive administration of protein antigen: Transdermal immunization with bovine serum albumin in transfersomes, *Vaccine Res.*, **4**, pp. 145–164.
6. Degim, T.I. (2006). New tools and approaches for predicting skin permeability, *Drug Discov. Today*, **11**, pp. 11–12.
7. Hadgraft, J. (2001). Skin, the final frontier, *Int. J. Pharm.*, **224**, pp. 1–18.
8. Moser, K., Kriwet, K., Naik, A., Kalia, Y.N., and Guy, R.H. (2001). Passive skin penetration enhancement and its quantification *in vitro*, *Eur. J. Pharm. Biopharm.*, **52**, pp. 103–112.
9. Groen, D., Poole, D.S., Gooris, G.S., and Bouwstra, J.A. (2011). Investigating the barrier function of skin lipid models with varying compositions, *Eur. J. Pharm. Biopharm.*, **79**, pp. 334–342.

10. Janusová, B., Zbytovská, J., Lorenc, P., Vavrysová, H., Palát, K., Hrabálek, A., and Vávrová, K. (2011). Effect of ceramide acyl chain length on skin permeability and thermotropic phase behavior of model stratum corneum lipid membranes, *Biochim. Biophys. Acta*, **1811**, pp. 129–137.
11. Brandner, J.M., Poetzl, C., Schmage, P., Hauswirth, U., and Moll, I. (2008). A (leaky?) barrier: Tight junction proteins in skin diseases, *Drug Discov. Today: Dis. Mech.*, **5**, pp. 39–45.
12. Muizzuddin, N., Hellemans, L., Overloop, L.V., Corstjens, H., Declercq, L., and Maes, D. (2010). Structural and functional differences in barrier properties of African American, Caucasian and East Asian skin, *J. Dermatol. Sci.*, **59**, pp. 123–128.
13. Buck, P. (2004). Skin barrier function: Effect of age, race and inflammatory disease, *Int. J. Aromather.*, **14**, pp. 70–76.
14. Konda, S., Meier-Davis, S., Cayme, B., Shudo, J., and Maibach, H.I. (2012). Age-related percutaneous penetration part 1: Skin factors, *Skin Therapy Lett.*, **17**, pp. 1–5.
15. Hadgraft, J., and Lane, M.E. (2005). Skin permeation: The years of enlightenment, *Int. J. Pharm.*, **305**, pp. 2–12.
16. Maibach, H.I., Feldmann, R.J., Milby, T.H., and Serat, W.F. (1971). Regional variation in percutaneous penetration in man, *Arch. Env. Health,* **23**, pp. 208–211.
17. Selzer, D., Abdel-Mottaleb, M.M.A., Hahn, T., Schaefer, U.F., and Neumann, D. (2013). Finite and infinite dosing: Difficulties in measurements, evaluations and predictions, *Adv. Drug Deliv. Rev.*, **65**, pp. 278–294.
18. OECD Guidance notes on dermal absorption. (2010). OECD Guidance Notes. Paris, Organization for Economic Co-operation and Development.
19. Karande, P., and Mitragotri, S. (2009). Enhancement of transdermal drug delivery via synergistic action of chemicals, *Biochim. Biophys. Acta*, **1788**, pp. 2362–2373.
20. Willams, A.C., and Barry, B.W. (2012). Penetration enhancers, *Adv. Drug Deliv. Rev.*, **64**, pp. 128–137.
21. Lane, M.E. (2013). Skin penetration enhancers, *Int. J. Pharm.*, **447**, pp. 12–21.
22. Tanner, T., and Marks, R. (2008). Delivering drugs by the transdermal route: Review and comment, *Skin Res. Tech.*, **14**, pp. 249–260.
23. Femenía-Font, A., Balaguer-Fernández, C., Merino, V., and López-Castellano, A. (2006). Combination strategies for enhancing transdermal absorption of sumatriptan through skin, *Int. J. Pharm.*, **323**, pp. 125–130.

24. Arora, A., Kisak, E., Karande, P., Newsam, J., and Mitragotri, S. (2010). Multicomponent chemical enhancer formulations for transdermal drug delivery: More is not always better, *J. Control. Release*, **144**, pp. 175–180.
25. Loftsson, T., and Masson, M. (2001). Cyclodextrins in topical drug formulations: Theory and practice, *Int. J. Pharm.*, **225**, pp. 15–30.
26. Alexander, A., Dwivedi, S., Ajazuddin, Giri, T. K., Saraf, S., Saraf, S., and Tripathi, D.K. (2012). Approaches for breaking the barriers of drug permeation through transdermal drug delivery, *J. Control. Release*, **164**, pp. 26–40.
27. Kogan, A., and Nissim, G. (2006). Microemulsions as transdermal drug delivery vehicles, *Adv. Colloid Interface Sci.*, **123–126**, pp. 369–385.
28. Nolan, L.M.A., Corish, J., Corrigan, O.I., and Fitzpatrick, D. (2007). Combined effects of iontophoretic and chemical enhancement on drug delivery. II. Transport across human and murine skin, *Int. J. Pharm.*, **341**, pp. 114–124.
29. Pillai, O., and Panchagnula, R. (2003). Transdermal delivery of insulin from poloxamer gel: *Ex vivo* and *in vivo* skin permeation studies in rat using iontophoresis and chemical enhancers, *J. Control. Release*, **89**, pp. 127–140.
30. Zhang, L.-W., Fang, Y.-P., and Fang, J.-Y. (2011). Enhancement techniques for improving 5-aminolevulinic acid delivery through the skin, *Dermatol. Sin.*, **29**, pp. 1–7.
31. Bounoure, F., Skiba, M.L., Besnard, M., Arnaud, P., Mallet, E., and Skiba, M. (2008). Effect of iontophoresis and penetration enhancers on transdermal absorption of metopimazine, *J. Dermatol. Sci.*, **52**, pp. 170–177.
32. Becker, S. (2012). Transport modeling of skin electroporation and the thermal behavior of the stratum corneum, *Int. J. Therm. Sci.*, **54**, pp. 48–61.
33. Lavon, I., and Kost, J. (2004). Ultrasound and transdermal drug delivery, *Drug Discov. Today*, **9**, pp. 670–676.
34. Bhalla, M., and Thami Gurvinder, P. (2006). Microdermabrasion: Reappraisal and brief review of literature, *Dermatol. Surg.*, **32**, pp. 809–814.
35. Murthy, S.N., Sammeta, S.M., and Bowers, C. (2010). Magnetophoresis for enhancing transdermal drug delivery: Mechanistic studies and patch design, *J. Control. Release*, **148**, pp. 197–203.

36. Wong, T.W., and Khaisan, A.N. (2013). Physicochemical modulation of skin barrier by microwave for transdermal drug delivery, *Pharm. Res.*, **30**, pp. 90–103.
37. Contri, R., Fiel, L.A., Pohlmann, A.R., Guterres, S.S., and Beck, R.C.R. (2011). Transport of substances and nanoparticles across the skin and *in vitro* models to evaluate skin permeation and/or penetration. In *Nanocosmetics and Nanomedicines: New Approaches for Skin Care*, 1st Edition, Beck, R.C.R, Guterres, S., and Pohlman, A. (Eds.), Springer, Berlin, pp. 3–36.
38. Prow, T.W., Grice, J.E., Lin, L.L., Faye, R., Butler, M., Becker, W., Wurm, E.M.T., Yoong, C., Robertson, T.A., Soyer, H.E., and Roberts, M.S. (2011). Nanoparticles and microparticles for skin drug delivery, *Adv. Drug Deliv. Rev.*, **63**, pp. 470–491.
39. Müller, R.H., Radtke, M., and Wissing, S.A. (2002). Solid lipid nanoparticles (SLN) and nanostructured lipid carriers (NLC) in cosmetic and dermatological preparations, *Adv. Drug Deliv. Rev.*, **54** Suppl. 1, pp. S131–S155.
40. Chen, M., Liu, X., and Fahr, A. (2011). Skin penetration and deposition of carboxyfluorescein and temoporfin from different lipid vesicular systems: *In vitro* study with finite and infinite dosage application, *Int. J. Pharm.*, **408**, pp. 223–234.
41. Fang, J.-Y., Hwang, T.-L., and Huang, Y.L. (2006). Liposomes as vehicles for enhancing drug delivery via skin routes, *Curr. Nanosci.*, **2**, pp. 55–70.
42. Maali, A., and Mosavian, M.T.H. (2013). Preparation and application of nanoemulsions in the last decade (2000–2010), *J. Dispersion Sci. Tech.*, **34**, pp. 92–105.
43. Mehnert, W., and Mäder, K. (2012). Solid lipid nanoparticles: Production, characterization and applications, *Adv. Drug Deliv. Rev.*, **64**, pp. 83–101.
44. Baspinar, Y., and Borchert, H.-H. (2012). Penetration and release studies of positively and negatively charged nanoemulsions: Is there a benefit of the positive charge? *Int. J. Pharm.*, **430**, pp. 247–252.
45. Kuchler, S., Radowski, M.R., Blaschke, T., Dathe, M., Plendl, J., Haag, R., Schäfer-Korting, M., and Kramer, K.D. (2009). Nanoparticles for skin penetration enhancement: A comparison of a dendritic core-multishell-nanotransporter and solid lipid nanoparticles, *Eur. J. Pharm. Biopharm.*, **71**, pp. 243–250.
46. Jeon, H.S., Seo, J.E., Kim, M.S., Kang, M.H., Oh, D.H., Jeon, S.O., Jeong, S.H., Choi, Y.W., and Lee, S. (2013). A retinyl palmitate-loaded solid lipid nanoparticle system: Effect of surface modification with dicetyl

phosphate on skin permeation *in vitro* and anti-wrinkle effect *in vivo*, *Int. J. Pharm.*, **452**, pp. 311-320.

47. Montenegro, L., Sinico, C., Castangia, I., Carbone, C., and Puglisi, G. (2012). Idebenone-loaded solid lipid nanoparticles for drug delivery to the skin: *In vitro* evaluation, *Int. J. Pharm.*, **434**, pp. 169–174.
48. Muller, R.H., Mader, K., and Gohla, S. (2000). Solid lipid nanoparticles (SLN) for controlled drug delivery: A review of the state of the art, *Eur. J. Pharm. Biopharm.*, **50**, pp. 161–177.
49. Das, S., Ng, W.K., and Tan, R.B.H. (2012). Are nanostructured lipid carriers (NLCs) better than solid lipid nanoparticles (SLNs): Development, characterizations and comparative evaluations of clotrimazole-loaded SLNs and NLCs? *Eur. J. Pharm. Sci.*, **47**, pp. 139–151.
50. Müller, R.H., Petersen, R.D., Hommoss, A., and Pardeike, J. (2007). Nanostructured lipid carriers (NLC) in cosmetic dermal products, *Adv. Drug Deliv. Rev.*, **59**, pp. 522–530.
51. Borgia, S.L., Regehly, M., Sivaramakrishnan, R., Mehnert, W., Korting, H.C., Danker, K., Röder, B., Kramer, K.D., and Schäfer-Korting, M. (2005). Lipid nanoparticles for skin penetration enhancement-correlation to drug localization within the particle matrix as determined by fluorescence and parelectric spectroscopy, *J. Control. Release*, **110**, pp. 151–163.
52. Teeranachaideeku, V., Boonme, P., Souto, E.B., Müller, R.H., and Junyaprasert, V.B. (2008). Influence of oil content on physicochemical properties and skin distribution of Nile red-loaded NLC, *J. Control. Release*, **128**, pp. 134–141.
53. Couvreur, P., Barratt, G., Fattal, E., Legrand, P., and Vauthier, C. (2002). Nanocapsule technology: A review, *Crit. Rev. Ther. Drug*, **19**, pp. 99–134.
54. Jäger, E., Venturini, C.G., Poletto, F.S., Colomé, L.M., Pohlmann, J.P.U., Bernardi, A., Battastini, A.M.O., Guterres, S.S., and Pohlmann, A.R. (2009). Sustained release from lipid-core nanocapsules by varying the core viscosity and the particle surface area, *J. Biomed. Nanotechnol.*, **5**, pp. 130–140.
55. Banerjee, R. (2013). Overcoming the stratum corneum barrier: A nano approach, *Drug Deliv. Transl. Res.*, **3**, pp. 205–208.
56. Siqueira, N.M., Contri, R.V., Paese, K., Beck, R., Pohlmann, A.R., and Guterres, S.S. (2011). Innovative sunscreen formulation based on benzophenone-3-Loaded chitosan-coated polymeric nanocapsules, *Skin Pharmacol. Physiol.*, **24**, pp. 166–174.

57. Guterres, S.S., Alves, M.P., and Pohlmann, A.R. (2007). Polymeric nanoparticles, nanospheres and nanocapsules for cutaneous applications, *Drug Target Insights*, **2**, pp. 147–157.
58. Silva, A.L.M., Contri, R.V., Jornada, D.S., Pohlmann, A.R., and Guterres, S.S. (2013). Vitamin K1-loaded lipid-core nanocapsules: Physicochemical characterization and *in vitro* skin permeation, *Skin Res. Technol.*, **19**, pp. 223–230.
59. Luengo, J., Weiss, B., Schneider, M., Ehlers, A., Stracke, F., König, K., Kostha, K.-H., Lehr, C.-M., and Schaefer, U.F. (2006). Influence of nanoencapsulation on human skin transport of flufenamic acid, *Skin Pharmacol. Physiol.*, **19**, pp. 190–197.
60. Tomoda, K., Terashima, H., Suzuki, K., Inagi, T., Terada, H., and Makino, K. (2012). Enhanced transdermal delivery of indomethacin using combination of PLGA nanoparticles and iontophoresis *in vivo*, *Colloid. Surf. B*, **92**, pp. 50–54.
61. Miyazaki, S., Takahashi, A., Kubo, W., Bachynsky, J., and Lobenberg, R. (2003). Poly *n*-butylcyanoacrylate (PNBCA) nanocapsules as a carrier for NSAIDs: *In vitro* release and *in vivo* skin penetration, *J. Pharm. Pharm. Sci.*, **6**, pp. 238–245.
62. Rastogi, R., Anand, S., and Koul, V. (2009). Flexible polymerosomes: An alternative vehicle for topical delivery, *Colloid. Surf. B*, **72**, pp. 161–166.
63. Lademann, J., Richter, H., Schanzer, S., Knorr, F., Meinke, M., Sterry, W., and Patzelt, A. (2011). Penetretion and storage of particles in human skin: Perspectives and safety aspects, *Eur. J. Pharm. Biopharm.*, **77**, pp. 465–468.

Chapter 6

Mucosal Barriers

Michelle R. Dawson and Deepraj Ghosh

School of Chemical and Biomolecular Engineering, Georgia Institute of Technology, 311 Ferst Dr. NW Atlanta, GA 30332-0100, USA

mdawson@gatech.edu

Mucus lining epithelial surfaces throughout the body poses a critical barrier to the transport of nanoparticle (NP) drug and gene carriers. Quantitative studies of the barrier properties of physiological mucus secretions are important in the development of new strategies for targeting the mucosal epithelium. This chapter summarizes the physiochemical properties of the mucus barrier, techniques used to characterize particle transport in mucus, and recent advancements in this field.

6.1 Introduction

The mucus layer in the mucosal epithelium acts as a semipermeable barrier allowing the exchange of water, gases, and small molecules, while preventing the transport of bacteria, pathogens, and particulates. Drug and gene carriers targeting epithelial cells in these tissues must penetrate this transport barrier before they are removed by mucosal clearance or digestion; however, mucus

Drug Delivery Across Physiological Barriers

Edited by Silvia Muro

ISBN 978-981-4669-40-5 (Hardcover), 978-981-4669-41-2 (eBook)

www.panstanford.com

is highly viscoelastic and sticky, which severely limits particle transport. Yet, virus particles, with diameters similar to drug and gene carriers, often diffuse as readily in mucus as in water [4]. Though virus particles have sufficient surface area to form multivalent interactions with mucus, they largely avoid sticking to mucus by displaying a virus surface that is densely coated (charge separation distance ~5 Å) with both positive and negative charges (for net neutral surface) [6].

The effects of size, surface charge, and hydrophobicity of standard surface-modified polystyrene particles on their permeation through mucus have been studied [1, 7]. These observations suggest that the rate of transport of drug and gene carriers, whose size, surface charge, and surface chemistry can be manipulated, is affected by their interactions with mucus components. Particles that can overcome the mucoadhesive forces and translocate to the epithelial cells may still have reduced therapeutic efficacy due to degradation by mucus components [8]. In addition, interactions with mucus may lead to particle aggregation or disintegration, thereby motivating the development of carriers that exhibit minimal interaction with mucus components. These interactions can also be exploited in the development of mucoadhesive particles that stick to mucus secretions for local delivery of therapeutics [9, 10].

Quantitative microscopy techniques, such as multiple particle tracking (MPT), have been used to analyze the transport of particles in physiological mucus samples, which has greatly expanded our knowledge of the dynamic transport of particulate drug and gene carriers in mucus [1, 11–14]. These studies have provided crucial information on mucus rheology and the physiology of the mucosal epithelia. This information has also guided the development of mucus-penetrating [15] and mucoadhesive [16] particles.

Future directions may be focused on tracking the path of particulates in the mucosal epithelium *in vivo*. Fluorescent labeling of particulates is still a major hurdle in the analysis of drug and gene carrier transport; therefore, advancements in labeling techniques are essential in the application of particle-tracking technology to the analysis of real drug and gene carriers, not standard particles. Even with standard fluorescent latex or polystyrene particles, which encapsulate a large amount of fluorescent material, the minimum particle diameter, due to the fluorescence intensity required for tracking experiments, is approximately 60 nm [17] (alternatively, techniques such as fluorescence recovery after photobleaching have been used to probe at smaller length scales). Drug and gene carriers, tagged with fluorescent molecules, often have lower fluorescence intensities, resulting in further limitations in the minimum size for tracking experiments (~100 nm).

6.2 Mucus Structure

Mucus is a dynamic and sticky gel that coats epithelial surfaces throughout the body (i.e., respiratory, gastrointestinal, and cervicovaginal tracts and the surface of the eye) protecting them from environmental invaders, such as bacteria, dirt, particles, and viruses (Fig. 6.1). After these particles are trapped in mucus, they are rapidly cleared through mucociliary transport in respiratory airways and by luminal mucus movement in other tracts. The thickness of the mucus layer, which varies from 10 to 450 μm in airway to 50–600 μm in gastrointestinal tract, depends on the rate at which mucin is secreted from goblet cells in the airways and other cells in the epithelium and submucosal glands, along with the rate of degradation and shedding [6]. The physiological parameters of mucus may also vary, for example the pH of mucus can form a gradient varying from acidic in the lumen to neutral at the epithelial cell surface [18–20], which also leads to differences in mucus viscoelasticity [21, 22]. Neves et al. have shown that a change in the pH of mucus can alter transport properties of charged NPs [23]. Irritating substances or toxins stimulate mucin secretion, which also results in mucus thickening and abnormal mucus transport (i.e., similar to what happens in the asthmatic airways). Even in the absorptive epithelium of the small intestine, where water is transported toward the cell surface, particle drug and gene carriers must cross a thick mucus gel, which is being transported away from the cell surface, to reach the epithelium.

The mucus layer in the airways is uniquely regulated by ciliated epithelial cells. Viscous shear stress from ciliary motion, coughing, or muscle contraction results in the formation of two mucus layers, an unstirred thicker gel layer (at the top of the mucus layer) and a thin more hydrated periciliary sol layer (near the cell surface) (Fig. 6.1A) [24, 25]. Volume and height of these layers are very closely controlled by the underlying epithelia by adjusting the mucin secretion rate via a mechanosensitive ATP-release mechanism [26, 27]. In diseases such as cystic fibrosis (CF), the mucin secretion rate is aberrant and mucus becomes poorly hydrated, which leads to airway obstruction.

The barrier properties of mucus stem from the large concentrations of high molecular weight mucin glycoproteins, which constitute 80% of the mucus dry weight [13]. Mucin glycoproteins, which include sialomucins, fucomucins, and sulfomucins, are composed of four to five mucin subunits, each with molecular weight of ~500 kDa [28, 29]. These subunits, which have highly glycosylated protein backbones and non-glycosylated protein ends, are attached at their non-glycosylated regions by disulfide bonds and crosslinked, bundled, and entangled with other mucin glycoproteins through carbohydrate interactions to form diverse mucin fibers (2–40 MDa in size, 3–10 nm in diameter, and 0.5–10 μm in length) (Fig. 6.2) [24, 28, 29]. These glycoproteins belong to a large family of mucins, which encompasses 16 different genes in humans with the following subfamilies: secreted gel forming, secreted gel non-forming, and cell-surface mucins [30, 31]. The expression of secreted gel forming mucins, important in maintaining mucus viscoelasticity, varies based on the anatomical location of the mucus layer (e.g., MUC5AC and MUC5B are predominant in respiratory pathway, whereas MUC2, MUC6 along with MUC5AC form the gastrointestinal tract mucus layer). Gel non-forming secreted mucin (MUC7), present in saliva, does not contribute to viscoelastic properties of mucus. Cell-surface mucins (e.g., MUC1 expressed across all epithelia, MUC16 expressed in stomach) are tethered to the cell membrane and form a closely knit structure above the epithelial cells known as glycocalyx, which may act as final layer of defense against foreign materials [32]. Several other constituents interact with mucus glycoproteins to form a dense network, including phospholipids, cellular and serum macromolecules, cell-derived DNA and F-actin, alginate, electrolytes, microorganisms, and sloughed cells [33]. Concentrations of various constituents are dependent on the anatomical location, as well as the

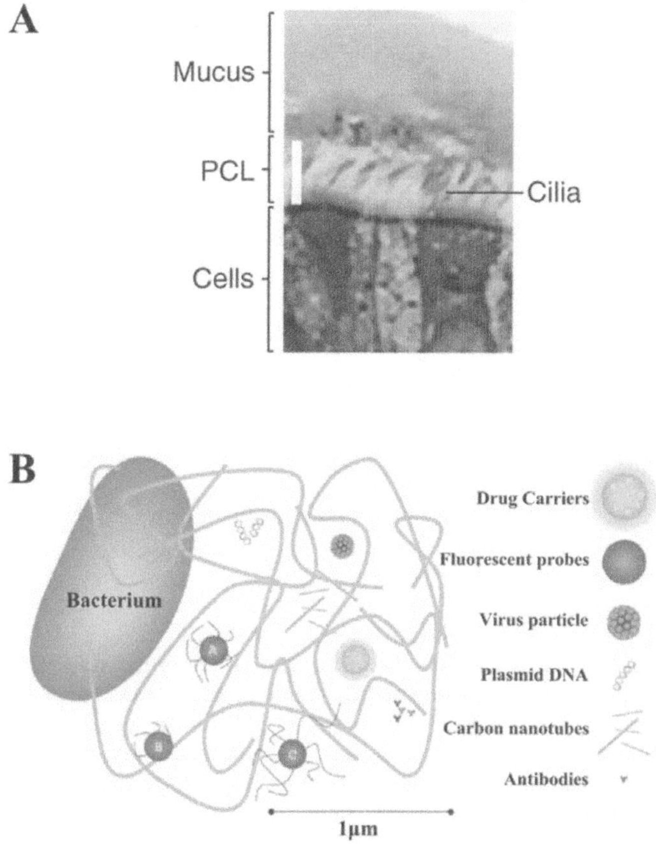

Figure 6.1 Mucus structure. (A) Light microscopy view of the airway surface layer, comprising the mucus layer and the periciliary layer (PCL), from fixed human bronchial epithelial cultures. Scale bar = 7 μm. Reprinted from Ref. 3, Copyright 2012, with permission from AAAS. (B) Schematic representation of interaction between mucus and foreign elements. Scale bar = 1 μm. Reprinted by permission from Macmillan Publishers Ltd: Ref. 5.

physiological and pathophysiological conditions of the mucus donor [33].

Interactions of mucin glycoproteins with various constituents result in the formation of porous mucus networks. Different techniques have been used to characterize pore size in mucus. Imaging methods such as scanning electron microscopy (SEM) with fixed and dehydrated mucus samples may lead to erroneous

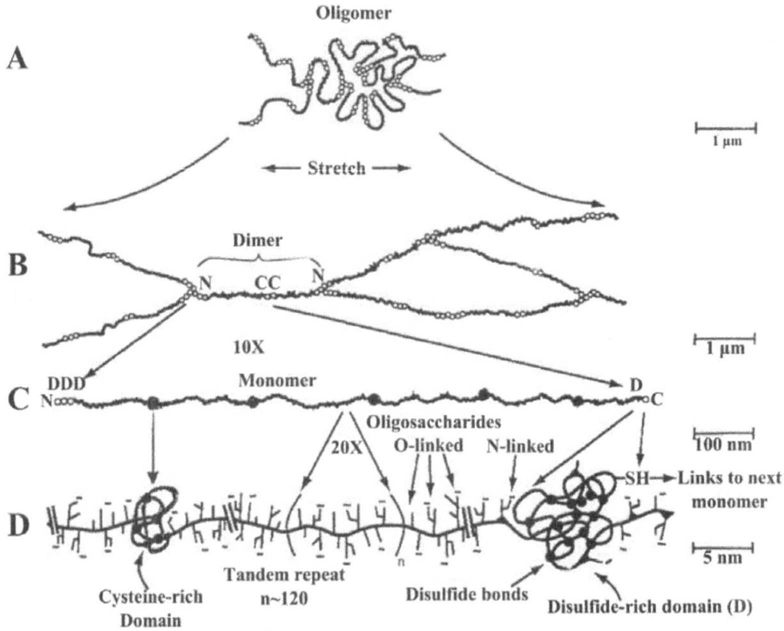

Figure 6.2 **Scaled representation of mucin chemical structure.** (A, B) Oligomeric mucin consisting multiple monomers linked via disulfide bonds (labeled "D"). (C, D) Mucin monomer (~500 nm in length and 5 nm in width) comprises an amino acid backbone with highly glycosylated regions and non-glycosylated cysteine-rich regions. Reprinted from Ref. 2, Copyright 2009, with permission from Elsevier.

results. Olmsted et al. found cervicovaginal mucus (CVM) mesh size distribution range of 10–200 nm using SEM [4], whereas Sanders et al. found pore diameter of CF sputum ranging from 100 to 400 nm [33]. In contrast, MPT probes into particle transport in a dynamic three-dimensional native mucus state. In CVM, this technique suggests that the mucus network consists of a highly heterogeneous (approximately 50–1800 nm) microporous mesh with average mesh spacing $\sim$340 $\pm$ 70 nm [34], whereas a similar heterogeneous network structure, albeit with reduced average mesh spacing of $\sim$145 $\pm$ 50 nm, has been observed in CF sputum [35] and in synthetic mucus formulated to model CF sputum. Increased understanding of the unique microstructure of physiological mucus secretions from different anatomical locations and donors will aid in the development of drug and gene carriers capable of passing through the mucosal barrier.

6.3 Mucus Viscoelasticity

Mucus is a viscoelastic gel, which means that it exhibits viscous and elastic properties upon deformation. When small strains are applied, mucus fibers can be stretched elastically without severing chemical bonds or mucus entanglements; however, large strains result in viscous deformations, which are important in forming the mucus sol layer. The viscoelasticity of mucus is critical for mucociliary transport in the respiratory tract, since mucus must be watery enough to allow for ciliary motion yet elastic enough for cilia to engage it and sweep it from the body [24].

The viscoelasticity of mucus can be characterized using bulk rheological techniques, such as cone and plate rheometry (Fig. 6.3A,B) [1, 11]. Bulk characterization of mucus rheology (i.e., cone and plate rheometry of CF mucus) has shown that mucus is a thixotropic gel, which means that it undergoes shear thickening at low shear rates (corresponding with physiological shear rates in the airways) and shear thinning at high shear rates (Fig. 6.3C) [1, 11]. Thus, changes in physiological shear rates, resulting from disease or therapeutic intervention, may alter mucus viscoelasticity and the rate of mucus clearance. The viscoelasticity of mucus can also vary dramatically depending on the anatomical location in the same way as mucus structure and mesh spacing. Recent investigation by Schuster et al. found CVM to have bigger mesh spacing and higher viscoelastic properties compared to normal airway mucus [36–39]. The observed difference was attributed to hydrophobic bundling of filamentous mucins (Fig. 6.4A), leading to the formation of thicker mucin fibers with high porosity and higher viscoelastic properties. In diseases such as chronic rhinosinusitis (CRS) or CF, the mucus layer becomes highly viscoelastic [40, 41] probably due to enhanced concentration of cell-secreted mucins, actin, and DNA materials, leading to smaller average mesh size distribution.

Changes in mucus viscoelasticity can also affect the rate of particle transport. Increased viscoelasticity results in slower mucus

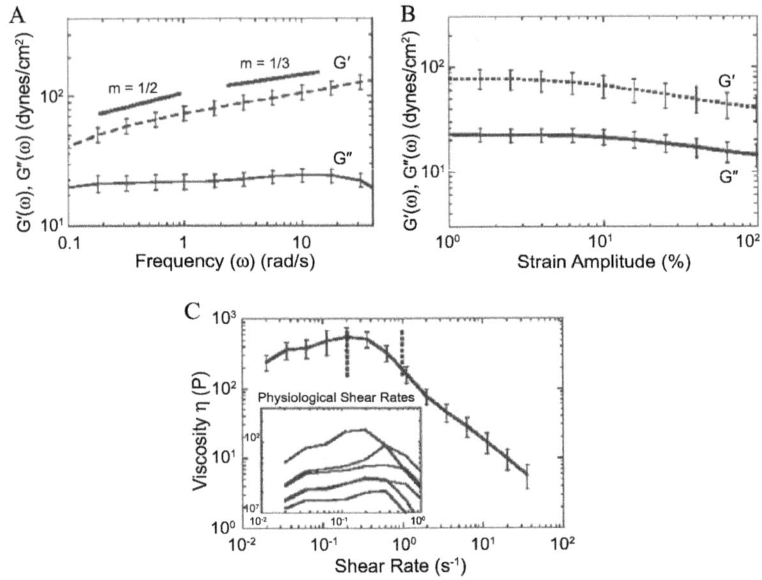

Figure 6.3 Macro-rheology of human cystic fibrotic (CF) sputum. (A) The frequency-dependent elastic ($G'(\omega)$) and viscous ($G''(\omega)$) moduli of CF samples ($n = 6$) recorded at a constant strain amplitude of 1%. (B) Strain-dependent elastic and viscous moduli from 0.1% to 100% strain amplitude. (C) The steady-state viscosity of CF sputum shows the shear thickening behavior at shear rates between 10^{-2} and 10^{2} s^{-1}. Physiological rates in the normal lung are 0.2–0.9 s^{-1}. *Inset,* viscosities of individual CF sputum samples at physiological shear rates. Adapted from Ref. [1].

clearance [42, 43], which may promote particle transport by increasing particle residence times; however, increased viscoelasticity also results in thickening of airway secretions, which may lead to steric obstruction of particles in mucus and mucus–particle interactions that limit mucus permeability (Fig. 6.4) [24, 44, 45]. Interestingly, Sanders et al. found the correlation between particle transport and elasticity of mucus to be counterintuitive as they reported that with increasing elasticity ($G' > 100$ Pa), the percentage of particles (124–270 nm) transported through mucus increased [35]. Particle-tracking experiments later revealed that this unexpected finding could be attributed to an increase in the heterogeneity of the mucus

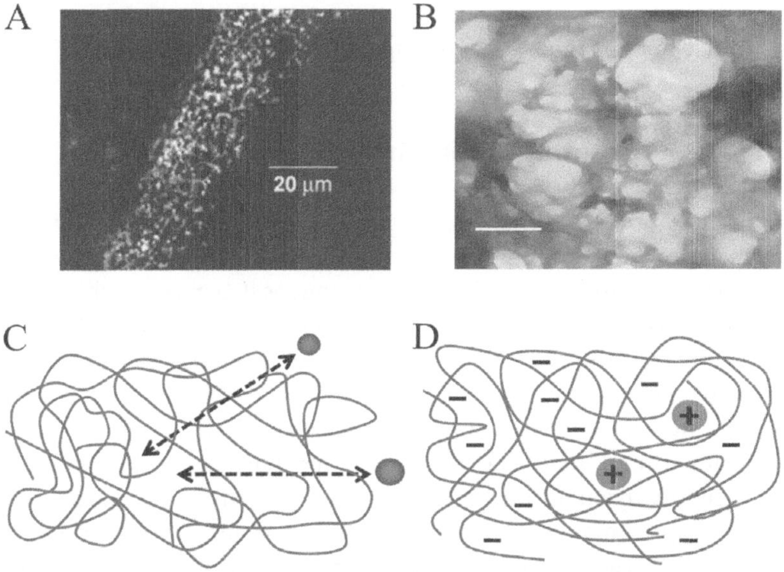

Figure 6.4 **Adhesive and steric interactions in mucus networks.** (A) Fluorescent image of 200 nm carboxylated polystyrene microspheres in CF mucus. The mucus fibers are bundled together, forming a thick mucus cord, which is illuminated by fluorescent microspheres that have adhered to mucus fibers coating their surface. (B) Transmission electron micrograph of CF sputum diluted 1:20 in sputum buffer and adsorbed to ionized formvar grid, which was negatively stained with uranyl acetate. The mucus fibers formed a mesh-like structure with tiny pores that had diameters between 100 nm and 500 nm. Previous studies showed similar mesh structure in CVM. Length scale bar corresponds to a distance of 200 nm. The adhesion of particles with mucus may result from (C) steric (entropic) interactions of particles with mucus components present in the crowded mesh-like structure of mucus or (D) enthalpic barriers, which result from the interactions of glycosylated and negatively charged mucus glycoproteins with hydrophobic or charged nanoparticles.

mesh (Fig. 6.4B), allowing larger pores for particle transport as elasticity increased [1].

Bulk rheology is critical for mucus clearance; however, this information may be less relevant to particle transport, since bulk viscosities of mucus may be up to 2000-fold higher than water [1], yet select proteins and viruses transport through mucus at rates similar to their theoretical transport in water [24].

6.4 Particle Transport in Mucus

Investigations of diffusion through mucus gels demonstrated that small molecules, such as testosterone with a molecular weight of 401 Da, diffuse rapidly through mucosal barriers, while large molecules are often trapped due to steric hindrance (Fig. 6.4C) [4, 35]. In fact, the diffusivities of particles in mucus can be related to the pore sizes of the mucus mesh and the concentration of high molecular weight mucin glycoproteins [24]. Previous studies demonstrated that the cutoff size for particles able to diffuse efficiently through concentrated mucus gels, such as colonic mucus, was approximately 100 nm [24]. NP drug and gene carriers are typically larger than 100 nm, which may significantly limit their ability to permeate mucus to reach the underlying epithelial cell targets.

If the translocation of particles through mucus were only a function of particle size, virus-sized latex particles with diameters less than 100 nm would diffuse through mucus readily. However, particles small enough to diffuse through the microscopic pores may still be unable to overcome this barrier due to adhesive interactions, which limit their transport in mucus [1, 46]. In previous studies, we showed that amine-modified and carboxylated polystyrene particles permanently adhere to mucin fibers via hydrophobic and electrostatic interactions (Fig. 6.4A), which cause mucin fibers in the mesh to collapse resulting in the formation of more heterogeneous pore structure (Fig. 6.4B) [4, 7].

MPT has been used to improve our understanding of the dynamic transport of particulate drug and gene carriers in mucus [1]. This approach involves simultaneously tracking the microscopic displacements of dozens of individual particles embedded in mucus gels using real-time video microscopy technology. This is a powerful technique for analyzing drug and gene carrier transport in mucus, since the physical and chemical properties of the tracked particles, including their size, shape, composition, and surface charge, can be directly related to their transport rates and the number of mucus–particle interactions. Since data are obtained from hundreds of individual particles, this technique also provides statistical insights on the heterogeneity in particle transport [1]. This information can also be used to quantify important properties of the mucus mesh, such as viscoelasticity [47] and porosity [48]. Furthermore, the individual particle transport rates can be averaged to determine the bulk transport properties of the embedded particles, along with the bulk-fluid rheology.

6.4.1 *Multiple Particle Tracking in Mucus*

The theory of Brownian motion and the concept of a random walk are applied in MPT experiments, which involve tracking the spontaneous displacements of thermally excited particles in fluid-like environments. Hundreds of images are collected over a short time period using high-resolution time-resolved fluorescence microscopy. The coordinates of the intensity-weighted centroids of the individual particles are then determined from these images using tracking software, often with 5 nm spatial and 33 ms temporal resolution. The spatial coordinates of the particle centroids, recorded in the two-dimensional plane of focus, are then converted to families of time-averaged mean squared displacements (MSDs):

$$<\Delta r^2(\tau)> = <[x(t+\tau) - x(t)]^2 + [y(t+\tau) - y(t)]^2>$$

where τ is the time scale or time lag. From this distributions of MSDs, time-dependent particle diffusion coefficients, and local rheological parameters of the mucus network can be determined [47, 49–51]. Previous studies have shown that tracking of particles in two dimensions can be used to characterize three-dimensional particle motion in isotropic, not necessarily homogeneous, environments [1]. Though fluorescent polystyrene NPs of controlled size and surface charge are often used in these experiments, homogeneous populations of fluorescently labeled drug and gene carriers can be used in particle-tracking experiments, provided that the particle size and surface charge are well characterized. For tracking experiments, particle size generally varies from 100 to 500 nm, which is the relevant size range for adenovirus, liposomal, and polymer-mediated transport in mucus [35].

In the simplest case, when the tracked particles are rigid spheres that move by Brownian diffusion, the fluid viscosity can be calculated directly from the Stokes–Einstein equation ($D = k_B T/6\pi\eta a$); however, when the size of tracked particles approaches the mucus pore size, particle motion is restricted by the viscosity and elasticity of the surrounding fluid. In this regime of particle transport, the viscous ($G''(\omega)$) and elastic ($G'(\omega)$) moduli are determined from the amplitude and slope of the time-dependent MSD [49]. Assuming the viscoelasticity of the fluid surrounding the tracked particle is continuous, which occurs when the particle radius is much larger than the size of the fluid pore, the Stokes–Einstein equation can be used to determine the frequency-dependent viscoelasticity of a complex fluid.

MPT was previously used to characterize the time-dependent diffusion coefficients of hundreds of individual 100, 200, and 500 nm carboxylated polystyrene particles embedded in CF sputum [1]. Mean diffusivity and diffusivity distributions of NPs dispersed in CF sputum were measured (Fig. 6.5). NP transport rates are generally slow in CF sputum; however, a small percentage of 100–500 nm particles has significantly higher transport rates than their respective averages (inset Fig. 6.5B). These fast-moving "outlier" particles contribute proportionally more than slow particles to the mean diffusivity, which is dominated by slow-moving particles. The increased mobility of outlier particles, which account for approximately 1%, 4%, and 6% of 500, 200, and 100 nm particles (inset Fig. 6.5B), respectively, makes them more likely to overcome mucus transport barriers. Recently, Suk et al. compared the viscoelastic properties of CF sputum using MPT with data obtained from conventional cone and plate rheometer [41] and found that bulk viscosity of CF sputum was 30,000 times higher at low shear rate and 300 times higher at high shear rate compared to water when probed using cone and plate rheometer. However, MPT data suggested heterogeneous regions with only fivefold higher nanoscale viscosity than water. Comparison of results between cone and plate rheometer and MPT highlights the necessity of probing at different length scales to understand relevant processes such as the

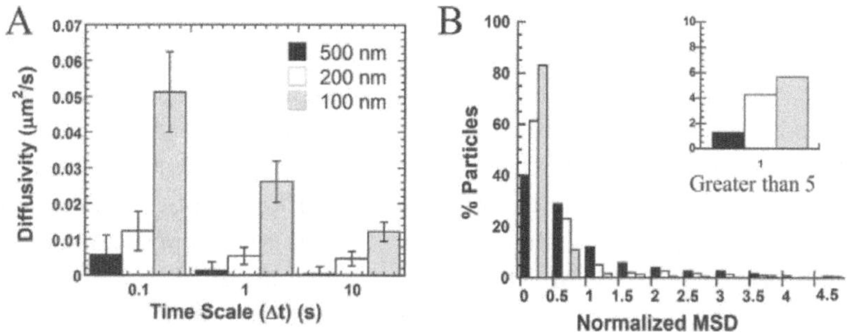

Figure 6.5 Characterization of nanoparticle transport in untreated CF sputum. (A) Effective (average) diffusion coefficients of 100, 200, and 500 nm particles in CF sputum. (B) Distribution of 100, 200, and 500 nm particle mean squared displacements (MSDs) at a time scale of 0.1 s normalized with their respective averages. The percentage of particles with MSDs greater than fivefold the average included in the inset. The trajectories of polystyrene particles were recorded using a silicon-intensified target camera mounted on an inverted epifluorescence microscope equipped with an oil-immersion objective. All MPT measurements were conducted in a Bioptechs FCS2 thermal-regulated microscope chamber (Butler, PA) maintained at 37°C. Adapted from Ref. [1].

relatively unhindered diffusion of viruses across the mucus layer, which agrees well with nanoscale rheological data. This approach has been applied to the analysis of other mucus barriers, including gastrointestinal [11] and vaginal [38] mucus.

6.5 Mucoadhesive Particles

Various mucosal routes of drug delivery provide certain advantages over other conventional routes such as avoiding the metabolism in the gastrointestinal tract. One of the primary challenges in mucosal drug delivery is to enhance particle retention times to allow for continuous release of drugs, transport across mucus barriers, and uptake in target tissues. To overcome this limitation, mucoadhesive NPs are widely investigated for applications in drug delivery. Mucoadhesive systems can interact with both mucus and underlying epithelial cells to increase residence time and longevity of drugs. Mucosal adhesion of NPs to the mucus layer consists of two stages: the contact stage and the consolidation stage [52]. During the contact stage, particles approach mucosal surfaces to establish physical interactions with mucus, and parameters such as hydration and crosslinking density play a key role in this process. During the consolidation stage, surface chemical groups interpenetrate the mucus layer to form both covalent and noncovalent bonds, which increase the strength of mucus adhesions. Surface properties of NPs are primarily designed and modified using various polymeric groups to form specific bonds with mucus layers [53–55]. Polymeric groups tailored on the surface of NPs include hydrophilic groups such as carboxyls and amides, lectin and thiolated groups, and charged or nonionic groups. The NP surface can establish highly heterogeneous interactions with mucus such as hydrogen bonds, hydrophobic or electrostatic interactions, and van der Waals forces, which are based on the interaction of functional groups present on the particle surface with mucus. For example, chitosan is used as a coating material since it has a large number of cationic groups that are adhesive with mucus in part to electrostatic interactions [56]; whereas the presence of polymeric thiomer or acrylate on the particle surface enables the formation of stable covalent bonds (e.g., disulfide) with mucins [57, 58]. Other parameters that can contribute to particle adhesion with mucus include molecular weight, polymer chain flexibility, pH, and excipient drug concentration. Mucoadhesive particles are currently applied to various drug delivery routes, including oral, buccal, ocular, nasal, vaginal, and rectal.

The primary goal of developing mucoadhesive particles is to provide sufficient time for drug release and uptake by the underlying epithelia before particle clearance. However, mucoadhesion may reduce the rate of particle uptake by cells since mucoadhesive particles may be permanently immobilized in mucus. To resolve this issue, second-generation mucoadhesive particles with surfaces conjugated with lectins have emerged [59]. Lectin-based mucoadhesive NPs can directly bind to receptors on the cell surface or mucus layer, increasing their specificity and uptake efficiency. This approach has also found some critical drawbacks such as premature adsorption of particles to the mucus gel layer, which can lead to mucociliary clearance. Chemical permeation enhancers (CPEs) are chemical agents that interact with the epithelial cells to modify their structure and increase their permeability. CPEs have also been utilized to increase the bioavailability of therapeutic drugs delivered via mucoadhesive particles by modifying the underlying epithelial cells or cell–cell junctions [60–63]. The efficacy of mucoadhesive drugs is largely dependent on their retention in the mucus layer, and consequently it is directly correlated with mucus clearance rates. Such properties can severely limit the effective drug delivery across rapidly clearing mucus layers at certain anatomical locations. Subsequently, another class of particles that has been proposed to travel across mucus barriers more efficiently to reach the epithelia is discussed next.

6.6 Mucus-Penetrating Nanoparticles

To increase the bioavailability of specific drugs diffusing through the mucus layer, it is desirable that particles can transport across mucus barriers relatively unhindered. In nature, capsid viruses exhibit high diffusion rates through mucus barriers due to their unique surface properties, including the dense packing of charged functional groups and the net neutral charge [64]. Dense presentation of charged groups on the virus surface allows hydrophilic interaction while minimizing chances of hydrophobic entrapment in mucus [65]. Furthermore, the net neutral charge of the surface allows particles to overcome electrostatic binding forces to move across mucus more freely. The surface properties of NPs are often manipulated to mimic the virus surface by controlling both net surface charge and distribution of exposed functional groups. However, dense packing of oppositely charged groups to create neutral surface on particles has proven to be extremely challenging. Alternatively, researchers have proposed that muco-inert surfaces with neutrally charged individual functional groups can be sufficient to minimize electrostatic interactions while the surface remains hydrophilic. This approach also promises to be less immunogenic than the virus surface. Polyethylene glycol (PEG) is a Food and Drug Administration–approved polymer, which is neutral and hydrophilic. Lai et al. found that dense coverage of latex particle surface with

PEG made their surfaces neutral and increased their diffusion rate through mucus compared to uncoated latex particles, which were often immobilized in mucus [38]. Dense coating with low molecular weight PEG minimized interaction between particles and mucin by limiting exposure of the hydrophobic core. This concept was also extended to biodegradable polymer poly sebacic acid (PSA), where copolymers of PSA-PEG were engineered with dense hydrophilic PEG on their surface [66]. PSA-PEG particles diffused at a much faster rate in fresh undiluted human CVM compared to equivalently sized PSA alone. Since previous reports on PEG-coated NPs were attributed to being both mucoadhesive and mucus penetrating, Wang et al. clarified that while dense coating with low molecular weight PEG could achieve successful mucus penetration, surface coating with higher molecular weight PEG or lower coverage with low molecular weight PEG could lead to increased mucoadhesion [67]. The PEGylated NPs also perform better than their uncoated counterparts against enzymatic degradation and renal clearance.

Polystyrene NPs coated with low molecular weight PEG (PS-PEG) have been shown to rapidly penetrate both airway mucus and undiluted human CVM [36, 37]. However, the cutoff sizes of the particles that can freely move across these two mucus layers are different, with 500 nm particles that are immobilized in respiratory tract mucus freely moving in CVM due to its inherent larger mesh spacing. MPT experiments also revealed that mucus thickening in pathological conditions, such as CF or CRS, may result in smaller mesh sizes and reduced transport of mucus-penetrating particles (MPPs) [40, 41]. PS-PEG particles exhibited high effective diffusivity in the following order: 200 nm >100 nm >500 nm particles, where 500 nm particles become immobilized in CF sputum as compared to remaining motile in CVM [41]. Possible discrepancy found with 100 nm particle transport was due to the unsaturated coating of NPs at lower sizes, which was later resolved by Schuster et al. while probing normal airway mucus. Interestingly, MPPs could not pass through CRS mucus from patients with nasal polyps, which was attributed to the high concentrations of DNA and actin (released by dying neutrophils) and increased mucin content due to goblet cell hyperplasia. These studies suggest that an efficient design of MPPs would require two main characteristics: (1) small particle size of $\leq$200 nm to overcome steric hindrance and (2) particle surface densely coated with neutral hydrophilic PEG to minimize adhesive interactions.

MPP formulations were also extended to use biodegradable materials for successful transition to *in vivo* studies. Recently, studies with biodegradable MPP formulations using poly(lactic-co-glycolic acid) (PLGA) copolymerized with PEG to form PLGA-PEG-diblock copolymers have been reported. PLGA-PEG NPs with near-neutral charge diffused much more freely compared to uncoated negatively charged similar sized PLGA particles in human CVM [68, 69]. Furthermore, MPP formulations using covalent and noncovalent PEG coatings have been proposed [68, 70], which can lead to noncovalent coating of PEGs using commercial pluronics (for the formation of triblock copolymers with a hydrophobic core), which are generally recognized as safe materials. Yang et al. demonstrated that pluronic molecules containing ampiphilic PEG-PPO(poly-propylene oxide)-PEG triblock enhanced particle transport in CVM, which was directly correlated with the molecular weight of the central PPO [70]. Higher molecular weight of PPO segments ($\geq$3 kDa) resulted in close to neutral surface charge (by zetapotential), and subsequent enhanced transport in CVM.

In part to the tremendous progress in the design of MPPs that can cross mucus barriers, recent studies have focused on investigating the *in vivo* transport of MPPs using animal models. Woodrow et al. used biodegradable PLGA NPs (less than 200 nm in diameter with zetapotential of −10 mV) loaded with small-interfering RNA targeted against mitogen-activated protein kinase 1 for sustained gene silencing in mice [71]. In a subsequent study, two formulations of PEG-coated NPs were investigated for intravaginal delivery, including PEG-coated and avidin-coated PLGA NPs. Retention in the mucus layers was similar for PLGA-PEG and PLGA-avidin NPs; however, penetration of the underlying cells was significantly higher for the PLGA-PEG NPs (found in epithelial cells and fibroblasts) [72]. In a similar study, PEG-based MPPs were designed with both biodegradable (PLGA) and nonbiodegradable (polystyrene) core for vaginal drug delivery [73, 74]. Both particles demonstrated better distribution, including in vaginal folds compared to conventional mucoadhesive particles. Finally, a protective microbicide drug (acyclovir monophosphate) when delivered using MPPs performed and protected better against herpes simplex virus-2 (HSV-2) compared to conventional soluble drug. Additionally, delivery of MPPs in hypotonic solution was shown to enhance the distribution and retention in vagina via advection as compared to simple diffusion. Osmotically induced absorption augmented the distribution of the drug via simple diffusion to enhance its efficiency [75]. Suk et al. has recently reported successful delivery of highly compact dense PEG-coated mucus-penetrating DNA NPs carrying wild-type *CFTR* gene in mice [76]. Intranasal delivery of the NPs significantly improved the gene transfer compared to other conventional delivery vehicles (including cationic polymer gene carriers). More importantly, inflammation was not observed in the lungs.

6.7 Mucolytic Agents

The prolonged retention of mucus in diseased tissues (e.g., CF and chronic obstructive pulmonary disease) can lead to recurring bacterial infections, resulting in a viscous, more purulent mucus [77]. Increased mucus viscoelasticity may be attributed to extensive disulfide and lectin bonding, poor hydration, and/or excess concentrations of extracellular DNA or actin [77]. In these situations, therapeutics have been used to reduce the viscosity of mucus to improve the rate of mucociliary clearance.

Mucolytic agents such as N-acetylcysteine [78, 79], guaifenesin [80], and dithiothreitol are used clinically and/or *in vitro* to reduce mucus viscosity. Lytic agents, such as recombinant human Deoxynuclease I (rhDNAse I or Pulmozyme®) and gelsolin, have also been effective in aiding clearance of mucus secretions [77]. Pulmozyme® enzymatically degrades extracellular DNA in human sputum [81]. Gelsolin severs noncovalent bonds between polymerized monomers of actin filaments. Gelsolin may scavenge actin filaments released during the inflammation process, increasing the effectiveness of DNAse I [82]. Finally, the application of pulmonary surfactant to canine airways increased the mucus transport velocity by nearly 400% in one study [83]. The effects of using mucus-altering agents as adjuncts in gene delivery have also been investigated: the addition of mucus-altering agents produced similar transgene expression to mechanical mucus depletion [84].

6.8 Conclusions and Perspectives

Inefficient particle transport in mucus is a major limitation in the therapeutic delivery of genes or drugs to the mucosal epithelium. Efficient drug delivery across the mucus layer has the potential to improve the treatment for diseases such as CF and cervical cancers, or to protect against viruses such as HSV. These surfaces are coated with thick and sticky mucus, which limits the transport of NPs to these otherwise accessible tissues. MPT has been used to determine the physiochemical properties of particle gene and drug carriers that more rapidly move through these barriers. The transport of particles in mucus is greatly limited by the adhesive and viscoelastic properties of mucus secretions, which can be exploited in the delivery of mucoadhesive particles. Thorough characterization of NP transport through the mesh structure of mucus has led to the engineering of next-generation MPPs that diffuse through the mucus barrier more readily. Extension of such techniques to more complex *in vitro* systems with different classes of drug carriers and polymers and correlating these data with other *in vivo* results may provide new avenues for the development of the next generation of mucus-targeted gene and drug delivery vectors. Current efforts have been made to develop drug delivery vehicles that are mucoadhesive for improved retention or muco-inert for improved penetration. At this time, no single particle system has been formulated to accomplish both feats and thus represents a major area of opportunity for future research.

References

1. Dawson, M., Wirtz, D., and Hanes, J. (2003). Enhanced viscoelasticity of human cystic fibrotic sputum correlates with increasing microheterogeneity in particle transport, *J. Biol. Chem.*, **50**, pp. 50393–50401.

2. Lai, S. K., Wang, Y.-Y., Wirtz, D., and Hanes, J. (2009). Micro- and macrorheology of mucus, *Adv. Drug Deliv. Rev.*, **2**, pp. 86–100.
3. Button, B., Cai, L.-H., Ehre, C., Kesimer, M., Hill, D. B., Sheehan, J. K., Boucher, R. C., and Rubinstein, M. (2012). A periciliary brush promotes the lung health by separating the mucus layer from airway epithelia, *Science*, **6097**, pp. 937–941.
4. Olmsted, S. S., Padgett, J. L., Yudin, A. I., Whaley, K. J., Moench, T. R., and Cone, R. A. (2001). Diffusion of macromolecules and virus-like particles in human cervical mucus, *Biophys. J.*, **4**, pp. 1930–1937.
5. Cu, Y., and Saltzman, W. M. (2009). Drug delivery: Stealth particles give mucus the slip, *Nat. Mater.*, **1**, pp. 11–13.
6. Cone, R. A. (2009). Barrier properties of mucus, *Adv. Drug Deliv. Rev.*, **2**, pp. 75–85.
7. Norris, D., and Sinko, P. (1996). Effect of size, surface charge, and hydrophobicity on the translocation of polystyrene microspheres through gastrointestinal mucin, *J. Appl. Polym. Sci.*, pp. 1481–1492.
8. Sanders, N. N., de Smedt, S. C., and Demeester, J. (2003). Mobility and stability of gene complexes in biogels, *J. Control. Release*, pp. 117–129.
9. Maurya, S. K., Pathak, K., and Bali, V. (2010). Therapeutic potential of mucoadhesive drug delivery systems: An updated patent review, *Recent Pat. Drug Deliv. Formul.*, **3**, pp. 256–265.
10. Peppas, N. A., and Sahlin, J. J. (1996). Hydrogels as mucoadhesive and bioadhesive materials: A review, *Biomaterials*, **16**, pp. 1553–1561.
11. Dawson, M., Krauland, E., Wirtz, D., and Hanes, J. (2004). Transport of polymeric nanoparticle gene carriers in gastric mucus, *Biotechnol. Prog.*, **3**, pp. 851–857.
12. Ensign, L. M., Cone, R., and Hanes, J. (2012). Oral drug delivery with polymeric nanoparticles: The gastrointestinal mucus barriers, *Adv. Drug Deliv. Rev.*, **6**, pp. 557–570.
13. Hanes, J., and Demeester, J. (2009). Drug and gene delivery to mucosal tissues: The mucus barrier. Preface, *Adv. Drug Deliv. Rev.*, **2**, pp. 73–74.
14. Lai, S. K., and Hanes, J. (2008). Real-time multiple particle tracking of gene nanocarriers in complex biological environments, *Methods Mol. Biol.*, **434**, pp. 81–97.
15. Ensign, L. M., Schneider, C., Suk, J. S., Cone, R., and Hanes, J. (2012). Mucus penetrating nanoparticles: Biophysical tool and method of drug and gene delivery, *Adv. Mater.*, **28**, pp. 3887–3894.

16. Wang, Y. Y., Lai, S. K., So, C., Schneider, C., Cone, R., and Hanes, J. (2011). Mucoadhesive nanoparticles may disrupt the protective human mucus barrier by altering its microstructure, *PLoS One*, **6**, pp. e21547.
17. Wereley, S. T., and Meinhart, C. D. (2010). Recent advances in micro-particle image velocimetry, *Annu. Rev. Fluid. Mech.*, **42** pp. 557–576.
18. Justin-Temu, M., Damian, F., Kinget, R., and Mooter, G. V. D. (2004). Intravaginal gels as drug delivery systems, *J. Womens Health*, **7**, pp. 834–844.
19. Livingston, E. H., and Engel, E. (1994). Modeling of the gastric gel mucus layer: Application to the measured pH gradient, *J. Clin. Gastroenterol.*, **1**, pp. S120–124.
20. Chu, S., Tanaka, S., Kaunitz, J. D., and Montrose, M. H. (1999). Dynamic regulation of gastric surface pH by luminal pH, *J. Clin. Invest.*, **5**, pp. 605–612.
21. Cao, X., Bansil, R., Bhaskar, K. R., Turner, B. S., LaMont, J. T., Niu, N., and Afdhal, N. H. (1999). pH-dependent conformational change of gastric mucin leads to sol-gel transition, *Biophys. J.*, **3**, pp. 1250–1258.
22. Celli, J. P., Turner, B. S., Afdhal, N. H., Ewoldt, R. H., McKinley, G. H., Bansil, R., and Erramilli, S. (2007). Rheology of gastric mucin exhibits a pH-dependent sol-gel transition, *Biomacromolecules*, **5**, pp. 1580–1586.
23. das Neves, J., Rocha, C. M. R., Gonçalves, M. P., Carrier, R. L., Amiji, M., Bahia, M. F., and Sarmento, B. (2012). Interactions of microbicide nanoparticles with a simulated vaginal fluid, *Mol. Pharm.*, **11**, pp. 3347–3356.
24. Cone, R. A. (1999). Mucus. In *Handbook of Mucosal Immunology*, Ogra, P. L. (Ed.), Academic Press, San Diego, pp. 43–64.
25. Tarran, R., Grubb, B. R., Gatzy, J. T., Davis, C. W., and Boucher, R. C. (2001). The relative roles of passive surface forces and active ion transport in the modulation of airway surface liquid volume and composition, *J. Gen. Physiol.*, **2**, pp. 223–236.
26. Button, B., Picher, M., and Boucher, R. C. (2007). Differential effects of cyclic and constant stress on ATP release and mucociliary transport by human airway epithelia, *J. Physiol.*, **2**, pp. 577–592.
27. Button, B., Okada, S. F., Frederick, C. B., Thelin, W. R., and Boucher, R. C. (2013). Mechanosensitive ATP release maintains proper mucus hydration of airways, *Sci. Signal.*, **279**, pp. ra46.
28. Silberberg, A. (1988). Models of mucus structure. In *Methods in Bronchial Mucology*, Braga, P., and Allegra, L. (Eds.), Raven Press, New York, pp. 51–62.

29. Khanvilkar, K., Donovan, M. D., and Flanagan, D. R. (2001). Drug transfer through mucus, *Adv. Drug Del. Rev.*, **2–3**, pp. 173–193.
30. Linden, S., Sutton, P., Karlsson, N., Korolik, V., and McGuckin, M. (2008). Mucins in the mucosal barrier to infection, *Mucosal Immunol.*, **3**, pp. 183–197.
31. McGuckin, M. A., Lindén, S. K., Sutton, P., and Florin, T. H. (2011). Mucin dynamics and enteric pathogens, *Nat. Rev. Microbiol.*, **4**, pp. 265–278.
32. Kesimer, M., Ehre, C., Burns, K., Davis, C., Sheehan, J., and Pickles, R. (2013). Molecular organization of the mucins and glycocalyx underlying mucus transport over mucosal surfaces of the airways, *Mucosal Immunol.*, **2**, pp. 379–392.
33. Sanders, N. N., de Smedt, S. C., and Demeester, J. (1999). The physical properties of biogels and their permeability for macromolecular drugs and colloidal drug carriers, *J. Pharm. Sci.*, **7**, pp. 835–849.
34. Willits, R. K., and Saltzman, W. M. (2001). Synthetic polymers alter the structure of cervical mucus, *Biomaterials*, **5**, pp. 445–452.
35. Sanders, N. N., de Smedt, S. C., van Rompaey, E., Simoens, P., de Baets, F., and Demeester, J. (2000). Cystic fibrosis sputum: A barrier to the transport of nanospheres, *Am. J. Respir. Crit. Care Med.*, **5**, pp. 1905–1911.
36. Schuster, B. S., Suk, J. S., Woodworth, G. F., and Hanes, J. (2013). Nanoparticle diffusion in respiratory mucus from humans without lung disease, *Biomaterials*, **13**, pp. 3439–3446.
37. Lai, S. K., O'Hanlon, D. E., Harrold, S., Man, S. T., Wang, Y.-Y., Cone, R., and Hanes, J. (2007). Rapid transport of large polymeric nanoparticles in fresh undiluted human mucus, *Proc. Natl. Acad. Sci. U. S. A.*, **5**, pp. 1482–1487.
38. Lai, S. K., Wang, Y.-Y., Hida, K., Cone, R., and Hanes, J. (2010). Nanoparticles reveal that human cervicovaginal mucus is riddled with pores larger than viruses, *Proc. Natl. Acad. Sci. U. S. A.*, **2**, pp. 598–603.
39. Lai, S. K., Wang, Y.-Y., Cone, R., Wirtz, D., and Hanes, J. (2009). Altering mucus rheology to "solidify" human mucus at the nanoscale, *PLoS One*, **1**, pp. e4294.
40. Lai, S. K., Suk, J. S., Pace, A., Wang, Y.-Y., Yang, M., Mert, O., Chen, J., Kim, J., and Hanes, J. (2011). Drug carrier nanoparticles that penetrate human chronic rhinosinusitis mucus, *Biomaterials*, **26**, pp. 6285–6290.
41. Suk, J. S., Lai, S. K., Wang, Y.-Y., Ensign, L. M., Zeitlin, P. L., Boyle, M. P., and Hanes, J. (2009). The penetration of fresh undiluted sputum

expectorated by cystic fibrosis patients by non-adhesive polymer nanoparticles, *Biomaterials*, **13**, pp. 2591–2597.

42. Zahm, J. M., Galabert, C., Chaffin, A., Chazalette, J. P., Grosskopf, C., and Puchelle, E. (1998). Improvement of cystic fibrosis airway mucus transportability by recombinant human DNase is related to changes in phospholipid profile, *Am. J. Respir. Crit. Care Med.*, **6 Pt 1**, pp. 1779–1784.
43. Quraishi, M. S., Jones, N. S., and Mason, J. (1998). The rheology of nasal mucus: A review, *Clin. Otolaryngol.*, **5**, pp. 403–413.
44. De Smedt, S., Meyvis, E., van Oostveldt, P., Blonk, J., Hennink, W., and Demeester, J. (1997). Diffusion of macromolecules in dextran methacrylate solutions and gels as studied by confocal scanning laser microscopy, *Macromolecules*, **30**, pp. 4863–4870.
45. Bhat, P. G., Flanagan, D. R., and Donovan, M. D. (1996). Drug diffusion through cystic fibrotic mucus: Steady-state permeation, rheologic properties, and glycoprotein morphology, *J. Pharm. Sci.*, **6**, pp. 624–630.
46. Hanes, J., Dawson, M., Har-el, Y., Suh, J., and Fiegel, J. (2003). Gene therapy in the lung. In *Pharmaceutical Inhalation Aerosol Technology*, Hickey, A. J. (Eds.), Marcel Dekker Inc., New York, pp. 489–539.
47. Tseng, Y., An, K. M., and Wirtz, D. (2002). Microheterogeneity controls the rate of gelation of actin filament networks, *J. Biol. Chem.*, **20**, pp. 18143–18150.
48. Valentine, M. T., Kaplan, P. D., Thota, D., Crocker, J. C., Gisler, T., Prud'homme, R. K., Beck, M., and Weitz, D. A. (2001). Investigating the microenvironments of inhomogeneous soft materials with multiple particle tracking, *Phys. Rev. E*, **6**, pp. 061506.
49. Tseng, Y., Kole, T. P., and Wirtz, D. (2002). Micromechanical mapping of live cells by multiple-particle-tracking microrheology, *Biophys. J.*, **6**, pp. 3162–3176.
50. Mason, T. G., and Weitz, D. A. (1995). Optical measurements of frequency-dependent linear viscoelastic moduli of complex fluids, *Phys. Rev. Lett.*, **7**, pp. 1250–1253.
51. Mason, T. G., and Weitz, D. A. (1995). Linear viscoelasticity of colloidal hard sphere suspensions near the glass transition, *Phys. Rev. Lett.*, **14**, pp. 2770–2773.
52. Smart, J. D. (2005). The basics and underlying mechanisms of mucoadhesion, *Adv. Drug Del. Rev.*, **11**, pp. 1556–1568.
53. Grabovac, V., Guggi, D., and Bernkop-Schnürch, A. (2005). Comparison of the mucoadhesive properties of various polymers, *Adv. Drug Del. Rev.*, **11**, pp. 1713–1723.

54. Salamat-Miller, N., Chittchang, M., and Johnston, T. P. (2005). The use of mucoadhesive polymers in buccal drug delivery, *Adv. Drug Del. Rev.*, **11**, pp. 1666–1691.

55. Khutoryanskiy, V. V. (2011). Advances in mucoadhesion and mucoadhesive polymers, *Macromol. Biosci.*, **6**, pp. 748–764.

56. Bernkop-Schnürch, A., and Dünnhaupt, S. (2012). Chitosan-based drug delivery systems, *Eur. J. Pharm. Biopharm.*, **3**, pp. 463–469.

57. Bernkop-Schnürch, A. (2005). Thiomers: A new generation of mucoadhesive polymers, *Adv. Drug Del. Rev.*, **11**, pp. 1569–1582.

58. Davidovich-Pinhas, M., and Bianco-Peled, H. (2010). Novel mucoadhesive system based on sulfhydryl-acrylate interactions, *J. Mater. Sci. Mater. Med.*, **7**, pp. 2027–2034.

59. Lehr, C.-M. (2000). Lectin-mediated drug delivery: The second generation of bioadhesives, *J. Control. Release*, **1**, pp. 19–29.

60. Nicolazzo, J. A., Reed, B. L., and Finnin, B. C. (2005). Buccal penetration enhancers: How do they really work? *J. Control. Release*, **1**, pp. 1–15.

61. Whitehead, K., and Mitragotri, S. (2008). Mechanistic analysis of chemical permeation enhancers for oral drug delivery, *Pharm. Res.*, **6**, pp. 1412–1419.

62. Whitehead, K., Karr, N., and Mitragotri, S. (2008). Safe and effective permeation enhancers for oral drug delivery, *Pharm. Res.*, **8**, pp. 1782–1788.

63. Gupta, V., Hwang, B. H., Doshi, N., and Mitragotri, S. (2013). A permeation enhancer for increasing transport of therapeutic macromolecules across the intestine, *J. Control. Release*, **2**, pp. 541–549.

64. Olmsted, S. S., Padgett, J. L., Yudin, A. I., Whaley, K. J., Moench, T. R., and Cone, R. A. (2001). Diffusion of macromolecules and virus-like particles in human cervical mucus, *Biophys. J.*, **4**, pp. 1930–1937.

65. Lai, S. K., Wang, Y.-Y., and Hanes, J. (2009). Mucus-penetrating nanoparticles for drug and gene delivery to mucosal tissues, *Adv. Drug Del. Rev.*, **2**, pp. 158–171.

66. Tang, B. C., Dawson, M., Lai, S. K., Wang, Y.-Y., Suk, J. S., Yang, M., Zeitlin, P., Boyle, M. P., Fu, J., and Hanes, J. (2009). Biodegradable polymer nanoparticles that rapidly penetrate the human mucus barrier, *Proc. Natl. Acad. Sci. U. S. A.*, **46**, pp. 19268–19273.

67. Wang, Y. Y., Lai, S. K., Suk, J. S., Pace, A., Cone, R., and Hanes, J. (2008). Addressing the PEG mucoadhesivity paradox to engineer nanoparticles that "slip" through the human mucus barrier, *Angew. Chem. Int. Ed. Engl.*, **50**, pp. 9726–9729.

68. Yu, T., Wang, Y.-Y., Yang, M., Schneider, C., Zhong, W., Pulicare, S., Choi, W.-J., Mert, O., Fu, J., and Lai, S. K. (2012). Biodegradable mucus-penetrating nanoparticles composed of diblock copolymers of polyethylene glycol and poly(lactic-co-glycolic acid), *Drug. Deliv. Transl. Res.*, **2**, pp. 124–128.
69. Cu, Y., and Saltzman, W. M. (2008). Controlled surface modification with poly(ethylene) glycol enhances diffusion of PLGA nanoparticles in human cervical mucus, *Mol. Pharm.*, **1**, pp. 173–181.
70. Yang, M., Lai, S. K., Wang, Y. Y., Zhong, W., Happe, C., Zhang, M., Fu, J., and Hanes, J. (2011). Biodegradable nanoparticles composed entirely of safe materials that rapidly penetrate human mucus, *Angew. Chem. Int. Ed. Engl.*, **11**, pp. 2597–2600.
71. Woodrow, K. A., Cu, Y., Booth, C. J., Saucier-Sawyer, J. K., Wood, M. J., and Saltzman, W. M. (2009). Intravaginal gene silencing using biodegradable polymer nanoparticles densely loaded with small-interfering RNA, *Nat. Mater.*, **6**, pp. 526–533.
72. Cu, Y., Booth, C. J., and Saltzman, W. M. (2011). *In vivo* distribution of surface-modified PLGA nanoparticles following intravaginal delivery, *J. Control. Release*, **2**, pp. 258–264.
73. Ensign, L. M., Tang, B. C., Wang, Y.-Y., Terence, A. T., Hoen, T., Cone, R., and Hanes, J. (2012). Mucus-penetrating nanoparticles for vaginal drug delivery protect against herpes simplex virus, *Sci. Transl. Med.*, **138**, pp. 138ra179.
74. Ensign, L. M., Schneider, C., Suk, J. S., Cone, R., and Hanes, J. (2012). Mucus penetrating nanoparticles: Biophysical tool and method of drug and gene delivery, *Adv. Mater.*, **28**, pp. 3887–3894.
75. Ensign, L. M., Hoen, T. E., Maisel, K., Cone, R. A., and Hanes, J. S. (2013). Enhanced vaginal drug delivery through the use of hypotonic formulations that induce fluid uptake, *Biomaterials*, **28**, pp. 6922–6929.
76. Suk, J. S., Kim, A. J., Trehan, K., Schneider, C. S., Cebotaru, L., Woodward, O. M., Boylan, N. J., Boyle, M. P., Lai, S. K., Guggino, W. B., and Hanes, J. (2014). Lung gene therapy with highly compacted DNA nanoparticles that overcome the mucus barrier, *J. Control. Release*, **178**, pp. 8–17.
77. Mrsny, R., Daugherty, A., Short, S., Widmer, R., Siegel, M., and Keller, G. (1996). Distribution of DNA and alginate in purulent cystic fibrosis sputum: Implications to pulmonary targeting strategies, *J. Drug Target.*, **4**, pp. 233–243.
78. Tomkiewicz, R. P., App, E. M., de Sanctis, G. T., Coffiner, M., Maes, P., Rubin, B. K., and King, M. (1995). A comparison of a new mucolytic

N-acetylcysteine L-lysinate with N-acetylcysteine: Airway epithelial function and mucus changes in dog, *Pulm. Pharmacol.*, **6**, pp. 259–265.

79. Sun, F., Tai, S., Lim, T., Baumann, U., and King, M. (2002). Additive effect of dornase alfa and Nacystelyn on transportability and viscoelasticity of cystic fibrosis sputum, *Can. Respir. J.*, **6**, pp. 401–406.
80. Sisson, J. H., Yonkers, A. J., and Waldman, R. H. (1995). Effects of guaifenesin on nasal mucociliary clearance and ciliary beat frequency in healthy volunteers, *Chest*, **3**, pp. 747–751.
81. Gonda, I. (1996). Inhalation therapy with recombinant human deoxyribonuclease I, *Adv. Drug Del. Rev.*, **1**, pp. 37–46.
82. Vasconcellos, C. A. (2002). Reduction in viscosity of cystic fibrosis sputum *in vitro* by gelsolin, *Science*, **5149**, pp. 969–971.
83. Rubin, B. (1996). Therapeutic aerosols and airway secretions, *J. Aerosol Med.*, **1**, pp. 123–130.
84. Ferrari, S., Kitson, C., Farley, R., Steel, R., Marriott, C., Parkins, D. A., Scarpa, M., Wainwright, B., Evans, M. J., Colledge, W. H., Geddes, D. M., and Alton, E. W. (2001). Mucus altering agents as adjuncts for nonviral gene transfer to airway epithelium, *Gene Ther.*, **18**, pp. 1380–1386.

Chapter 7

Transport of Therapeutics across Gastrointestinal Epithelium

Frederic Lagarce and Emilie Roger
University of Angers, UMR-S 1066, 16 Boulevard Daviers, 49045 Angers Cedex 01, France
frederic.lagarce@univ-angers.fr, emilie.roger@univ-angers.fr

This chapter focuses on the oral administration of therapeutics. Patients prefer oral administration because it is painless and simple to implement. However, oral administration remains a challenge since it is limited by various physiological barriers. The first part of this chapter describes the physiological features of gastrointestinal (GI) tract related to drug absorption. Drug solubility and permeability across the GI epithelium are the two main features involved in drug bioavailability. To overcome the barriers to these features, many different technological solutions are proposed to increase the bioavailability of poorly water-soluble and/or poorly permeable drugs. The second part of this chapter presents various techniques developed to enhance drug solubility, such as physical (size reduction) and chemical modifications or lipid formulations, and illustrates them by marketed products. Finally, the chapter describes approaches to improve drug permeability, more specifically novel drug delivery systems such as nanoparticles.

Drug Delivery Across Physiological Barriers
Edited by Silvia Muro

ISBN 978-981-4669-40-5 (Hardcover), 978-981-4669-41-2 (eBook)
www.panstanford.com

7.1 Introduction

Oral delivery is the main route of administration for drugs because it has many advantages. In fact, the oral route, if efficient, allows drug delivery very easily and safely. The acceptance and compliance of treatment are high; thus, oral delivery is well adapted to chronic treatments. The oral route can be used anywhere, anytime, and in most situations. The cost of manufacturing is often very low in comparison to injectable drugs, which have to be sterile. Injectable drugs are mostly used in the case of emergency care and if the oral route is impossible. The oral route is then the first choice in the pharmaceutical industry when a formulation of a new active pharmaceutical ingredient (API) is sought. Unfortunately, not all APIs are adapted to the oral route: one of the main limitations is the lack of bioavailability for drugs that are not able to cross the gastrointestinal (GI) epithelium or are not soluble in the GI tract. As proposed by Amidon et al. in 1995, solubility and permeability are the two main API features involved in its bioavailability after oral administration [1]. This reasoning leads to the famous biopharmaceutical classification scheme (BCS) in which drugs are separated in four classes based on the ability of the therapeutic dose to be transported across the intestinal epithelium and its solubility in a pH range of 1–8. In this classification, class I drugs display good solubility and permeability, whereas class IV drugs display limited permeability and solubility. Class II and III drugs represent intermediate situations with a lack of solubility for class II drugs and a lack of permeability for class III drugs. Unfortunately, many important therapeutic drugs belong to classes II, III, or IV and thus have often limited and unreliable bioavailability. Between 2000 and 2011 in the USA, 42% of approvals for generic drugs were for class I drugs and 0% for class IV drugs because the latter are too difficult to formulate for the oral route [2]. Moreover, newly discovered APIs belong mostly to BCS class II and class IV. Thus, there is an important need to find efficient formulations that will enhance permeability and apparent solubility of these compounds if oral route is the target. This is the topic of this chapter, but before exploring the design of new oral drug delivery systems, it is mandatory to review the physiology of the GI tract with an emphasis on the physiological barrier to drug absorption.

7.2 Physiological Features of GI Tract Related to Drug Absorption

After oral administration, the drug and its formulation have to follow a complex journey in the body (Fig. 7.1A). This trip begins in the mouth and ends in the GI tract. Many physiological barriers have to be overcome to reach drug absorption. In this chapter, we will focus on the small intestine because this is where the absorption phenomenon mainly occurs. However, it is worth to note that before this step the drug can be degraded by enzymes in the mouth or in the stomach, where amylases, lipases, and peptidases are found. The drug can also be degraded in the strong acidic conditions found in the stomach. In fact, the efficacy to reach the intestine in an intact manner is very different between different drugs. For an accurate overview of this phenomenon, the reader is advised to read the book from Neena Washington [3]. In order to be absorbed, the therapeutic dose of the drug should reach the intestinal tract without being degraded and remain in a soluble state. Then, there are three types of barriers to cross: a chemical barrier, a physical barrier, and a biological barrier (Fig. 7.1B).

7.2.1 *Chemical Barriers to Drug Absorption*

7.2.1.1 pH, solubility, and absorption

The concentration of H^+, represented by pH, has an influence on the solubility of the majority of therapeutic drugs, which are weak acids or bases. The Henderson–Haselbach equation (Eq. 7.1) allows to calculate the ratio of the ionic over the unionized moieties as a function of pH and pK_a of the molecules.

$$pH - pK_a = \log[\text{basic form}]/[\text{acidic form}] \tag{7.1}$$

The acidic form is unionized for weak acids and ionized for weak bases. For example, being a weak acid and having a pK_a of 3.5, aspirin

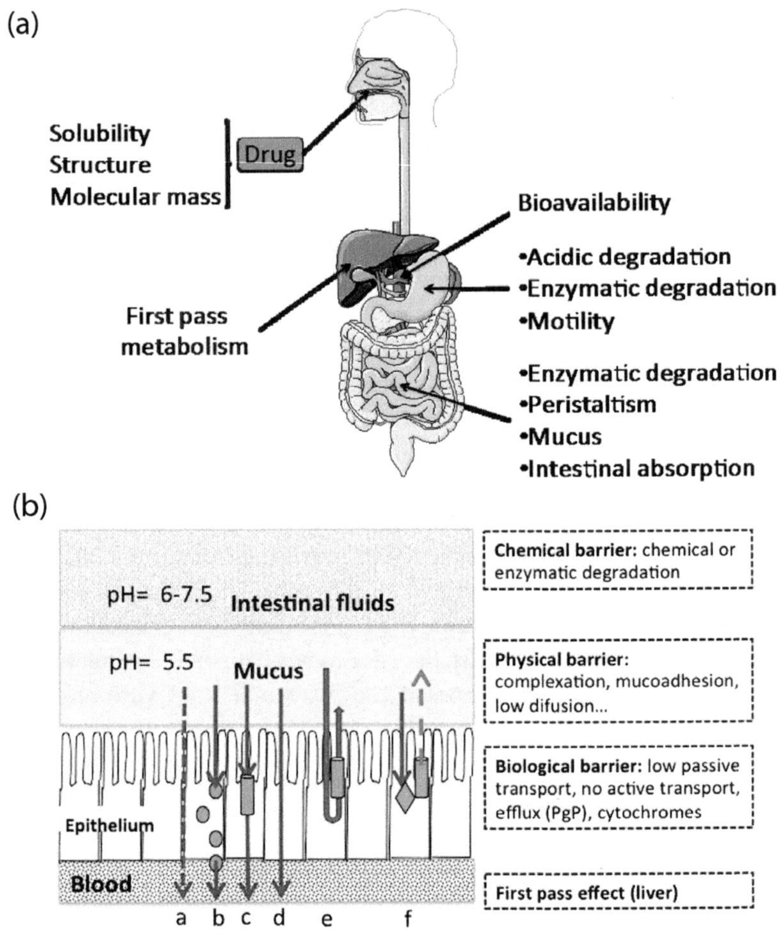

Figure 7.1 (A) Different barriers encountered after the oral administration of drug. (B) Intestinal barriers to drug absorption and mechanism of transport across epithelium: (a) paracellular transport, (b) transcytosis, (c) active transport, (d) passive transport, (e) efflux by transporter such as PgP, and (f) metabolization by cytochrome and efflux. Circles represent the vesicles that cross the cells via transcytosis. Cylinders represent membrane transporters. The diamond represents enzymes such as cytochrome p450.

is mostly in its ionized form in the duodenum at pH 6.5 (from Eq. 7.1 one can calculate the ratio between the basic form and the acidic form, which is 1000 in this case). Local pH thus has an important impact on the solubility of the compounds because the ionic form is the most soluble. The pH varies along the GI tract and is the most acidic in the stomach (pH range 1.5 to 4.5). The GI tract becomes slightly acidic at the entrance of the duodenum (pH range 5.5 to 6.5) and then neutral and slightly basic in the jejunum, ileum, and colon. From the above explanations, one can deduce that weak bases are more soluble in acidic conditions, that is, in the stomach and at the entrance of small intestine, and that weak acids are more soluble in basic conditions, that is, after the duodenum.

Solubility is mandatory for absorption, but the drug also has to be absorbed, and this is mainly due to passive diffusion. To diffuse across cell membrane, which are composed mainly of phospholipids, the drug should also be lipophilic. The unionized form is, in this case, the better state for the molecule to diffuse. To be absorbed, the molecule should be lipophilic enough but not too much to remain soluble. The ideal drug partition coefficient has been found to be around 100 in favor of cellular phospholipids over water to maximize the possibilities for passive diffusion. This corresponds to a log P of the concentrations in octanol versus concentrations in water of 2. Octanol is the lipophilic phase used because it mimics the phospholipids of the cellular membrane. One should be aware that the molecular mass of the drug also has a role for passive diffusion. A molecular mass below 500 is considered a favorable factor for absorption across the small intestine. If the drug can be absorbed using active transport, these conditions (log P and molecular weight) are not required as soon as there is a transporter for the drug helping it to cross the cell membrane of the enterocytes. Hydrophilic drugs are not able to diffuse passively and have to use active transport or paracellular transport to be absorbed. For paracellular transport, the most important parameter besides solubility is the molecular weight of the drug, which should be below 200 Da. This allows for the diffusion and convection processes between cells and across tight junction that limit their passage.

7.2.1.2 Enzymes and other chemical interactions in small intestine

Glands found in the small intestine secrete bicarbonate, mucus, and the intestinal juice *succus entericus*, which contains an enteropeptidase that transforms trypsinogen into trypsin. Secretions in the small intestine also come from the pancreas.

Pancreatic secretion (1 L/day) contains proteases, which are converted to their active form in the lumen of the small intestine. It also contains lipases and amylase. These enzymes, used physiologically to digest food, can have a very significant impact on drug stability and can modify the formulations. This is why the importance of the use of biomimetic media is often raised in order to evaluate drug solubility and permeability *in vitro*.

Besides pancreatic secretion, the lumen of the small intestine also receives bile coming from the liver (600 mL/day). Bile contains ions, bile acids, phospholipids, cholesterol, and bilirubin. The role of bile salts is to emulsify the fat contained in food and help the absorption of fatty acids and cholesterol. In the same way, bile salts can also help the emulsifications of lipophilic drugs and thus allow a better absorption.

Chemical interactions can also be observed with food in the stomach or small intestine. Food can change the pH. Food components such as fiber can entrap drugs, for example steroids. Calcium present in the food can be complexed with drugs such as tetracyclin, which can limit their absorption. The presence of fat in the food can help lipophilic drugs dissolve in the fat and then stimulate the secretion of bile salts leading to better absorption. One should note that food also has nonchemical consequences on drug absorption because it impacts gastric emptying, blood flow, and lymphatic secretions.

The interaction with food is very complex to study. In a first *in vitro* approach, biomimetic media are used [4]. Those media have a composition close to gastric fluid or intestinal fluid at the fasted and fed states. Dissolution (using USP apparatus) and diffusion studies (using cell cultures) in these biomimetic media allow to discriminate the effect of food on absorption in most situations [5]. For the majority of APIs, these biomimetic media (Fa-SGF, Fe-SGF,

Fassif-V2, Fessif-V2) [5] have much more predictive *in vivo* features in comparison to ancient standard USP media such as simulated intestinal fluid and simulated gastric fluid.

7.2.2 *Physical Barriers to Drug Absorption: Mucus*

Besides viscosity of the small intestine content, which is related to the presence and nature of food, the physical barrier to drug diffusion preceding absorption is mucus. Mucus is a hydrogel composed of proteins called mucins, lipids, carbohydrates, cellular fragments, and salts. This hydrogel protects many epitheliums of the body against physical or chemical aggression (see Section 2.3 of this book). It is found all along the GI tract and has to be crossed before absorption through the enterocytes. On the GI tract, mucus is composed of two layers: a firmly adherent layer and a loosely adherent layer with a thickness that varies throughout the GI tract. Total thickness of both layers goes from 150 μm to 800 μm. In the duodenum and jejunum, where most drug absorption occurs, the firmly adherent layer is very thin (<20 μm) and the loosely adherent layer is around 150–200 μm, which are important distances for diffusion considering the size of the drug molecule [6]. The structure of mucus is complex, which arises from the properties of mucins. Mucins are large glycoproteins composed of more than 800 amino acids, also containing cysteine-rich and disulfide-rich domains. Mucins have long flexible proline, threonine, and serine (PTS) domains that are glycosylated. The glycans terminate with negatively charged carboxylic groups. Diffusion in the mucus structure depends on the charge of the molecules. Mucus contains pores that are 200–400 nm diameter, thus allowing diffusion of many APIs. If APIs are encapsulated in nano- or microcarriers (as we will see later in this chapter), the size of the carrier can preclude diffusion in mucus. Neutral molecules are thought to diffuse better. Positively charged molecules can be trapped in mucus by electrostatic interactions, but because the surfaces of the enterocytes are also negatively charged, the molecules can also be attracted by cell surfaces and then be absorbed. *In vitro* studies should not be limited to mucus diffusion studies and should include cellular uptake studies with Caco-2 and/or HT29 cells. At least the source of mucus should be chosen wisely because diffusion properties differ significantly if one uses only reconstructed mucus from purified mucins diluted in water or crude mucus extracted from animals. The literature is full of good review papers on mucus barrier properties and diffusion for molecules or nanoparticles [7–10]. The conditions for mucus diffusion studies have an important impact on the results found: membrane thickness should be the lowest and the experimentation duration should be high enough in order to overlook the time of flow establishment. If the membrane volume is negligible versus the volume of diffusion chambers, one can use the following diffusion equation to calculate the permeation coefficient P_{app}:

$$P_{\mathrm{app}} = dQ/AC_0dt \tag{7.2}$$

where dQ/dt is the rate of drug appearance on the received side ($\mu\mathrm{gs}^{-1}$), C_0 is the initial concentration over the donor side ($\mu\mathrm{gmL}^{-1}$), and A is the surface area (cm^2).

7.2.3 *Biological Barriers to Drug Absorption*

7.2.3.1 Transport across cell membrane

Because many chapters of this book are dedicated to cell transport, we will not detail this part in the present chapter. As described in Fig. 7.1B, there are many possible mechanisms to cross the intestinal epithelium: passive diffusion, which depends on the concentration difference between apical and basal sides of the enterocyte; active transport, which is specific and energy dependent; paracellular diffusion, which is limited to small and hydrophilic molecules; and endocytosis. However, it is important to note that drug absorption in the GI tract is mostly linked to the passive diffusion across the cell membrane. This hypothesis was verified using parallel artificial membrane permeability assay (PAMPA). PAMPA measures passive transport and compares the absorption through this artificial membrane with diffusion across Caco-2 cells, the reference cellular assay for drug absorption, as well as passive and active transport. In fact, in some cases, absorption can be predicted only by performing diffusion studies across phospholipid bilayers. In this case, Avdeef et al. found a correlation between cellular models and PAMPA [11].

Table 7.1 Examples of drugs actively transported in the intestinal tract

Active transport	Drug substrate
Amino acid transporters	Baclofen, L-dopa, gabapentin, methyldopa
Peptide transporters	β—lactams antibiotics, ACE inhibitors, methyldopa
Nucleoside transporters	Zidovudin, zalcitabin
Organic anion transporters	Ceftriaxone, methotrexate, pravastatin
Organic cation transporters	Quinidine, verapamil, midazolam

This suggests that for some drugs (weakly basic drugs in the case of Avdeef study), passive diffusion is the limiting factor for drug absorption. Besides passive diffusion, drugs can use active transport if their structures mimic endogenous molecules (Table 7.1). Drugs can also enter in the enterocyte via endocytosis. This is particularly the case if the drug is associated to micelles or into nanocarriers.

Diffusion across intestinal epithelium allows the drug to reach the blood stream but also possibly the lymphatic system. In fact, diffusion across M-cells, which belong to the gut-associated lymphoid tissue and are situated in the ileum and jejunum, can target the drug into lymph. In this case, the first pass effect is avoided: the drug reaches lymphatic thoracic duct and then the left jugular vein. APIs with high log P (log P>3.5) are likely to reach the lymphatic circulation. Unfortunately, the output of lymphatics is only 1/500 of the blood's output, thus limiting the importance of this pathway.

7.2.3.2 Efflux pumps as a barrier to intestinal absorption

After cellular uptake, hydrophilic or lipophilic drugs can be transported out of the enterocytes by ATP-binding cassette (ABC) transporters. ABC transporters are a large family of proteins, some of which are involved in drug transport in the intestine: ABCB1 (MDR1), ABCC1-6 (MRP1-6), and ABCG2 (BCRP). ABCB1, ABCC2, and ABCG2 localized at the apical side of the enterocyte are responsible for drug efflux from the enterocyte to the intestinal lumen. On the contrary, ABCC1 or ABBCC3 are located on the basal membrane of the enterocyte and help the drug to be excluded from the enterocyte to the blood stream. ABCC4-6 has been found in the enterocyte, but its role remains unknown. The most described efflux transporter in the enterocyte is ABCB1, also called P-glycoprotein (PgP).

More than 40 therapeutic drugs of primary importance, including taxanes and other anticancer drugs, antiretroviral drugs, morphine, or dexamethasone, are substrates of PgP. For a complete list of transporters and substrates, please refer to Chapter 24 of *Oral Bioavailability* edited by Ming Hu and Xiaoling Li [12]. PgP can be inhibited by therapeutic drugs such as amiodarone, chloroquine, verapamil, and cyclosporine A, but also by synthetic specific second- and third-generation inhibitors such as Elacridar or Zosuquidar. Third-generation PgP inhibitors are highly specific and do not suffer limitations of previous inhibitors, which are substrates of CYP 3A4 enzymes or other ABC transporters. Moreover, they are not as pharmacologically active as first-generation inhibitors. It is important to note that surfactants such as poloxamers (or fatty acid ester surfactants such as Cremophor EL and Solutol HS 15), which are often used to obtain drug delivery systems, can inhibit PgP and thus provide an enhancement of drug absorption [13]. Vitamin E-TPGS has also been shown to inhibit PgP [14]. Inhibiting PgP can lead to a great improvement in drug absorption. This was, for example, shown with paclitaxel given orally to mdr1a (-/-) mice. In this group, the oral bioavailability was 35% compared to 11% found in wild-type mice [15].

ABCG2, also named breast cancer cell resistant protein (BCRP), is also an important efflux protein with many substrates in common with PgP. Inhibiting BCRP also has important consequences on the bioavailability of its substrate. Thus, irinotecan bioavailability was enhanced by 63% if given with the inhibitor gefitinib [16]. The active metabolite of topotecan, Sn-38, is also a substrate to BCRP. This case is not the only one. In fact, if the parent drugs are substrates of efflux transporters, their metabolites are likely to be also substrates of these proteins. An interplay between metabolic enzymes and efflux transporters has been demonstrated. Phase I presystemic metabolism induced by cytochrome P450 (CYP) located in the endoplasmic reticulum of enterocytes is connected with efflux transporters. The double jeopardy theory proposed by L. Benet and his team explains that the API, when absorbed in the enterocyte, can be substrate of the CYP and metabolized or effluxed by the transporters (eventually, it can reach the blood stream) [17]. If effluxed, it can re-enter the cell and have a second "chance" to be metabolized by CYP. Thus, the efficiency of CYP metabolism is enhanced by BCRP or PgP. Inhibition of efflux transporters has, in this case, two consequences: CYP is more likely to be saturated and some API can avoid this enzyme and more easily reach the systemic circulation [18].

The communication and interplay between enzymes and transporters with the physiological goal to protect our organism against xenobiotics are very complex. The ensemble of these phenomena is called the bioavailability barrier matrix (BBM), a term proposed by Wang [19]. The theory is that too many mechanisms are involved to limit the blood access from molecules that are strangers to our body (xenobiotics). The challenge of drug discovery and pharmaceutical technology is to overcome the BBM, as we will see in the following text of this chapter.

7.3 Drug Delivery Solutions for Drug Bioavailability Enhancement

The oral administration of drugs can be a great challenge in the function of the physicochemical properties of drugs. For drugs characterized as BCS class II, III, or IV, which all together represent approximately 90% of pipeline drugs, strategies are needed to administer them by the oral route (Fig. 7.2) [20]. There are two main strategies to increase absorption: modification of the physicochemical properties of a drug or drug encapsulation into the delivery system.

7.3.1 *Drug Delivery Solutions to Improve Drug Solubility*

BCS class II and IV drugs are identified to have low solubility, so various approaches to overcome this problem have been investigated. The aqueous solubility of drug is a critical determinant of its dissolution rate [20]. Increasing the solubility of drug increases the

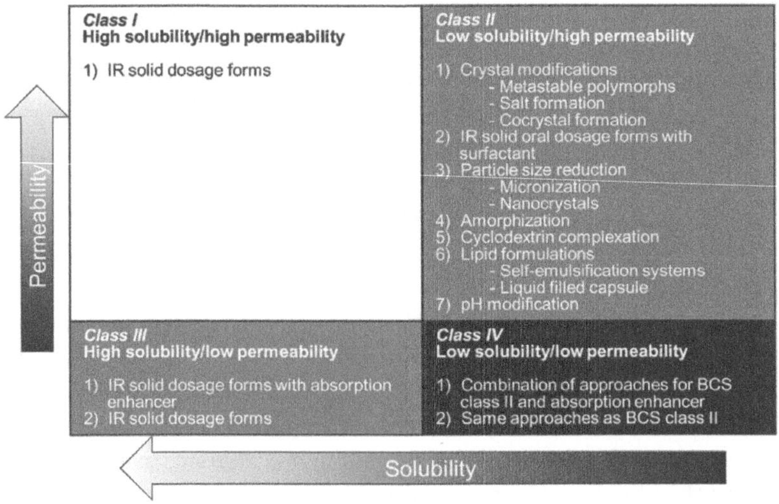

Figure 7.2 Biopharmaceutics classification system (BCS) and viable formulation options based on BCS. Reprinted from Ref. [20], Copyright 2011, with permission from Elsevier.

rate of dissolution and by consequence, the bioavailability of drug. According to the Noyes–Whitney equation, several physicochemical factors influence the drug dissolution rate such as the effective area of the particles to dissolve, the diffusion coefficient of the drug at molecular state, the diffusion thickness of the layer at saturation concentration around the particle, the saturation concentration (solubility), the amount of dissolved drug, and the volume of dissolution media [21]. More specifically, the dissolution rate is directly proportional to the difference between the solubility and the solution concentration.

$$dm/dt = D/V \cdot S/h(C_s - C_t) \tag{7.3}$$

where dm/dt is the dissolution rate, expressed as the change in the amount of drug dissolved (m) per unit time (t); D is the diffusion coefficient calculated from Fick's first law; S is the specific surface area of the solid particles; h is the thickness of the diffusion film adjacent to the dissolving surface; C_s is the saturation solubility of the drug molecule; and C_t is the concentration of the dissolved molecule at time t.

Different approaches to improve solubility classified in the forms of physical and chemical modifications or formulation techniques are presented as follows.

7.3.1.1 Modification of chemical form: salt form, amorphous form, prodrug

First, changing the chemical structure, or more specifically the use of "amorphous" or "salt" formulations, is an option to increase the solubility of a drug.

Crystal modification can be an effective approach. Polymorphism in crystalline (or amorphous) form consists of the same chemical composition of the drug with a different arrangement and/or different molecular conformations that can modify the physicochemical properties such as solubility, melting point, or stability [22]. Typically, amorphous form represents the most energetic solid state of a pharmacologically active material but is thermodynamically unstable. Thus, the amorphous form can change into various crystalline forms over time, which are more stable but less soluble. Therefore, the amorphous form of a drug is utilized accompanied by the required excipients and a suitable manufacturing process to limit the transformation of the solid-state form (see Section 7.3.1.3 on solid dispersion). Amorphous active pharmaceuticals marketed as drug products include Ceftin® (cefuroxime axetil), Accupril® (quinapril hydrochloride), or viracept® (nelfinavir).

Another commonly used approach in the pharmaceutical industry is the ionization of drug by salt formation. Salt forms are typically more soluble than the free bases. The salt form of the drug is often made from the free acidic functional group, most commonly the carboxylic acid group, or from the free basic group, most commonly the amines. Because the majority of adsorption occurs in the small intestine where the pH is typically in the range of 6 to 8, it is preferable to have the maximum solubility of the drug at neutral or slightly basic pH values. In general, the acidic form, which is ionized at pH values of intestine and consequently more soluble, is suitable (see Eq. 7.1). Many drugs are marketed in salt forms [23]. Some examples are Levothyroxine Sodium, Pravastatin Sodium, Diclofenac

Sodium, Naproxen, Metformin HCl, Hydrocodone, and Metoprolol Tartrate.

Finally, the prodrug is another chemical approach to improve the solubility of a drug. A prodrug is a chemical modification of the parent drug with an additional functional group (e.g., phosphate ester group), which undergoes enzymatic and/or chemical transformation prior to exhibiting its pharmacological activity. Different prodrugs are present on the market: fosphenytoin (prodrug of phenytoin), fosampremavir (prodrug of amprenavir), estramustine phosphate (prodrug of estramustine), predinsolone phosphate (prodrug of prednisolone), and parecoxib (prodrug of valdecoxib). The formation of a prodrug is useful to enhance solubility but also stability or transport across biological membrane [24].

7.3.1.2 Modification of physical form: particles size reduction

As shown in the Noyes–Whithey law, particle size and the resulting surface area could have an effect on the rate of dissolution of a drug. A strategy to improve the bioavailability of a drug is to reduce the particle size (micronization or nanocrystals). Micronization is a milling process allowing to obtain particle size below 10 μm. The term nanocrystals is used for particles of nanometer range (<1 μm). Indeed, when the size of a drug decreases, the surface specific area increases and the dissolution rate is expected to increase proportionally. Moreover, a relationship between the diffusion layer thickness (happ) and particle size was reported: reducing particle size decreases happ, which also increases the dissolution rate. Therefore, an increase in the dissolution rate resulting from a particle size reduction process is attributed not only to an increase in the surface-area-to-volume ratio but also to a decrease in happ. In the case of nanocrystals, there is also an enhancement of the saturation solubility, which allows increasing the amount of drug available for absorption. The common method of obtaining micronized drug particles is mechanical pulverization using milling techniques (jet mill, rotor stator colloid mills, ball mill, and pin mill) [20, 25]. Nanocrystals are produced by bead milling, high-pressure homogenization, controlled precipitation, supercritical fluid process, or combination technologies (e.g., precipitation with high-pressure homogenization) [26, 27]. Numerous products based on micronization or nanocrystal are actually found on the pharmaceutical market: micronized spirolactone tablets, micronized Glyburide tablets, Estrace® (micronized estradiol), Prometrium® (micronized progesterone), Tricor® and Triglide® (nanocrystal of fenofibrate), Rapamune® (nanocrystal of sirolimus), Emend® (nanocrystal of aprepitant), and Ritalin® (nanocrystal of methylphenidate) [27].

7.3.1.3 Solid dispersion

Solid dispersion consists of two different components, which are generally a hydrophilic matrix at solid state and a hydrophobic drug. The drug can be dispersed molecularly in amorphous or crystalline form. Different types of solid dispersions are described. Eutectic mixtures consist of two compounds completely miscible in the liquid state but only to a very limited extend in the solid state. Solid solutions are comparable to liquid solutions, where all components are totally miscible and soluble to form one phase irrespective of the number of components. Solid solutions can be classified according to their miscibility as continuous, where all components are miscible in all proportions, or discontinuous, where the solubility of a drug in the carrier is limited (practically, the term solid solution should be applied when the mutual solubility of the two components exceeds 5%). They also can be classified according to the way in which the solvate molecules are distributed in the solvendum: substitutional (the solid solution has crystalline structure and the solute molecules are substituted by solvent molecules in the crystal lattice), interstitial (the dissolved molecules occupy the interstitial spaces between the solvent in the crystal lattice), or amorphous (the solute molecules are dispersed molecularly but irregularly within the amorphous solvent) [28]. There are two main processes for preparing solid dispersions: melting method (hot melt extrusion method, melt agglomeration) and solvent evaporation method (rotor evaporator, spray drying, freeze drying, coprecipitation, and supercritical fluid) [29]. Various carriers are used in solid dispersion: polyethylene glycol (PEG); polyvinylpyrrolidone (PVP); polyvinylalcohol (PVA); cellulose derivatives such as hydroxypropylmethylcellulose (HPMC); polyacrylate and polymethacrylate; urea; sugars such as mannitol, sorbitol, and chitosan; emulsifiers such as Tween® 80; and sodium lauryl sulfate (SLS). Thus, particles in solid dispersion experience a size reduction and a higher degree of porosity, and they improve the wetting characteristic of the drug surface. This altogether improves the apparent solubility of a drug under physiologically relevant conditions and, consequently, its bioavailability. Some drugs are already used in the form of solid dispersion and marketed: Tegretol® (carbamazepine and PEG), Sporanox® (itraconazole and HPMC), Karetra® (lopinavir/Ritonavir and PEG), or Nimotop® (nimodipine and HPMC, PVP/VA, PEG).

7.3.1.4 Cyclodextrin complexes

Drug-cyclodextrin complexes correspond to the inclusion of poorly water-soluble drugs in the core of cyclodextrins, which increases the apparent solubility as well as the dissolution rate and the bioavailability of drug. Cyclodextrins are cyclic oligosaccharides containing six, seven, or eight glucopyranose units called, respectively, α, β, or γ cyclodextrins. These compounds are hydrogen bonded internally to create three-dimensional truncated cones with a hydrophilic outer surface and a lipophilic central cavity. The most commonly used cyclodextrins in pharmaceutical formulation are derivatives of β-cyclodextrin such as hydroxypropyl-β-cyclodextrin [30]. Drugs can be included in one or two cyclodextrins (Fig. 7.3). Different techniques are reported to prepare cyclodextrin complexes: kneading method, spray drying method, freeze drying method, microwave irradiation method, or grinding method. Currently there are about 35 different pharmaceutical products containing drug/cyclodextrin complexes in the market [31], as for oral administration [30, 32]: Sporanox® (itraconazole/hydroxypropyl-β-cyclodextrin complexe), Brexin® (piroxicam/β-cyclodextrin complexe), Albenza® (albendazole/β-cyclodextrin complexe), or Abilify® (Aripiprazole/complexe sulfobutyl ether β-cyclodextrin). In addition, cyclodextrin also has the ability to enhance permeability of drugs [33].

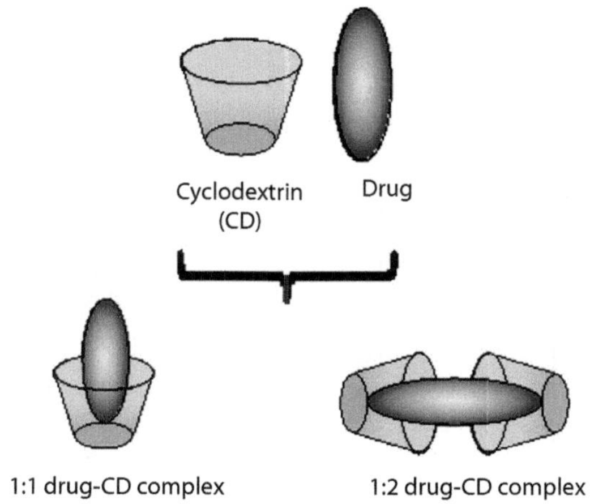

Figure 7.3 Schematic structure of cyclodextrin.

7.3.1.5 Lipid formulations

The use of lipid excipient allows to increase the solubilization of drug and its absorption. Indeed, the presence of lipids in the GI tract stimulates and increases the secretion of bile salts and endogenous lipids (phospholipids, cholesterol) that increase the capacity of solubilization of the GI tract. Moreover, these exogenous lipids reduce the gastric transit, which increases the time available for the dissolution. Lipid-based formulations include systems with pure oils (triglyceride or mixed mono- and diglycerides) or combination of oils with cosolvent and/or surfactants to form micellar solutions, emulsions, and self-emulsifying systems. Pouton [34] introduced a classification of these systems (Table 7.2). The uses of solutions or emulsions are limited due to the low stability of these formulations. The self-emulsifying systems called SEDDS (self-emulsifying drug delivery systems) are isotropic mixtures of oil, surfactant, cosolvent, and soubilized drug, which form oil-in-water fine emulsions when in contact with aqueous phase, such as the GI fluid [35]. According to the size range of the oil droplets, SMEDDS (self-microemulsifying drug delivery systems) were obtained by ranging the droplet size from 100 to 200 nm or SNEDDS (self-nanoemulsifying drug delivery systems) were obtained by ranging the droplet size by less [20].

Table 7.2 Lipid formulation classification system: composition and characteristic feature

Excipients in formulation	Content of formulation (%, w/w)				
	Type I	Type II	Type IIIA	Type IIIB	Type IV
Oils: triglycerides or mixed mono- and diglycerides	100	40–80	40–80	<20	—
Water-insoluble surfactant (HLB <12)	—	20–60	—	—	0–20
Water soluble surfactants (HLB >12)	—	—	20–40	20–50	30–80
Hydrophilic cosolvent	—	—	0–40	20–50	0–50
Characteristics	Nondispersing; requires digestion	SEDDS without water-soluble components	SEDDS/SMEDDS with water-soluble components	SMEDDS with water-soluble components and low oil content	Formulation dispersion typically to form a micellar solution: oil-free formulation based on surfactant and cosolvents

Source: Ref. [34, 35]

Actually, different drug formulations of poorly water-soluble drugs using these systems are available and marketed: Tretinoin (SEDDS), Isotretinoin (SEDDS), Cyclosporin A (SMEDDS or SNEDDS), Lopinavir and ritonavir (SEDDS), Tipranavir (SEDDS), and Amprenavir (SEDDS) [36].

7.3.2 *Drug Delivery Solutions to Improve Drug Permeability*

Oral bioavailability of drugs is also limited due to low permeability across the intestinal barrier, particularly for drugs characterized as BCS class III or BCS class IV. The main approaches for enhancing drug permeability are currently under study and mainly based on the potential of novel drug delivery systems [37]. These systems include polymeric or lipid-based carriers such as liposomes (vesicles in which an aqueous volume is entirely surrounded by one or several phospholipid bilayers mimicking the cell membrane), polymeric micelles (self-assembled core-shell nanostructures formed in an aqueous solution consisting of amphiphilic block copolymers), dendrimers (polymeric complexes that comprise a series of well-defined branches around an inner core), nanoparticles (nanocapsules that are core/shell structure or nanospheres) (Fig. 7.4).

Properties of drug delivery systems that increase the permeability of drug could be explained by their small size (under 1 μm and more often under 200 nm), which allows diffusion within the mucus and transports the drug across the epithelial membrane in different ways. Moreover, some drug delivery systems have the ability to protect the drug against GI degradation.

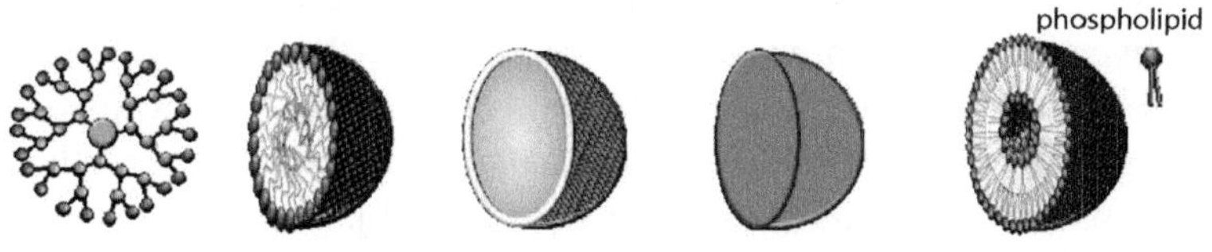

Figure 7.4 Schematic structure of nanocarriers used for oral delivery.

7.3.2.1 Drug delivery systems to improve GI stability

To reach the small intestine, where absorption essentially takes place, drugs must resist degradation from the acidic pH of the stomach and digestive enzymes as explained in the first part of this chapter. Thus, to protect encapsulated drug, some drug delivery systems are stable in GI tract that limit drug release. Liposomes structured in concentric bilayer of phospholipids are rapidly destabilized in the gastric and intestinal media. Similarly, nanoparticles formulated with insoluble polymers, polyelectrolyte such as chitosan or lipids are also destabilized by the pH or by enzymes (e.g., lipase) [38, 39]. Nevertheless, the addition of saturated phospholipids and cholesterol improves the rigidity of liposomes, and the presence of poloxamer or PEG at the surface stabilizes the drug delivery systems and decreases the possibility of enzymatic degradation.

7.3.2.2 Drug delivery systems to diffuse across mucus

To reach the intestinal barrier, drug delivery systems should adhere to the mucus and cross the mucus layer to avoid mucus clearance mechanisms. Mucoadhesive properties were obtained with positively charged drug delivery systems. Mucin is negatively charged due to the high number of sulfate sialic acid or sulfonic acid and sugar moieties. Consequently, electrostatic interactions are driven between mucin and positively charged drug delivery systems and the retention at mucosal surface could increase. Chitosan (cationic polysaccharide) and its derivatives are widely used to generate positive charges on the surface of drug delivery systems. Different studies with chitosan-modified liposomes [40] and chitosan nanoparticles [41–43] illustrate that chitosan-coated delivery systems improve penetration into the mucosa and, by consequence, improve the intestinal absorption of encapsulated drugs. Similar results were also obtained with particles coated with groups binding to mucin, such as thiol [44, 45].

Moreover, the penetration of drug delivery systems in the mucus is also a critical step. Mucus-penetrating particles were developed with hydrophilic nonionic long chains on their surface, such as polyethylene glycol (PEG), poloxamer 407 (pluronic® F127). Indeed, densely coated particles were able to rapidly penetrate mucus layers and approach the enterocyte by minimizing interactions between particles and mucins [10, 40, 46, 47]. The stability of lipid nanocapsules in the mucus and their ability to diffuse across this physical barrier have been demonstrated *in vitro* [48]. For some authors [8], negatively charged hydrophobic nanoparticles form low interactions with mucins and thus diffuse better in mucus thickness. Influences of charge and size are linked, but the literature is very controversial on that point. In consequence, a balance between cationic charge to facilitate mucoadhesion and hydrophilic coating to diffuse through the mucus is important for an efficient oral delivery of these systems. The diffusion system used to evaluate nanoparticle diffusion has a significant impact on the results. It is thus very important to choose this system wisely and to understand its limitation and its relevancy [49].

7.3.2.3 Drug delivery systems to improve intestinal permeability

Finally, drug delivery systems have the potential to enhance drug transport across intestinal barrier via different routes: paracellular transport, active or passive transcellular transport across enterocytes or M-cells. The paracellular pathway between the epithelial cells is both size- and charge-dependent. In general, large-size particles (>1 μm) are not efficiently absorbed due to the small available absorptive area and the restriction by tight junctions. In order to allow drug passage, the opening of tight junctions is required. Different polymers such as anionic polyacrylate derivatives, poly(methacrylic) derivatives, chitosan and its derivatives, or poly(amidoamine) used to formulate drug delivery systems are known to act directly or indirectly on tight junctions' components and their opening [50]. For example, quaternized and carboxylated chitosan-based nanoparticles [40, 51] or amine-terminated, hydroxyl-terminated, and carboxyl-terminated poly(amido amine) (PAMAM) dendrimers [52] enhance paracellular transport by opening tight junctions.

Transcellular transport occurs by a process by which drug delivery systems are taken by enterocytes or M-cells. Different mechanisms can be used by drug delivery systems such as diffusion (passive transport) or endocytosis/transcytosis (pinocytosis, phagocytosis, receptor-mediated endocytosis). To study the mechanism of transport of drug delivery systems across the intestinal barrier, a lot of studies are performed with *in vitro* models (Caco-2 cells, co-culture of Caco-2 cells with HT-29 or Caco-2 cells with Raji-cells). Transport of drug delivery systems by enterocyte can be mediated through different endocytic process: macropinocytosis, clathrin-mediated endocytosis, caveolae-mediated endocytosis, and/or clathrin- and caveolae-independent endocytosis. Incorporation of ligands onto nanoparticles that could bind to these cell-surface proteins could enable transcytosis of nanoparticles and enhance the oral absorption of encapsulated drugs. For example, several proteins are known to initiate caveolae-mediated endocytosis: folic acid receptor, glycophosphatidylinositol anchor, autocrine motility factor receptor, interleukin-2 receptor, GM1 gangliosides, integrins, platelet-derived growth factor receptor, epidermal growth factor receptor, and the CCK receptor [38, 39]. Thus, some results demonstrated the ability of drug delivery systems to enhance intestinal transport and to improve oral bioavailability. For example, N-trimethyl chitosan nanoparticles improve urease absorption and the immune response after oral administration [53]. These nanoparticles also improve insulin permeability via a caveolae- and clathrin-mediated endocytosis process and increased insulin bioavailability [54]. Lipid nanocapsules increase paclitaxel permeability via a clathrin- and caveolae-dependent endocytosis [55] leading to a threefold enhancement of oral bioavailability of paclitaxel [56]. Polymeric micelles increase oral bioavailability of salmon calcitonin and improve intestinal absorption by the transcellular pathway [57]. Wheat germ agglutinin-coated liposomes also improve salmon calcitonin efficacy and its oral absorption via clathrin-dependent transport [58]. Moreover, solid lipid nanoparticles improved the oral bioavailability and permeability of γ-Tocotrienol, and this enhancement was due to passive permeability [59]. Thus, polymeric micelles, solid lipid nanoparticles, lipids nanocapsules, liposomes, polymeric nanoparticles, or dendrimers improve drug permeability [38, 39, 60]. Likewise, grafting drug delivery systems with specific ligands has also shown to enhance nanocarrier transport by intestinal cells [61–65].

7.4 Conclusions and Perspectives

As seen in this chapter, oral drug delivery can be a challenge especially for BCS class II, III, and IV drugs. Because of the oral route's advantages over the other routes of administration, the game is worth the candle. Thus, many formulation strategies are sought to enhance the bioavailability of poorly soluble or fragile drugs across the intestinal epithelium. In this area, nanotechnology can be a great hope, but new drug delivery systems have to be easy to produce at a large scale and have to provide a real breakthrough over conventional formulations, which is yet to be realized. This is why, it is possible to state that the road is still long for the development of efficient oral delivery for many drugs.

References

1. Amidon, G.L., H. Lennernäs, V.P. Shah, and J.R. Crison. (1995). A theoritical basis for a biopharmaceutic drug classification: The correlation of *in vitro* drug product dissolution an *in vivo* bioavailability, *Pharm. Res.*, **12**, pp. 413–420.
2. Nair, A.K., O. Anand, N. Chun, D.P. Conner, M.U. Mehta, D.T. Nhu, J.E. Polli, L.X. Yu, and B.M. Davit. (2012). Statistics on BCS classification of generic drug products approved between 2000 and 2011 in the USA, *AAPS J.*, **14**, pp. 664–676.
3. Washington, N., C. Washington, and C.G. Wilson. (2001). *Physiological Pharmaceutics: Barriers to Drug Absorption*, 2nd Ed, CRC Press, London, New York.
4. Jantratid, E., N. Janssen, C. Reppas, and J.B. Dressman. (2008). Dissolution media simulating conditions in the proximal human gastrointestinal tract: An update, *Pharm. Res.*, **25**, pp. 1663–1676.
5. Dressman, J.B., K. Thelen, and E. Jantratid. (2008). Towards quantitative prediction of oral drug absorption, *Clin. Pharmacokin.*, **47**, pp. 655–667.
6. Atuma, C., V. Strugala, A. Allen, and L. Holm. (2001). The adherent gastrointestinal mucus gel layer: Thickness and physical state *in vivo*, *Am. J. Physiol.*, **280**, pp. G922–G929.

7. Cone, R.A. (2009). Barrier properties of mucus, *Adv. Drug Deliv. Rev.*, **61**, pp. 75–85.
8. Crater, J.S., and R.L. Carrier. (2010). Barrier properties of gastrointestinal mucus to nanoparticle transport, *Macromol. Biosci.*, **10**, pp. 1473–1483.
9. Ensign, L.M., R. Cone, and J. Hanes. (2012). Oral drug delivery with polymeric nanoparticles: The gastrointestinal mucus barriers, *Adv. Drug Deliv. Rev.*, **64**, pp. 557–570.
10. Lai, S.K., Y.-Y. Wang, and J. Hanes. (2009). Mucus-penetrating nanoparticles for drug and gene delivery to mucosal tissues, *Adv. Drug Deliv. Rev.*, **61**, pp. 158–171.
11. Avdeef, A., P. Artursson, S. Neuhoff, L. Lazorova, and S. Tavelin. (2005). Caco-2 permeability of weakly basic drugs predicted with the double-sink PAMPA pKa(flux) method, *Eur. J. Pharm. Sci.*, **24**, pp. 333–349.
12. Hu, M., and X. Li. (2011). Barriers to oral bioavailability: An overview. In *Oral Bioavailability: Basic Principles, Advanced Concepts, and Applications*, Hu, M., and Li, X. (Eds.), John Wiley & Sons, Inc., New Jersey, USA.
13. Cornaire, G., J. Woodley, P. Hermann, A. Cloarec, C.C. Arellano, and G. Houin. (2004). Impact of excipients on the absorption of P-glycoprotein substrates *in vitro* and *in vivo*, *Int. J. Pharm.*, **278**, pp. 119–131.
14. Bogman, K., F. Erne-Brand, J. Alsenz, and J. Drewe. (2003). The role of surfactants in the reversal of active transport mediated by multidrug resistance proteins, *J. Pharm. Sci.*, **92**, pp. 1250–1261.
15. Sparreboom, A., J. van Asperen, U. Mayer, A.H. Schinkel, J.W. Smit, D.K. Meijer, P. Borst, W.J. Nooijen, J.H. Beijnen, and O. van Tellingen. (1997). Limited oral bioavailability and active epithelial excretion of paclitaxel (Taxol) caused by P-glycoprotein in the intestine, *Proc. Nat. Acad. Sci.* **94**, pp. 2031–2035.
16. Stewart, C.F., M. Leggas, J.D. Schuetz, J.C. Panetta, P.J. Cheshire, J. Peterson, N. Daw, J.J. Jenkins, 3rd, R. Gilbertson, G.S. Germain, F.C. Harwood, and P.J. Houghton. (2004). Gefitinib enhances the antitumor activity and oral bioavailability of irinotecan in mice, *Cancer Res.*, **64**, pp. 7491–7499.
17. Benet, L.Z., and C.L. Cummins. (2001). The drug efflux-metabolism alliance: Biochemical aspects, *Adv. Drug Deliv. Rev.*, **50 Suppl 1**, pp. S3–S11.
18. Jeong, E.J., X. Liu, X. Jia, J. Chen, and M. Hu. (2005). Coupling of conjugating enzymes and efflux transporters: Impact on bioavailability and drug interactions, *Curr. Drug Metab.*, **6**, pp. 455–468.

19. Wang, S. (2011). Interplay between efflux transporters and metabolic enzymes. In *Oral Bioavailability: Basic Principles, Advanced Concepts, and Applications,* Hu, M., and Li, X. (Eds.), John Wiley & Sons, Inc., New Jersey, USA.
20. Kawabata, Y., K. Wada, M. Nakatani, S. Yamada, and S. Onoue. (2011). Formulation design for poorly water-soluble drugs based on biopharmaceutics classification system: Basic approaches and practical applications, *Int. J. Pharm.*, **420**, pp. 1–10.
21. Hörter, D., and J.B. Dressman. (2000). Influence of physicochemical properties on dissolution of drugs in the gastrointestinal tract, *Adv. Drug Deliv. Rev.*, **46**, pp. 75–87.
22. Gu, C.H., and D.J. Grant. (2001). Estimating the relative stability of polymorphs and hydrates from heats of solution and solubility data, *J. Pharm. Sci.*, **90**, pp. 1277–1287.
23. Prohotsky, D.L., and F. Zhao. (2012). A survey of top 200 drugs-inconsistent practice of drug strength expression for drugs containing salt forms, *J. Pharm. Sci.*, **101**, pp. 1–6.
24. Rautio, J., H. Kumpulainen, T. Heimbach, R. Oliyai, D. Oh, and J. Savolainen. (2008). Prodrugs: Design and clinical applications, *Nat. Rev. Drug Discov.*, **7**, pp. 255–270.
25. Savjani, K.T., A.K. Gajjar, and J.K. Savjani. (2012). Drug solubility: Importance and enhancement techniques, *ISRN Pharm.*, **2012**, pp. 195727–195727.
26. Müller, R.H., S. Gohla, and C.M. Keck. (2011). State of the art of nanocrystals: Special features, production, nanotoxicology aspects and intracellular delivery, *Eur. J. Pharm. Biopharm.*, **78**, pp. 1–9.
27. Shegokar, R., and R.H. Muller. (2010). Nanocrystals: Industrially feasible multifunctional formulation technology for poorly soluble actives, *Int. J. Pharm.*, **399**, pp. 129–139.
28. Leuner, C., and J. Dressman. (2000). Improving drug solubility for oral delivery using solid dispersions, *Eur. J. Pharm. Biopharm.*, **50**, pp. 47–60.
29. Vasconcelos, T.F., B. Sarmento, and P. Costa. (2007). Solid dispersions as strategy to improve oral bioavailability of poor water soluble drug, *Drug Discov. Today*, **12**, pp. 1068–1075.
30. Davis, M.E., and M.E. Brewster. (2004). Cyclodextrin-based pharmaceutics: Past, present and future, *Nat. Rev. Drug Discov.*, **3**, pp. 1023–1035.
31. Kurkov, S.V., and T. Loftsson. (2013). Cyclodextrins, *Int. J. Pharm.*, **453**, pp. 167–180.

32. Rasheed, A., A. Kumar, and V.V.N.S.S. Sravanthi. (2008). Cyclodextrins as drug carrier molecule: A review, *Sci. Pharm.*, **76**, pp. 567–598.
33. Carrier, R.L., L.A. Miller, and I. Ahmed. (2007). The utility of cyclodextrins for enhancing oral bioavailability, *J. Cont. Rel.*, **123**, pp. 78–99.
34. Pouton, C.W. (2000). Lipid formulations for oral administration of drugs: Non-emulsifying, self-emulsifying and self-microemulsifying drug delivery systems, *Eur. J. Pharm. Sci.*, **11, Supp**, pp. S93–S98.
35. Pouton, C.W., and C.J.H. Porter. (2008). Formulation of lipid-based delivery systems for oral administration: Materials, methods and strategies, *Adv. Drug Deliv. Rev.*, **60**, pp. 625–637.
36. Kohli, K., S. Chopra, D. Dhar, S. Arora, and R.K. Khar. (2010). Self-emulsifying drug delivery systems: An approach to enhance oral bioavailability, *Drug Discov. Today.*, **15**, pp. 958–965.
37. Bernkop-Schnurch, A. (2013). Reprint of: Nanocarrier systems for oral drug delivery: Do we really need them? *Eur. J. Pharm. Sci.*, **50**, pp. 2–7.
38. Plapied, L., N. Duhem, A. des Rieux, and V. Preat. (2011). Fate of polymeric nanocarriers for oral drug delivery, *Curr. Opin. Colloid Interface Sci.*, **16**, pp. 228–237.
39. Roger, E., F. Lagarce, E. Garcion, and J.-P. Benoit. (2010). Biopharmaceutical parameters to consider in order to alter the fate of nanocarriers after oral delivery, *Nanomed.*, **5**, pp. 287–306.
40. Chen, D., D. Xia, X. Li, Q. Zhu, H. Yu, C. Zhu, and Y. Gan. (2013). Comparative study of Pluronic® F127-modified liposomes and chitosan-modified liposomes for mucus penetration and oral absorption of cyclosporine A in rats, *Int. J. Pharm.*, **449**, pp. 1–9.
41. Behrens, I., A.I. Vila Pena, M.J. Alonso, and T. Kissel. (2002). Comparative uptake studies of bioadhesive and non-bioadhesive nanoparticles in human intestinal cell lines and rats: The effect of mucus on particle adsorption and transport, *Pharm. Res.* **19**, pp. 1185–1193.
42. Bravo-Osuna, I., C. Vauthier, A. Farabollini, G.F. Palmieri, and G. Ponchel. (2007). Mucoadhesion mechanism of chitosan and thiolated chitosan-poly(isobutyl cyanoacrylate) core-shell nanoparticles, *Biomaterials*, **28**, pp. 2233–2243.
43. Pan, Y., Y.J. Li, H.Y. Zhao, J.M. Zheng, H. Xu, G. Wei, J.S. Hao, and F.D. Cui. (2002). Bioadhesive polysaccharide in protein delivery system: Chitosan nanoparticles improve the intestinal absorption of insulin *in vivo*, *Int. J. Pharm.*, **249**, pp. 139–147.
44. Dufresne, M.H., M.A. Gauthier, and J.C. Leroux. (2005). Thiol-functionalized polymeric micelles: From molecular recognition to improved mucoadhesion, *Bioconjug. Chem.*, **16**, pp. 1027–1033.

45. Kalarickal, N.C., S. Rimmer, P. Sarker, and J.C. Leroux. (2007). Thiol-functionalized poly(ethylene glycol)-b-polyesters: Synthesis and characterization, *Macromolecules*, **40**, pp. 1874–1880.
46. Lai, S.K., D.E. O'Hanlon, S. Harrold, S.T. Man, Y.-Y. Wang, R. Cone, and J. Hanes. (2007). Rapid transport of large polymeric nanoparticles in fresh undiluted human mucus, *Pro. Nat. Acad. Sci.*, **104**, pp. 1482–1487.
47. Yang, M., S.K. Lai, Y.-Y. Wang, W. Zhong, C. Happe, M. Zhang, J. Fu, and J. Hanes. (2011). Biodegradable nanoparticles composed entirely of safe materials that rapidly penetrate human mucus, *Angew. Chem.*, **50**, pp. 2597–2600.
48. Groo, A.C., P. Saulnier, J.C. Gimel, J. Gravier, C. Ailhas, J.P. Benoit, and F. Lagarce. (2013). Fate of paclitaxel lipid nanocapsules in intestinal mucus in view of their oral delivery, *Int. J. Nanomed.*, **8**, pp. 4291–4302.
49. Groo, A., and F. Lagarce. (2014). Mucus models to evaluate nanomedicines for diffusion, *Drug Discov. Today.*, **19**, pp. 1097–1108.
50. Salamat-Miller, N., and T.P. Johnston. (2005). Current strategies used to enhance the paracellular transport of therapeutic polypeptides across the intestinal epithelium, *Int. J. Pharm.*, **294**, pp. 201–216.
51. Sadeghi, A.M., F.A. Dorkoosh, M.R. Avadi, M. Weinhold, A. Bayat, F. Delie, R. Gurny, B. Larijani, M. Rafiee-Tehrani, and H.E. Junginger. (2008). Permeation enhancer effect of chitosan and chitosan derivatives: Comparison of formulations as soluble polymers and nanoparticulate systems on insulin absorption in Caco-2 cells, *Eur. J. Pharm. Biopharm.*, **70**, pp. 270–278.
52. Sadekar, S., and H. Ghandehari. (2012). Transepithelial transport and toxicity of PAMAM dendrimers: Implications for oral drug delivery, *Adv. Drug Deliv. Rev.*, **64**, pp. 571–588.
53. Chen, F., Z.R. Zhang, F. Yuan, X. Qin, M. Wang, and Y. Huang. (2008). *In vitro* and *in vivo* study of N-trimethyl chitosan nanoparticles for oral protein delivery, *Int. J. Pharm.*, **349**, pp. 226–233.
54. Jin, Y., Y. Song, X. Zhu, D. Zhou, C. Chen, Z. Zhang, and Y. Huang. (2012). Goblet cell-targeting nanoparticles for oral insulin delivery and the influence of mucus on insulin transport, *Biomaterials*, **33**, pp. 1573–1582.
55. Roger, E., F. Lagarce, E. Garcion, and J.P. Benoit. (2009). Lipid nanocarriers improve paclitaxel transport throughout human intestinal epithelial cells by using vesicle-mediated transcytosis, *J. Cont. Rel.*, **140**, pp. 174–181.

56. Peltier, S., J.M. Oger, F. Lagarce, W. Couet, and J.P. Benoit. (2006). Enhanced oral paclitaxel bioavailability after administration of paclitaxel-loaded lipid nanocapsules, *Pharm. Res.*, **23**, pp. 1243–1250.

57. Li, N., X.-R. Li, Y.-X. Zhou, W.-J. Li, Y. Zhao, S.-J. Ma, J.-W. Li, Y.-J. Gao, Y. Liu, X.-L. Wang, and D.-D. Yin. (2012). The use of polyion complex micelles to enhance the oral delivery of salmon calcitonin and transport mechanism across the intestinal epithelial barrier, *Biomaterials*, **33**, pp. 8881–8892.

58. Makhlof, A., S. Fujimoto, Y. Tozuka, and H. Takeuchi. (2011). *In vitro* and *in vivo* evaluation of WGA-carbopol modified liposomes as carriers for oral peptide delivery, *Eur. J. Pharm. Biopharm.*, **77**, pp. 216–224.

59. Abuasal, B.S., C. Lucas, B. Peyton, A. Alayoubi, S. Nazzal, P.W. Sylvester, and A. Kaddoumi. (2012). Enhancement of intestinal permeability utilizing solid lipid nanoparticles increases γ-tocotrienol oral bioavailability, *Lipids*, **47**, pp. 461–469.

60. Mei, L., Z. Zhang, L. Zhao, L. Huang, X.L. Yang, J. Tang, and S.S. Feng. (2013). Pharmaceutical nanotechnology for oral delivery of anticancer drugs, *Adv. Drug Deliv. Rev.*, **65**, pp. 880–890.

61. Yin, Y., D. Chen, M. Qiao, X. Wei, and H. Hu. (2007). Lectin-conjugated PLGA nanoparticles loaded with thymopentin: *Ex vivo* bioadhesion and *in vivo* biodistribution, *J. Cont. Rel.*, **123**, pp. 27–38.

62. Roger, E., S. Kalscheuer, A. Kirtane, B.R. Guru, A.E. Grill, J. Whittum-Hudson, and J. Panyam. (2012). Folic acid functionalized nanoparticles for enhanced oral drug delivery, *Mol. Pharm.*, **9**, pp., 2103–2110.

63. Mishra, N., S. Tiwari, B. Vaidya, G.P. Agrawal, and S.P. Vyas. (2011). Lectin anchored PLGA nanoparticles for oral mucosal immunization against hepatitis B, *J. Drug Target.*, **19**, pp. 67–78.

64. Liu, Y., P. Wang, C. Sun, N. Feng, W. Zhou, Y. Yang, R. Tan, Z. Chen, S. Wu, and J. Zhao. (2010). Wheat germ agglutinin-grafted lipid nanoparticles: Preparation and *in vitro* evaluation of the association with Caco-2 monolayers, *Int. J. Pharm.*, **397**, pp. 155–163.

65. Francis, M.F., M. Cristea, and F.M. Winnik. (2005). Exploiting the vitamin B12 pathway to enhance oral drug delivery via polymeric micelles, *Biomacromolecules*, **6**, pp. 2462–2467.

Chapter 8

Crossing the Endothelial Barrier

Bhawani Aryasomayajula, Shravan Kumar Sriraman, and Vladimir P. Torchilin

Center for Pharmaceutical Biotechnology and Nanomedicine, Northeastern University, Boston, MA 02115, USA

v.torchilin@neu.edu

The endothelium is often regarded as a gatekeeper, playing an important role in separating blood from the surrounding tissues and regulating the passage of proteins and other molecules across the blood vessels. This transport can either be transcellular or paracellular. The cytoskeletal structures of the endothelial cells play a pivotal role in regulating this transport. This chapter attempts to discuss in brief these two types of transport and the components and factors governing them, an understanding of which is essential for designing more efficient drugs that can traverse the endothelium more effectively for increased therapeutic benefits. Novel approaches for enhancing the penetration of drugs are also reviewed here.

8.1 Introduction

The endothelium is described as the monolayer of cells that forms the inner lining of the circulatory system, that is, blood vessels

Drug Delivery Across Physiological Barriers

Edited by Silvia Muro

ISBN 978-981-4669-40-5 (Hardcover), 978-981-4669-41-2 (eBook)

www.panstanford.com

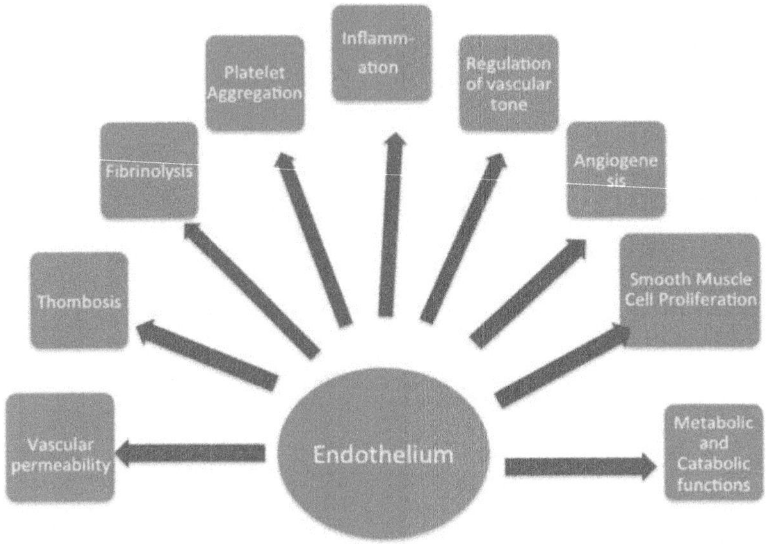

Figure 8.1 **Functions of the endothelial cells.** The endothelium is a dynamic barrier, forming the innermost lining of blood vessels in all tissues. Shown here are its important metabolic, secretory, and regulatory functions.

and the lymphatic system [1]. It can be regarded as the porous membrane through which fluids and solutes interchange between the plasma and tissue spaces [2]. It is a continuous, semipermeable barrier whose primary role is to regulate blood flow. It was initially believed that the endothelium is a passive, inert layer but now we know it to be metabolically and synthetically active. Some of its major functions include platelet activation, thrombosis, fibrinolysis, control of inflammation etc. (Fig. 8.1). Traversing the endothelium is a crucial step for many drugs to reach their target site and exert their intended effects. Being dynamic in nature, the endothelium has a plethora of mechanisms regulating its metabolism and transport functions, which is why targeting a medicine to the endothelium and designing it to cross the barrier efficiently can greatly improve the success of any therapy.

The endothelial layer in an average person covers the lumen of all organs and weighs about 1 kg due to the presence of 6–10 trillion (10^{13}) cells occupying an area of 4000–7000 m^2. This would mean that if the cells were lined next to each other, they would circle the circumference of the earth twice [3]! Before 1628, for a very long time, it was thought that the circulatory system consisted primarily of arteries and veins. Arteries were thought to contain air and vital spirits, and veins were perceived to carry blood. In 1628, an English physician William Harvey became the first person to hypothesize the existence of a closed network of arteries and veins, also suggesting the existence of capillaries. In 1961, Malpighi was the first to observe capillaries under the microscope. Although the term "endothelium" was first coined in 1865 by the Swiss anatomist Wilhelm His [4], its in-depth study became possible only after 1970 with the advent of *in vitro* cell culture. It was eventually discovered that the characteristics of the endothelial cells differ based on its location. Although there exists some heterogeneity in the endothelium from organ to organ, the basic structural elements of the endothelium, as well as its function, largely remain the same.

8.2 Endothelial Structural Elements

The endothelial cells are anchored to a basal lamina, which forms a scaffold for the endothelial cells and together they form a layer called the intima. Endothelial cells are about 50–70 μm in length, 10–30 μm in width, and 0.1–10 μm in thickness. They are typically mononuclear with a basophilic cytoplasm and contain intracellular organelles such as the Golgi complex, mitochondria, endoplasmic reticulum, etc. Endothelial cells are in direct contact with smooth muscle cells on the abluminal side and exert some control on their growth and regulation. On the luminal side, they are in direct contact with all the blood components, nutrients, and macromolecules and therefore, have the important responsibility of providing an anti-thrombotic surface for maintaining a smooth flow. The various structural elements of the endothelium play a vital role in achieving this function (Fig 8.2).

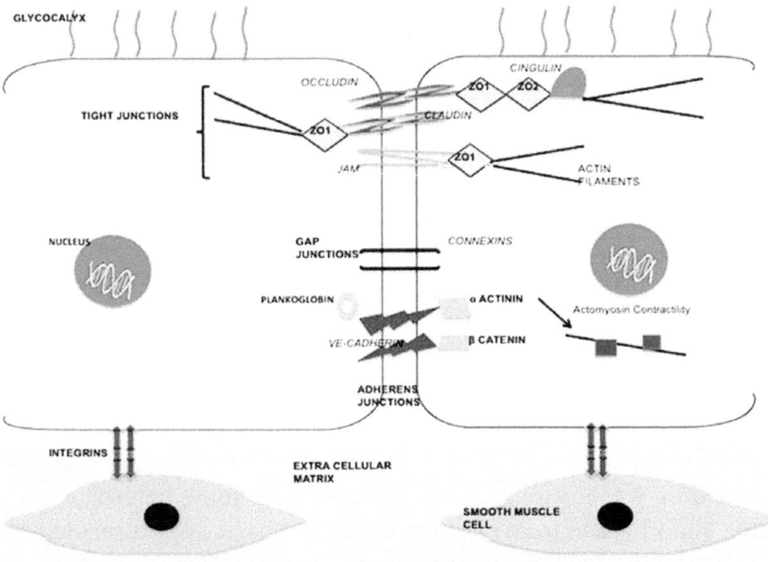

Figure 8.2 Endothelial cytoskeleton and cell junctions depicting the various junctional proteins. Elements such as glycocalyx and cytoskeleton actin filaments are depicted here associated with the downstream components of the gap, tight, and adherens junctions. Also shown in the figure are integrins, anchoring the cell to the extracellular matrix.

8.2.1 *Glycocalyx*

The glycocalyx is a uniform layer on the endothelium that is semipermeable in nature and composed of proteins, glycoproteins, glycolipids, and proteoglycans. It is 50–100 nm thick and holds a negative charge due to its composition [5–8]. Due to its negative charge and its anti-thrombogenic and anti-inflammatory surface, the glycocalyx repels the various cell components such as erythrocytes, cholesterol, leukocytes as well as lowdensity lipoproteins and maintains homeostasis [5, 6, 9]. Some pathological conditions such as salt-sensitive hypertention, diabetes, hypoxia, and hyperlipedemia disturb this negative charge or the structure of glycocalyx thereby increasing the cell permeability [6, 10, 11].

8.2.2 *Cytoskeleton*

An endothelial cell must at all times maintain their flat morphology and be able to migrate and multiply to perform its various functions. The cytoskeletal elements of the endothelium are responsible for achieving this. The cytoskeleton of a typical endothelial cell consists of microfilaments, intermediate filaments, and microtubules.

Microfilaments, also known as actin filaments, are about 5–8 nm in diameter and are composed primarily of a protein called actin, occurring either as G-actin (globular) or F-actin (filamentous) [12]. They arrange as bundles, close to the cell membrane and usually at particular attachment sites. Microfilaments play an important role in the contraction of endothelial cells, which has implications on the change in permeability of the endothelial barrier. Another important function of the microfilaments is to mediate the motility of endothelial cells. This is useful for the migration of endothelial cells in situations such as neovascularization of the cells and in response to tumor-derived growth factors. They are known to form stress fibers in the central portions of the cells and change their orientation and alignment during shear states. Microfilaments work in concert with integrins for maintaining cell–cell adhesion, cell differentiation, and cell proliferation [13, 14].

Intermediate filaments, also known as 10 nm filaments, as the name suggests are 10 nm in diameter. They are found in bundles containing up to 500–10000 filaments. Intermediate filaments, unlike microfilaments, are not localized just around the periphery of the cell but can be found around the nucleus as rings and extending through the cytoplasm, stretching up till the plasma membrane, and sometimes also the other cytoskeletal elements such as actin filaments and microtubules. The main function of the intermediate filaments is not contractility or motility but to act as a scaffold and provide mechanical support to the cell. Unlike actin filaments and microtubules, intermediate filaments vary with respect to their protein composition. More than 50 types of intermediate filament proteins such as keratin and vimentin are known, and hence the filaments are classified into six different types depending on their amino acid sequence [15].

Microtubules are about 24 nm in diameter and consist of a hollow center surrounded by 13 proto-filaments [16, 17], composed of α and β dimers of the protein tubulin. They are found throughout the cell and play a major role in the transport of intracellular vesicles. Their other important roles include moving chromosomes in mitosis, shuttling organelles and secretory products through the cytoplasm [17].

8.2.3 *Endothelial Junctions*

There are two types of junctions between the adjacent endothelial cells that help in the formation and maintenance of the monolayer: adherens junctions and tight junctions. Another type of junction known as the gap junction, facilitates the passage of small molecules, water, and ions between neighboring cells. The distribution of the adherens and tight junctions depends on the site of the endothelium and the functional need of the vascular network [18]. Adherens junctions are made up of clusters of calcium-dependent vascular endothelial (VE)-cadherin proteins [19, 20]. These VE-cadherins are formed earlier in microvasculature during embryogenesis compared to the tight junctions and are believed to play a more important role in maintaining the architectural integrity of the endothelium. The structure and the functioning mechanisms of the adherens junctions and tight junctions are discussed as follows.

Adherens junctions: As has been mentioned earlier in this chapter, adherens junctions are essentially clusters of transmembrane VE-cadherin proteins. In their cytoplasmic domain, they are associated with catenins, actin-binding proteins, Rho GTPases, kinases, and phosphatases. These molecules, in turn, are linked to the actin cytoskeleton, which is responsible for the contractile function.

Tight junctions: Tight junctions are made up of various trans-membrane proteins (which are cell specific), the most prominent of which are claudins; occludins; junctional adhesion molecules (JAMs) A, B, and C; and endothelial cell selective adhesion molecules (ESAMs) [21, 22]. These proteins are associated with their intracellular partners such as zona occludens (ZO1 and ZO-2), calcium/calmodulin-dependent serine protein kinase, and afadin.

ZO-1 can associate with cytoskeletal binding proteins such as α-catenin, cingulin, and cortactin.

Gap junctions: Gap junctions are composed of connexins (Cx37, Cx40, Cx43) and are involved in mediating cell migration and replication [23]. Gap junctions are responsible for transport of molecules and ions directly between the cytoplasm of two neighboring cells [24]. They are also involved in blood pressure regulation and vasodilation. Reduced connexin levels have been associated with phenotypic endothelial dysfunction [25, 26].

In addition to the three types of junctions just mentioned, another type of homophilic adhesion proteins belonging to the immunoglobulin family, known as platelet endothelial cell adhesion molecule (PECAM or CD-31), is also found at the endothelial cell junctions. These proteins play an important role in cell–cell recognition and cell junction assembly [18]. Proteins known as integrins are also important for anchoring the cell to the extracellular matrix, regulating cell permeability and mediating shear-induced signaling [27].

8.3 Endothelial Permeability and Transport of Substances Across the Barrier

The endothelium is regarded as a semipermeable barrier, meaning that the transport of substances across is regulated by various factors and involves different pathways depending on the physiological conditions and types of molecules. Under normal conditions, the basal permeability, which is the permeability across the capillaries, regulates the transport, whereas under disease states, induced permeability takes over. There are two major transport pathways: transcellular pathway (i.e., through the cell) and paracellular pathway (i.e., between the cells). The paracellular pathway comes into play while transporting molecules smaller than 3 nm, whereas transcellular (also known as transcytotic) pathway is used to transport macromolecules such as albumin. Majno and Palade are given the credit for the discovery of the paracellular pathway while studying the formation of cell gaps induced by histamine in endothelial cells [28]. This idea that the transport of molecules occurs through small or large cell gaps is now widely accepted. Transcytosis is the transport of macromolecular cargo from one side of a cell to the other within a membrane-bounded carrier vesicle. Until now it was widely believed that the two pathways were independent of each other. This theory is now being reevaluated as evidence is being uncovered about the two pathways being interdependent for maintaining tissue fluid homeostasis. The two pathways are discussed in brief in the next few paragraphs.

Another phenomenon that has been reviewed extensively, which we will not discuss in detail here, is the enhanced permeability and retention (EPR) effect. In certain disease states such as inflammation and especially in the case of tumors, abnormalities in vasculature are observed, such as hypervascularization, fenestrations, aberrant vascular architecture, and a certain amount of "leakiness" due to which the permeability of these blood vessels to cells and molecules is higher than in normal vessels. This is the result of factors such as a large gap between endothelial cells, abnormal basement membrane, and the production of a high amount of vascular endothelial growth factor (VEGF) [29–32]. This enhanced permeability ensures that abundant supply of nutrients is maintained for rapid tumor growth. The same phenomenon of EPR has been observed with macromolecules and even small particles such as lipid-based pharmaceutical nanocarriers (Fig. 8.3), while small molecules are able to return into the blood circulation via diffusion, thereby leading to tremendous application in cancer therapeutics [33–35]. Lately, it has been shown that the EPR effect is highly heterogeneous in human patients, just as tumors are heterogeneous in nature. This can be the result of varying sizes, normoxic to hypoxic conditions, varying interstitial pressure, partial to extensive necrosis, diverse genetic composition, etc. [36, 37]. Therefore, the clinical application of the EPR effect is not clear and is still a subject of extensive research.

8.3.1 *Paracellular Pathway*

Three major signaling mechanisms regulate the paracellular permeability: c-Src, Ca^{2+} signaling, and Rho-GTPase transduction pathways. A detailed description of these pathways is beyond

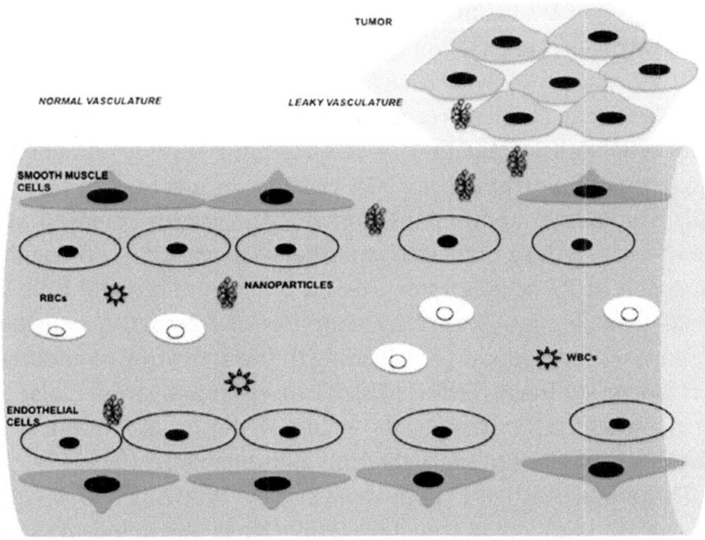

Figure 8.3 **EPR effect.** Blood vessel depicting the various blood components and nanoparticles. The vessels around the tumor area become permeable with large gaps between the cells leading to enhanced permeability and retention of the nanoparticles into the tumor tissues.

the scope of this chapter. However, the mediators that affect the paracellular permeability are essential for understanding the functioning of the barrier.

8.3.1.1 Mediators affecting endothelial permeability

Paracellular permeability of the endothelium changes with the change of environmental factors and external stimuli. In this section, we discuss the most well-studied permeability agents, many of which exert their effect by either destabilization of adherens junctions leading to VE-cadherin internalization or activation of acto-myosin contractility leading to the formation of stress fibers and breaking of the adherens junctions [38].

Inflammatory mediators such as histamine, thrombin, and bradykinin cause a transient increase in vascular permeability followed by barrier stabilization. Histamine causes a transient increase in permeability by inducing Src-mediated tyrosine phosphorylation of VE-cadherin and β–catenin and a transient increase in the intracellular Ca^{2+} ion concentration [39]. Thrombin does the same via the proteolytic cleavage and activation of receptor PAR-1, a G protein coupled receptor expressed on the surface of the endothelial cells. Activation of this receptor leads to downstream signaling and activation of PLCβ causing an increase in Ca^{2+} ions [40]. Bradykinin, another inflammatory mediator released at the site of tissue injury increases endothelial permeability through the activation of B2 receptors expressed on the endothelial cells [41]. Signal transduction occurs by coupling of B2 to Gα_q causing an increase in intracellular Ca^{2+} leading to the activation of endothelial nitric oxide synthase (eNOS) [42]. B1 receptor activation is also seen during tissue injury and leads to Src-mediated activation of ERK which in turn leads to the activation of inducible (i) NOS [43].

Therefore, both the pathways act by inducing NO synthesis but the B2 pathway is transient while B1 is sustained [41]. VEGF is a proangiogenic cytokine that is essential for developmental angiogenesis and has been implicated in pathological vessel growth in conditions such as inflammation, arthritis, and tumerogenesis [44]. VEGF induces permeability by several mechanisms, including PLC-dependent intracellular calcium release, src-kinase-mediated phosphorylation/internalization of junctional proteins, Rho-GTPase activation, cytoskeletal rearrangement, and NOS signaling. Platelet activating factor (PAF), a phospholipid produced by various cell types such as endothelial cells, is an important factor in regulating homeostasis [45]. PAF-R is a G protein coupled receptor that, upon activation, induces PLCβ_1 and PLCγ_1 leading to the activation of the downstream signaling cascade and ultimately causing the reorganization of the actin cytoskeleton associated with vascular leakiness [46, 47].

8.3.1.2 Mediators that enhance endothelial barriers

To understand the regulation of paracellular permeability, it becomes imperative that we also discuss the barrier-stabilizing agents that attenuate the response of the barrier to the permeability-enhancing agents discussed earlier and have an overall stabilizing effect on the endothelial barrier. These substances are essential for maintaining vascular integrity, and their suppression leads to the impairment of the barrier [48–50]. Fibroblast growth factor (FGF) is an angiogenic cytokine that plays an important role in adherens junction integrity [49, 50]. It binds to cell surface tyrosine kinase receptor FGFR and leads to the stabilization of VE-cadherin surface expression by modulating its interaction with p120 catenin [12]. Another potent barrier-stabilizing agent is sphingosine-1-phosphate (S1P). S1P circulates at high levels in the blood and signals through G protein coupled receptor S1P1. It brings about cortical actin reorganization via various pathways such as Rac-1, cortactin, FAK, paxillin, and actinin 1 and 4. S1P signals through G protein G_i, leading to PLC-dependent calcium release and FAK activation via phosphorylation [51–54]. Angiopoietin-1 (Ang-1) is a cytokine necessary for angiogenesis during development. It binds to the tyrosine kinase receptor Tie-2 and protects the mature vasculature from an increase in permeability due to VEGF, thrombin, bradykinin, and histamine [55]. Some of the mechanisms of action are inhibition of VEGF-induced Src activation thereby inhibiting the endocytosis of VE-cadherin, Src-mediated Ca^{2+} increase, and redistribution of Tie-1 to the sites on AJs where Tie-1 transinteraction between the opposite cells is mediated [56–58].

8.3.2 *Transcellular Pathway*

The transcellular pathway is defined as the energy-dependent trafficking of macromolecules through clathrin-coated pits, caveolae, vesiculo–vacuolar organelles, or transcellular channels [19]. Transcytosis may be receptor dependent or independent. The transcellular transport of albumin and albumin-bound ligands from vessel lumen into the interstitial space, the transport of hormones such as insulin, and lipid transport are all important examples of transcytosis. Of these, transport through vesicles called caveolae has been the best studied and is of great importance. Clathrin-mediated transport is a type of receptor-mediated endocytosis that is perhaps the most commonly observed transport in eukaryotes and is responsible for the internalization of nutrients, antigens, growth factors, pathogens, and recycling receptors such as transferrin and lowdensity lipoproteins.

8.3.2.1 Role of caveolae in transcytosis

Caveolae are cholesterol-rich non-clathrin-coated pits about 70 nm in diameter. They are made up of the structural protein caveolin-1 (Cav-1) that assembles into oligomers of 14–16 monomers, which cause the invagination of the plasma membrane of the endothelium [51]. A reduction of albumin transport in Cav-1 knockout mice was observed, indicating its importance in caveolae-mediated transport [59]. Caveolae make up about 20% of the cell volume in endothelial cells. They are cholesterol-rich regions coated with Cav-1, a 22 kDa protein [60, 61]. Cav-1 is associated with downstream signaling molecules such as G proteins, kinases, eNOS, or ion channels [62, 63]. The protein dynamin oligomerizes at the caveolae assembly and is followed by GTP hydrolysis and caveolar fission [64, 65]. The caveolar scission leads to actin reorganization and requires GTPase RhoG activity [66]. The released vesicles undergo vectorial transport to the albuminal side of the cell, where they fuse with the basolateral membrane, thus releasing albumin into interstitial space by exocytosis. Docking and fusion of caveolar vesicles involve the canonical membrane complex t-SNAREs (syntaxin and SNAP-23/25 family) on the basolateral membrane, which initiates the induction of the fission of the membrane-attached vesicle. Following this endocytosis, the transmigration of the vesicle across the endothelium, the fusion of the vesicle to the basolateral membrane, and finally exocytosis in the interstitial space can occur. However, the molecular events that determine whether a vesicle will be addressed to the basolateral membrane, completing the transcytosis process, or to other intracellular compartments are unknown.

8.3.2.2 Role of clathrin in endocytosis

Clathrin-mediated transcytosis occurs with the help of vesicles that are about 100 nm in diameter and characterized by the presence of polygonal lattices that serve as outer shell or scaffold, coated with a protein known as clathrin. It is perhaps the most widely observed pathway for the transport of molecules between eukaryotic cells. Clathrin triskelions are made up of subunits consisting of heavy and light chains and are linked with the membrane bilayer with the help of various adapter proteins, the most important of which is the AP-2 complex. AP-2 is made up of many subunits that are capable of recognizing and associating with the cytoplasmic domains of the cargo proteins and directing them into the invagination [67]. The invagination continues and proceeds onto scission, requiring the cytoplasmic GTPase dynamin. The uncoating of the vescicle is accomplished with the help of proteins such as 5-phosphoinositide phosphatase synaptojanin 1 and a chaperone complex known as Hsc70 and auxilin, after which the vescicle may either fuse with endosomes or recycled in the case of synaptic vesicles [68]. Many membrane proteins and lipids regulate the clathrin-coated pit formation, but a detailed summary of these factors and processes is beyond the scope of this chapter.

8.3.2.3 Role of vesiculo–vacuolar organelles in transcytosis

Vesiculo–vacuolar organelles (VVOs) are probably the least studied of all the transcytotic pathways, and the exact mechanism is still under investigation. They are clusters of interconnected, uncoated vesicles and vacuoles present in the endothelium lining venules, small veins, and tumor vessels [69, 70]. First described by Dvorak et. al. [69], they are substantially bigger (80–140 nm in diameter) than caveolae (70 nm) and occupy about 18% of the venular endothelium [71]. These assemble into transcellular channels and provide a transport pathway to macromolecules. Tracer ferritin has been shown to accumulate in the VVOs after administration of permeability-enhancing agents such as VEGF and histamine [71].

8.4 Recent Advances in Drug Delivery with a Focus on Overcoming the Challenges in Crossing the Endothelial Barrier

The study of the endothelium is of great importance with respect to drug delivery because of the fact that the endothelium forms a formidable barrier to drug penetration as it is located extensively throughout the body to form a physical barrier and is metabolically active. In addition to that, it performs many roles such as

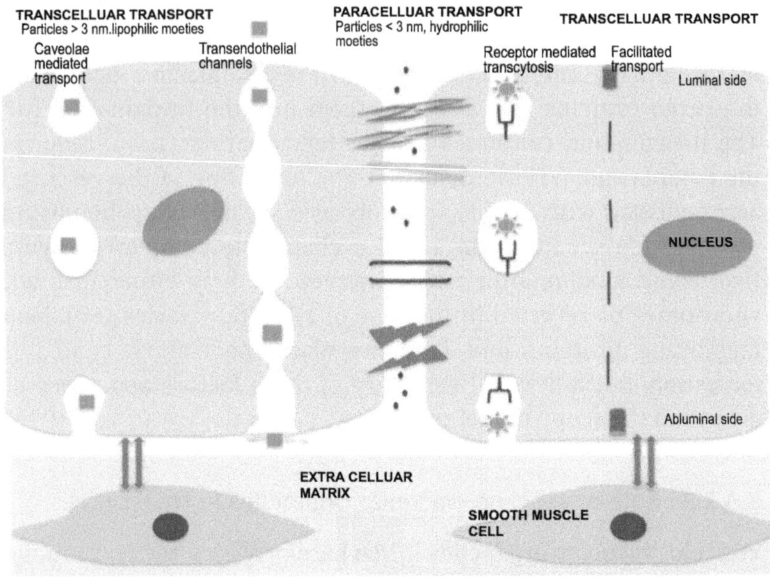

Figure 8.4 Various transport pathways across endothelial cells. The two different transendothelial pathways namely paracellular and transcellular, are shown here. The methods for transcellular transport such as caveolaemediated transport transendothelial channels, receptor-mediated transcytosis (eg., via clathrin pits) and facilitated transport are illustrated here.

regulating permeability, controlling the vasomotor tone, forming new blood vessels, and trafficking cells and nutrients across the barrier. Endothelial involvement is seen in most of the disease states spanning a vast number of disciplines such as neurology (stroke, multiple sclerosis), endocrinology (diabetes), pulmonary (asthma), cardiology (atherosclerosis), infectious diseases, cancer, rheumatology (arthritis), to name a few. This makes targeting the endothelial barrier for drug therapy, an option that holds immense promise. The regulation of endothelial permeability happens via various mechanisms (as has been discussed earlier in the chapter) and is a complicated process. This also provides the option of having ample sites or mechanisms for targeting. A myriad of techniques are being investigated today that target different approaches for drug therapy, the most important of which include pharmacological, genetic, and mechanical. In this section, we discuss some recent papers that explore novel approaches seeking to overcome the barrier and improve the effectiveness of the therapy.

8.4.1 *Formulation-Based Approaches*

The physicochemical properties of a formulation have a great bearing on the transport of the drug across the endothelial barrier, depending on the target tissue. Properties such as small size, the ability to remain unionized at physiological pH, and lipophilicity are some of the important criteria for the effective transport of a formulation [72]. As an example, when we consider the blood–brain barrier (BBB), the small and large hydrophilic compounds such as glucose, amino acids, and hormones use transporters to cross the barrier. The lipophilic moieties diffuse passively across the membrane, although they may be subjected to efflux pumps and some multidrug resistance proteins. The modification of drugs to decrease the polar groups or increase the nonpolar groups is one of the popular approaches [73, 74]. However, increasing the size of the molecule can, in some cases, result in a decrease in the transport of the drug [75], thereby making the process more complicated. Encapsulating the drug in a lipid vesicle is a popular practice where the unilamellar or multilamellar lipid vescicles of varying sizes such as micelles and liposomes are designed with a narrow size distribution and administered intravenously. These nanocarriers provide the advantage of being amenable to modifications to improve their targeting, uptake, and pharmacokinetics. They will be discussed in detail later in this section. Administering a prodrug is another popular approach, where the compound is modified chemically to give a pharmacologically inactive compound that usually has a useful property such as better solubility or lipophilicity. Once administered and after reaching the site of action, it gets converted into the active compound as a result of enzyme action resulting in its therapeutic effect. Esterification or amidation of hydroxy, amino, or carboxylic acid groups in a drug will enhance the lipid solubility. Once penetration is achieved, the hydrolysis of these groups will result in the generation of the active compound [76]. Coupling the drug to a lipophilic moiety has also been used for drugs such as levodopa, GABA, and valproate [77]. A possible disadvantage of this method is that increasing the lipophilicity may also result in increased efflux of the drug. This problem can be resolved by using chemical drug delivery systems, which originated from the prodrug concept but are considered to be more advanced. The active compound is rendered inactive by attaching various chemical moieties for site-specific targeting and lock-in, as well as for imparting favorable physicochemical properties. The molecule undergoes sequential metabolic conversions to disengage these moieties after reaching the target site. The loss of these moieties results in the loss of properties that are needed for transport, such as lipophilicity, and the molecule is now locked-in at the target site. This approach has been successfully used to deliver drugs to the brain, eyes, and lungs [78–80].

8.4.2 *Temporary Disruption of the Barrier*

Endothelial layer disruption is another popular approach for site-specific delivery of therapeutic agents. This can be achieved reversibly in several ways and has been extensively investigated in the case of the BBB.

Transcranial delivery of drugs can be achieved by three methods, namely, intracerebral implantation, intracerebroventricular (ICV) infusion, and convection-enhanced diffusion. The disadvantage of this method is that it is invasive and it is difficult to achieve optimal diffusion rates [74]. In solvent/adjuvant-mediated BBB disruption, the drug is dissolved in solvents such as DMSO, ethanol, or detergents such as Tween for administration. These agents are known to destabilize the BBB and improve penetration. Immune adjuvants such as Freun's complete or incomplete adjuvant have been used in cases such as Alzhimer's vaccine in transgenic mice with brain amyloid to increase the permeability of the barrier to IgG [81]. Osmotic BBB inhibition involves an intercarotid injection of an inert hypertonic solution of mannitol or arabinose to cause endothelial cell shrinkage [82]. This method, albeit controversial, is being actively employed by many researchers [83, 84]. Electroporation-based approaches rely on the formation of nanoscale defects in the membrane on application of an electric field [85, 86]. Ultrasound has also been employed to this effect by administering microbubbles containing an ultrasound contrast agent into the bloodstream [87, 88]. Administration of vasoactive amines such as bradykinin, histamine, and the bradykinin analog RMP-7 has been shown to selectively open the barrier in experimental animal models [89].

8.4.3 *Physiology-Based Approaches*

8.4.3.1 Receptor-mediated transcytosis

Although the endothelium impedes the free diffusion of large molecules, the endothelial cell surface contains a multitude of receptors for ensuring the transport of various proteins, hormones, and peptides. Most endocytic receptors mediate uptake via clathrin or caveolar pathways associated with 200 nm vesicles [90]. Drug molecules can be modified to become targets for these receptors for targeting and trafficking across the membrane. Monoclonal antibody (mAb) attachment is widely used to impart targeting property to a formulation. They are receptor specific and therefore, upon attachment, activate a signaling cascade, cytoskeletal rearrangement, membrane invagination, and internalization. One such example is the transferrin receptor. A peptidomimetic mAb against transferrin receptors was conjugated to BNDF for transport across the BBB for achieving a reduction in stroke volume [91]. Similarly, antibodies against the insulin receptor have also been developed and extensively studied [92]. A monoclonal antibody against lung caveolae has successfully been shown to increase the transcytosis across the lung endothelium [93]. Targeting the albumin transport protein gp60, which localizes in the caveolae, has been shown to be an efficient technique as well [94]. Apart from antibodies, one can use molecules essential for cell metabolism, which undergo receptor-mediated transport [95, 96], such as folate, leutinizing hormone–releasing hormone [95], etc. The various endocytic pathways such as caveolae and clathrincoated pits, however, limit the size of the drug that can be transported, although attempts are being made to overcome this problem, for example functionalizing the surface of the drugs or nanocarriers such as polystyrene microcarriers targeted to clathrin-coated mannose-6-phosphate receptor with sphingomyelinase. This mimics the ICAM-1 pathway of uptake and enables the uptake of larger particles [90].

8.4.3.2 Adsorptive-mediated transcytosis

This type of transcytosis is found mainly across the BBB. Cationic peptides and proteins as well as some lectins are internalized by endocytosis, but the initial binding to the plasma membrane is mediated by electrostatic interactions with the anionic site or by specific interactions with sugar residues, respectively. Several different synthetic moieties, such as polyethylenimine (PEI),2-diethylaminoether (DEAE)-dextran, artificial lipids, proteins and peptides which are used in transfection for stabilizing the DNA or in various nanoparticles, are known to be taken up by this mechanism [97]. Attachment of moieties known as cell-penetrating peptides (CPPs) is rapidly gaining popularity as another such approach. CPPs, as the name suggests, are cationic peptides (<20 amino acids) that can transport small molecules as well as large molecules (even greater than CP's own weight) such as proteins, DNA, antibodies, contrast (imaging) agents, toxins, and nanoparticular drug carriers among others [98]. Some examples of CPPs include TAT peptide, Antp, VP22, and Transportan. They have been extensively studied and used, and there is a plethora of literature documenting their use in endothelial targeting [99–101]. Another class of peptides smaller in size (3–10 amino acids), known as cell-targeting peptides, is now emerging as an alternative to the mAb approach. The most popular member of this class is the RGD peptide [102, 103]. This peptide contains RGD binding sequences recognized by integrins that are expressed on the endothelial cells, thereby making it a popular and widely studied agent for endothelial targeting [104, 105].

8.4.3.3 Transporter-facilitated transport

Drugs can be modified to mimic endogenous nutrients, peptides, and small molecules to exploit the barrier's transporter system. Numerous types of transporter systems have been identified and exploited for this purpose. A typical example is the drug L-Dopa, which is the metabolic precursor of the hormone dopamine. L-Dopa is transported across the BBB using the neutral amino acid transporter system for the treatment of Parkinsons [106]. Other BBB transporters that can be used for drug delivery include the GLUT1 transporter for glucose, the MCT1 lactate transporter, the CAT1 cationic amino acid transporter, etc.

The advantages and disadvantages of approaches discussed in this chapter are briefly summarized in Table 8.1.

8.5 What to Expect in the Coming Years: Challenges and Expectations

Although transport of substances across the endothelial barrier occurs via multiple pathways and molecules with a range of size and physicochemical properties, it is still extremely challenging to design drugs that exploit these pathways to achieve therapeutic concentrations. This has resulted in very few formulations targeted toward crossing endothelial barriers making it to the market. Comparatively, a lot more research is now being carried out to address this problem but is still fraught with a number of problems. The most prominent of those problems is the tendency of various pharmaceutical companies to adopt an oral delivery approach for a number of biological activities. This forces the drugs to come into contact with numerous biological barriers leading to a significant reduction in bioavailability. Taking the example of the BBB, though many novel delivery systems that target receptors such as transferrin, insulin, and glutathione have been investigated for over a decade in academia, this has not resulted in as many clinical drug candidates as one would expect from such vast amounts of research. The reason for this is the factors that govern penetration across the barrier. The molecules have to be lipophilic to be able to penetrate the barrier. Most of the small molecule research taking place is resulting in hydrophilic drug candidates [107, 108]. It is said that 98% of the discovered small molecules are not able to effectively cross the BBB [108]. Novel drug delivery systems offer a viable solution to this problem. Phospholipid and polymeric nanocarriers such as liposomes, micelles, and dendrimers offer various advantages to

Table 8.1 Approaches for targeting and penetrating the endothelial barrier

	Example	Advantages	Disadvantages
Formulation-based approaches	Increasing lipophilicity, administering prodrug, chemical drug delivery systems, encapsulation into nanocarriers	Easy approach for modifying drugs, applicable to a vast variety of molecules	Molecules may lose activity, increasing lipophilicity increases efflux and binding to non-target sites
Mechanical disruption, barrier disruption	ICV infusion, osmotic method, solvent/adjuvant-mediated method, electroporation, ultrasound, bradykinin analogs	Molecule transported without changes, one approach applicable to many molecules	Invasive, risky, some approaches need anesthesia and hospitalization and constant monitoring
Physiology-based approaches	Receptor-mediated transcytosis, adsorptive-mediated transcytosis, transporter-facilitated transport	Availability of a range of options for entry into the cell, exploits the cell's innate mechanisms thereby preventing any mechanical disruption	Receptors can be expressed in other tissues, reducing specificity; cationic lipids can be toxic; not all the pathways are well understood

drug delivery such as (1) the presence of lipophilic and hydrophilic regions for the encapsulation of a wide variety of drugs, (2) being amenable to surface modification for targeting the formulation to a number of targets (e.g., adhesion proteins such as ICAMs, PECAMs [109, 110]), (3) flexibility of modifying the pharmacokinetics of the formulation, (4) longer circulation time, and (5) protection from the mononuclear phagocyte system via PEGylation, among many other advantages.

A simple search of the word "endothelium" brings up around 950,000 journal articles. Tremendous progress has been made in the past few decades to better understand the endothelium, and cutting-edge breakthroughs have occurred in terms of endothelial biomedicine. In the near future, we hope this technology will overcome the bench to bedside barrier and translate into a more efficient therapy in clinical practice.

References

1. Cines, D.B., Pollak, E.S., Buck, C.A., Loscalzo, J., Zimmerman, G.A., McEver, R.P., Pober, J.S., Wick, T.M., Konkle, B.A., Schwartz, B.S., Barnathan, E.S., McCrae, K.R., Hug, B.A., Schmidt, A.M., Stern, D.M. (1998). Endothelial cells in physiology and in the pathophysiology of vascular disorders, *Blood,* **91**, pp. 3527–3561.
2. Malik, A.B., Lynch, J.J., Cooper, J.A. (1989). Endothelial barrier function, *J. Invest. Dermatol.,* **93**, pp. 62S–67S.
3. Aird, W.C. (2005). Spatial and temporal dynamics of the endothelium, *J. Thrombos. Haemost.: JTH,* **3**, pp. 1392–1406.
4. Aird, W.C. (2007). Phenotypic heterogeneity of the endothelium: I. Structure, function, and mechanisms, *Circ. Res.,* **100**, pp. 158–173.
5. Vink, H., Duling, B.R. (2000). Capillary endothelial surface layer selectively reduces plasma solute distribution volume, *Am. J. Physiol. Heart C.,* **278**, pp. H285–289.
6. Pelikan, P., Gimbrone, M.A., Jr., Cotran, R.S. (1979). Distribution and movement of anionic cell surface sites in cultured human vascular endothelial cells, *Atherosclerosis,* **32**, pp. 69–80.
7. Squire, J.M., Chew, M., Nneji, G., Neal, C., Barry, J., Michel, C. (2001). Quasi-periodic substructure in the microvessel endothelial glycocalyx:

A possible explanation for molecular filtering? *J. Struc. Biol.,* **136**, pp. 239–255.

8. Pries, A.R., Secomb, T.W., Gaehtgens, P. (2000). The endothelial surface layer, *Pflugers Arch.,* **440**, pp. 653–666.
9. Simionescu, M., Simionescu, N., Palade, G.E. (1982). Biochemically differentiated microdomains of the cell surface of capillary endothelium, *Ann. N.Y. Acad. Sci.,* **401**, pp. 9–24.
10. Raz, I., Havivi, Y., Yarom, R. (1988). Reduced negative surface charge on arterial endothelium of diabetic rats, *Diabetologia,* **31**, pp. 618–620.
11. Oberleithner, H., Peters, W., Kusche-Vihrog, K. Korte, S., Schillers, H., Kliche, K., Oberleithner, K. (2011). Salt overload damages the glycocalyx sodium barrier of vascular endothelium, *Pflugers Arch.,* **462**, 519–528.
12. Joris, I., Majno, G., Corey, E.J., Lewis, R.A. (1987). The mechanism of vascular leakage induced by leukotriene E4. Endothelial contraction, *Am. J. Pathol.,* **126**, pp. 19–24.
13. Lee, T.Y.J., Gotlieb, A.I. (2003). Microfilaments and microtubules maintain endothelial integrity, *Micros. Res. Tech.,* **60**, pp. 115–127.
14. Schoenwaelder, S.M., Burridge, K. (1999). Bidirectional signaling between the cytoskeleton and integrin, *Curr. Opin. Cell Biol.,* **11**, pp. 274–286.
15. Franke, W.W., Schmid, E., Osborn, M., Weber, K. (1979). Intermediate-sized filaments of human endothelial cells, *J. Cell Biol.* 81, pp. 570–580.
16. Dustin, P. (1984). Structure and chemistry of microtubules. In *Microtubules,* Springer Berlin Heidelberg, pp. 19–93.
17. Wang, E., Goldman, R.D. (1978). Functions of cytoplasmic fibers in intracellular movements in BHK-21 cells, *J. Cell Biol.,* **79**, pp. 708–726.
18. Dejana, E., Tournier-Lasserve, E., Weinstein, B.M. (2009). The control of vascular integrity by endothelial cell junctions: Molecular basis and pathological implications, *Dev. Cell,* **16**, pp. 209–221.
19. Goddard, L.M., Iruela-Arispe, M.L. (2013). Cellular and molecular regulation of vascular permeability, *Thromb. Haemost.,* **109**, pp. 407–415.
20. Dejana, E., Giampietro, C. (2012). Vascular endothelial-cadherin and vascular stability, *Curr. Opin. Hematol.,* **19**, pp. 218–223.
21. Muller, W.A. (2003). Leukocyte–endothelial cell interactions in leukocyte transmigration and the inflammatory response, *Trends Immunol.,* **24**, pp. 326–333.

22. Aurrand-Lions, M., Johnson-Leger, C., Wong, C., Du Pasquier, L., Imhof, B.A. (2001). Heterogeneity of endothelial junctions is reflected by differential expression and specific subcellular localization of the three JAM family members, *Blood,* **98**, pp. 3699–3707.
23. Hill, C., Rummery, N., Hickey, H., Sandow, S. (2002). Heterogeneity in the distribution of vascular gap junctions and connexins: Implications for function, *Clin. Exp. Pharmacol. Physiol.,* **29**, pp. 620–625.
24. Lampe, P.D., Lau, A.F. (2004). The effects of connexin phosphorylation on gap junctional communication, *Int. J. Biochem. Cell B.,* **36**, pp. 1171–1186.
25. Hammond, S., Mathewson, A.M., Baker, P.N., Mayhew, T.M., Dunn, W.R. (2011). Gap junctions and hydrogen peroxide are involved in endothelium-derived hyperpolarising responses to bradykinin in omental arteries and veins isolated from pregnant women, *Eur. J. Pharmacol.,* **668**, pp. 225–232.
26. Rummery, N.M., Hill, C.E. (2004). Vascular gap junctions and implications for hypertension, *Clin. Exp. Pharmacol. Physiol.,* **31**, pp. 659–667.
27. Langille, B.L. (2001). Morphologic responses of endothelium to shear stress: Reorganization of the adherens junction, *Microcirculation,* **8**, pp. 195–206.
28. Majno, G., Palade, G. (1961). Studies on inflammation I. The effect of histamine and serotonin on vascular permeability: An electron microscopic study, *J. Biophys. Biochem. Cytol.,* **11**, pp. 571–605.
29. Jain, R.K. (1987). Transport of molecules across tumor vasculature, *Cancer Metast. Rev.,* **6**, pp. 559–593.
30. Hashizume, H., Baluk, P., Morikawa, S., McLean, J.W., Thurston, G., Roberge, S., Jain, R.K., McDonald, D.M. (2000). Openings between defective endothelial cells explain tumor vessel leakiness, *Am. J. Path.,* **156**, pp. 1363–1380.
31. Chang, Y.S., Munn, L.L., Hillsley, M.V., Dull, R.O., Yuan, J., Lakshminarayanan, S., Gardner, T.W., Jain, R.K., Tarbell. J.M. (2000). Effect of vascular endothelial growth factor on cultured endothelial cell monolayer transport properties, *Microvasc. Res.,* **59**, pp. 265–277.
32. Roberts, W.G., Palade, G.E. (1995). Increased microvascular permeability and endothelial fenestration induced by vascular endothelial growth factor, *J. Cell Sci.,* **108**, pp. 2369–2379.
33. Maeda, H., Sawa, T., Konno, T. (2001). Mechanism of tumor-targeted delivery of macromolecular drugs, including the EPR effect in solid

tumor and clinical overview of the prototype polymeric drug SMANCS, *J. Control. Release,* **74**, pp. 47–61.

34. Torchilin, V. (2011). Tumor delivery of macromolecular drugs based on the EPR effect, *Adv. Drug Deliv. Rev.,* **63**, pp. 31–135.
35. Lukyanov, A.N., Gao, Z., Torchilin, V.P. (2003). Micelles from polyethylene glycol/phosphatidylethanolamine conjugates for tumor drug delivery, *J. Control. Release,* **91**, pp. 97–102.
36. Maeda, H. (2012). Macromolecular therapeutics in cancer treatment: The EPR effect and beyond, *J. Control. Release,* **164**, pp. 138–144.
37. Bae, Y.H. (2009). Drug targeting and tumor heterogeneity, *J. Control. Release,* **133**, pp. 2–3.
38. Mehta, D., Malik, A.B. (2006). Signaling mechanisms regulating endothelial permeability, *Physiol. Rev.,* **86**, pp. 279–367.
39. Guo, M., Breslin, J.W., Wu, M.H., Gottardi, C.J., Yuan, S.Y. (2008). VE-cadherin and beta-catenin binding dynamics during histamine-induced endothelial hyperpermeability, *Am. J. Physiol. Cell Physiol.,* **294**, pp. C977–984.
40. Coughlin, S.R. (2005). Proteaseactivated receptors in hemostasis, thrombosis and vascular biology, *J. Thrombos. Haemost.,* **3**, pp. 1800–1814.
41. Marceau, F., Regoli, D. (2004). Bradykinin receptor ligands: Therapeutic perspectives, *Nat. Rev. Drug Discovery,* **3**, pp. 845–852.
42. Leeb-Lundberg, L.M. (2004). Bradykinin specificity and signaling at GPR100 and B2 kinin receptors, *Br. J. Pharmacol.,* **143**, pp. 931–932.
43. Zhang, Y., Brovkovych, V., Brovkovych, S., Tan, F., Lee, B.S., Sharma, T., Skidgel, R.A. (2007). Dynamic receptor-dependent activation of inducible nitric-oxide synthase by ERK-mediated phosphorylation of Ser745, *J. Biol. Chem.,* **282**, pp. 32453–32461.
44. Dvorak, H.F. (2006). Discovery of vascular permeability factor (VPF), *Exp. Cell Res.,* **312**, pp. 522–526.
45. Marrache, A.M., Gobeil, Jr. F., Zhu, T., Chemtob, S. (2005). Intracellular signaling of lipid mediators via cognate nuclear G protein-coupled receptors, *Endothelium,* **12**, pp. 63–72.
46. Axelrad, T.W., deo, D.D., Ottino, P., Van Kirk, J., Bazan, N.G., Bazan, H.E., Hunt, J.D. (2004). Platelet-activating factor (PAF) induces activation of matrix metalloproteinase 2 activity and vascular endothelial cell invasion and migration, *FASEB J.,* **18**, pp. 568–570.
47. Bussolino, F., Camussi, G., Aglietta, M., Braquet, P., Bosia, A., Pescarmona, G., Sanavio, F., D'Urso, N., Marchisio, P.C. (1987). Human

endothelial cells are target for platelet-activating factor. I. Platelet-activating factor induces changes in cytoskeleton structures, *J. Immunol.,* **139**, pp. 2439–2446.

48. Suri, C., Jones, P.F., Patan, S., Bartunkova, S., Maisonpierre, P.C., Davis, S., Sato, T.N., Yancopoulos, G.D. (1996). Requisite role of angiopoietin-1, a ligand for the TIE2 receptor, during embryonic angiogenesis, *Cell,* **87**, pp. 1171–1180
49. Murakami, M., Nguyen, L.T., Zhang, Z.W., Moodie, K.L., Carmeliet, P., Stan, R.V., Simons, M. (2008). The FGF system has a key role in regulating vascular integrity, *J. Clin. Invest.,* **118**, pp. 3355–3366.
50. Rosen, H., Gonzalez-Cabrera, P.J., Sanna, M.G., Brown, S. (2009). Sphingosine 1-phosphate receptor signaling, *Annu. Rev. Biochem.,* **78**, pp. 743–768.
51. Komarova, Y., Malik, A.B. (2010). Regulation of endothelial permeability via paracellular and transcellular transport pathways, *Annu. Rev. Physiol.,* **72**, pp. 463–493.
52. Thennes, T., Mehta, D. (2012). Heterotrimeric G proteins, focal adhesion kinase, and endothelial barrier function, *Microvasc. Res.,* **83**, pp. 31–44
53. Sun, X., Shikata, Y., Wang, L., Ohmori, K., Watanabe, N., Wada, J., Shikata, K., Birukov, K.G., Makino, H., Jacobson, J.R., Dudek, S.M., Garcia, J.G. (2009). Enhanced interaction between focal adhesion and adherens junction proteins: Involvement in sphingosine 1-phosphate-induced endothelial barrier enhancement, *Microvasc. Res.,* **77**, pp. 304–313.
54. Shikata, Y., Birukov, K.G., Garcia, J.G. (2003). S1P induces FA remodeling in human pulmonary endothelial cells: Role of Rac, GIT1, FAK, and paxillin, *J. Appl. Physiol.,* **94**, pp. 1193–1203.
55. Thurston, G., Suri, C., Smith, K., McClain, J., Sato, T.N., Yancopoulos, G.D., McDonald, D.M. (1999). Leakage-resistant blood vessels in mice transgenically overexpressing angiopoietin-1, *Science,* **286**, pp. 2511–2514.
56. Hisatsune, C., Kuroda, Y., Nakamura, K., Inoue, T., Nakamura, T., Michikawa, T., Mizutani, A., Mikoshiba, K. (2004). Regulation of TRPC6 channel activity by tyrosine phosphorylation, *J. Biol. Chem.,* **279**, pp. 18887–18894.
57. Vazquez, G., Wedel, B.J., Kawasaki, B.T., Bird, G.S.J., Putney, J.W. (2004). Obligatory role of Src kinase in the signaling mechanism for TRPC3 cation channels, *J. Biol. Chem.,* **279**, pp. 40521–40528.

58. Fukuhara, S., Sako, K., Minami, T., Noda, K., Kim, H.Z., Kodama, T., Shibuya, M., Takakura, N., Koh, G.Y., Mochizuki, N. (2008). Differential function of Tie2 at cell–cell contacts and cell–substratum contacts regulated by angiopoietin-1, *Nat. Cell Biol.,* **10**, pp. 513–526.

59. Razani, B., Engelman, J.A., Wang, X.B., Schubert, W., Zhang, X.L., Marks, C.B., Macaluso, F., Russell, R.G., Li, M., Pestell, R.G., Di Vizio, D., Hou, H., Jr., Kneitz, B., Lagaud, G., Christ, G.J., Edelmann, W., Lisanti, M.P. (2001). Caveolin-1 null mice are viable but show evidence of hyperproliferative and vascular abnormalities, *J. Biol. Chem.,* **276**, pp. 38121–38138.

60. Predescu, D., Palade, G.E. (1993). Plasmalemmal vesicles represent the large pore system of continuous microvascular endothelium, *Am. J. Physiol. Heart Circ. Physiol.,* **265**, pp. H725–H733.

61. Murata, M., Peränen, J., Schreiner, R., Wieland, F., Kurzchalia, T.V., Simons, K. (1995). VIP21/caveolin is a cholesterol-binding protein, *Proc. Natl. Acad. Sci. U.S.A.,* **92**, pp. 10339–10343.

62. Absi, M., Burnham, M.P., Weston, A.H., Harno, E., Rogers, M., Edwards, G. (2007). Effects of methyl beta-cyclodextrin on EDHF responses in pig and rat arteries: Association between SK(Ca) channels and caveolin-rich domains, *Br. J. Pharmacol.,* **151**, pp 332–340.

63. Garcia-Cardena, G., Fan, R., Stern, D.F., Liu, J., Sessa, W.C. (1996). Endothelial nitric oxide synthase is regulated by tyrosine phosphorylation and interacts with caveolin-1, *J. Biol. Chem.,* **271**, pp. 27237–27240.

64. Minshall, R.D., Tiruppathi, C., Vogel, S.M., Malik, A.B. (2002). Vesicle formation and trafficking in endothelial cells and regulation of endothelial barrier function, *Histochem. Cell Biol.,* **117**, pp. 105–112.

65. Predescu, S.A., Predescu, D.N., Palade, G.E. (2001). Endothelial transcytotic machinery involves supramolecular protein–lipid complexes, *Mol. Biol. Cell,* **12**, pp. 1019–1033.

66. Prieto-Sánchez, R.M., Berenjeno, I.M., Bustelo, X.R. (2006). Involvement of the Rho/Rac family member RhoG in caveolar endocytosis, *Oncogene,* **25**, pp. 2961–2973.

67. Bonifacino, J.S., Dell'angelica, E.C. (1999). Molecular bases for the recognition of tyrosine-based sorting signals, *J. Cell Biol.,* **145**, 923–926.

68. Sorkin, A. (2004). Cargo recognition during clathrin-mediated endocytosis: A team effort, *Curr. Opin. Cell Biol.,* **16**, pp. 392–399.

69. Kohn, S., Nagy, J., Dvorak, H., Dvorak, A. (1992). Pathways of macromolecular tracer transport across venules and small veins. Structural basis for the hyperpermeability of tumor blood vessels, *Lab. Invest.*, **67**, pp. 596–607.

70. Lossinsky, A., Buttle, K., Pluta, R., Mossakowski, M., Wićniewski, H. (1999). Immunoultrastructural expression of intercellular adhesion molecule-1 in endothelial cell vesiculotubular structures and vesiculo-vacuolar organelles in blood-brain barrier development and injury, *Cell Tissue Res.*, **295**, 77–88.

71. Feng, D., Nagy, J.A., Hipp, J., Dvorak, H.F., Dvorak, A.M. (1996). Vesiculo-vacuolar organelles and the regulation of venule permeability to macromolecules by vascular permeability factor, histamine, and serotonin, *J. Exp. Med.*, **183**, pp. 1981–1986.

72. Lipinski, C.A., Lombardo, F., Dominy, B.W., Feeney, P.J. (2001). Experimental and computational approaches to estimate solubility and permeability in drug discovery and development settings, *Adv. Drug Deliv. Rev.*, **46**, 3–26.

73. Pardridge, W.M. (2003). Blood–brain barrier drug targeting: The future of brain drug development, *Mol. Interventions*, **3**, pp. 90–105.

74. Pardridge, W.M. (2005). The blood–brain barrier: Bottleneck in brain drug development, *NeuroRx.*, **2**, pp. 3–14.

75. Fischer, H., Gottschlich, R., Seelig, A. (1998). Blood–brain barrier permeation: Molecular parameters governing passive diffusion, *J. Membr. Biol.*, **165**, pp. 201–211.

76. Misra, A., Ganesh, S., Shahiwala, A., Shah, S.P. (2003). Drug delivery to the central nervous system: A review, *J. Pharm. Pharm. Sci.*, **6**, pp. 252–273.

77. Lambert, D.M. (2000). Rationale and applications of lipids as prodrug carriers, *Eu. J. Pharm. Sci.*, **11**, pp. S15–S27.

78. Bodor, N. (1997). Retrometabolic approaches for drug design and targeting, *Pharmazie*, **52**, pp. 491–494.

79. Wu, J., Yoon, S.H., Wu, W.M., Bodor, N. (2002). Synthesis and biological evaluations of brain-targeted chemical delivery systems of [Nva2]-TRH, *J. Pharm. Pharmacol.*, **54**, pp. 945–950.

80. Chen, P., Bodor, N., Wu, W.-M., Prokai, L. (1998). Strategies to target kyotorphin analogues to the brain, *J. Med. Chem.*, **41**, pp. 3773–3781.

81. Schenk, D., Barbour, R., Dunn, W., Gordon, G., Grajeda, H., Guido, T., Hu, K., Huang, J., Johnson-Wood, K., Khan, K., Kholodenko, D., Lee, M., Liao, Z., Lieberburg, I., Motter, R., Mutter, L., Soriano, F., Shopp, G.,

Vasquez, N., Vandevert, C., Walker, S., Wogulis, M., Yednock, T., Games, D., Seubert, P. (1999). Immunization with amyloid-beta attenuates Alzheimer-disease-like pathology in the PDAPP mouse, *Nature,* **400**, pp. 173–177.

82. Neuwelt, E.A., Frenkel, E.P., Diehl, J., Vu, L., Rapoport, S., Hill, S. (1980). Reversible osmotic blood–brain barrier disruption in humans: Implications for the chemotherapy of malignant brain tumors, *Neurosurgery,* **7**, pp. 44–52.
83. Hiesiger, E.M., Voorhies, R.M., Basler, G.A., Lipschutz, L.E., Posner, J.B., Shapiro, W.R. (1986). Opening the blood–brain and blood–tumor barriers in experimental rat brain tumors: The effect of intracarotid hyperosmolar mannitol on capillary permeability and blood flow, *Ann. Neurology,* **19**, pp. 50–59.
84. Boockvar, J.A., Tsiouris, A.J., Hofstetter, C.P., Kovanlikaya, I., Fralin, S., Kesavabhotla, K., Seedial, S.M., Pannullo, S.C., Schwartz, T.H., Stieg, P., Zimmerman, R.D., Knopman, J., Scheff, R.J., Christos, P., Vallabhajosula, S., Riina, H.A. (2011). Safety and maximum tolerated dose of superselective intraarterial cerebral infusion of bevacizumab after osmotic blood–brain barrier disruption for recurrent malignant glioma, *J. Neurosurg.,* **114**, pp. 624–632.
85. Weaver, J.C. (2000). Electroporation of cells and tissues, *IEEE Trans. Plasma. Sci.,* **28**, pp. 24–33.
86. Weaver, J.C., Chizmadzhev, Y.A. (1996). Theory of electroporation: A review, *Bioelectrochem. Bioener.,* **41**, pp. 135–160.
87. Hynynen, K., Mcdannold, N., Vykhodtseva, N., Raymond, S., Weissleder, R., Jolesz, F.A., Sheikov, N. (2006). Focal disruption of the blood–brain barrier due to 260-kHz ultrasound bursts: A method for molecular imaging and targeted drug delivery, *J. Neurosurg.,* **105**, pp. 445–454.
88. Kinoshita, M., Mcdannold, N., Jolesz, F.A., Hynynen, K. (2006). Targeted delivery of antibodies through the blood–brain barrier by MRI-guided focused ultrasound, *Biochem. Biophys. Res. Commun.,* **340**, pp. 1085–1090.
89. Matsukado, K., Nakano, S., Bartus, R.T., Black, K.L. (1997). Steroids decrease uptake of carboplatin in rat gliomas–uptake improved by intracarotid infusion of bradykinin analog, RMP-7, *J. Neurooncol.,* **34**, pp. 131–138.
90. Ansar, M., Serrano, D., Papademetriou, I., Bhowmick, T.K., Muro, S. (2013). Biological functionalization of drug delivery carriers to bypass size restrictions of receptor-mediated endocytosis independently from receptor targeting, *ACS nano.,* **7**, pp. 10597–10611.

91. Zhang, Y., Pardridge, W.M. (2006). Blood–brain barrier targeting of BDNF improves motor function in rats with middle cerebral artery occlusion, *Brain Res.*, **1111**, pp. 227–229.
92. Coloma, M.J., Lee, H.J., Kurihara, A., Landaw, E.M., Boado, R.J., Morrison, S.L., Pardridge, W.M. (2000). Transport across the primate blood–brain barrier of a genetically engineered chimeric monoclonal antibody to the human insulin receptor, *Pharm. Res.*, **17**, pp. 266–274.
93. Mcintosh, D.P., Tan, X.Y., Oh, P., Schnitzer, J.E. (2002). Targeting endothelium and its dynamic caveolae for tissue-specific transcytosis *in vivo*: A pathway to overcome cell barriers to drug and gene delivery, *Proc. Natl. Acad. Sci. U.S.A.*, **99**, pp. 1996–2001.
94. Malik, A.B. (2009). Targeting endothelial cell surface receptors: Novel mechanisms of microvascular endothelial barrier transport, *J. Med. Sci.*, **2**, pp. 13–17.
95. Dharap, S., Wang, Y., Chandna, P. Khandare, J.J., Qiu, B., Gunaseelan, S., Sinko, P.J., Stein, S., Farmanfarmaian, A., Minko, T. (2005). Tumor-specific targeting of an anticancer drug delivery system by LHRH peptide, *Proc. Natl. Acad. Sci. U.S.A.*, **102**, pp. 12962–12967.
96. Torchilin, V.P. (2007). Targeted pharmaceutical nanocarriers for cancer therapy and imaging, *AAPS J.*, **9**, pp. E128–E147.
97. Goula, D., Becker, N., Lemkine, G., Normandie, P., Rodrigues, J., Mantero, S., Levi, G., Demeneix, B.A. (2000). Rapid crossing of the pulmonary endothelial barrier by polyethylenimine/DNA complexes, *Gene Ther.*, **7**, pp. 499–504.
98. Gupta, B., Levchenko, T.S., Torchilin, V.P. (2005). Intracellular delivery of large molecules and small particles by cell-penetrating proteins and peptides, *Adv. Drug Deliv. Rev.*, **57**, pp. 637–651.
99. Lewin, M., Carlesso, N., Tung, C.-H., Tang, X.W., Cory, D., Scadden, D.T., Weissleder, R. (2000). Tat peptide-derivatized magnetic nanoparticles allow *in vivo* tracking and recovery of progenitor cells, *Nat. Biotechnol.*, **18**, pp. 410–414.
100. Torchilin, V.P. (2008). Tat peptide-mediated intracellular delivery of pharmaceutical nanocarriers, *Adv. Drug Deliv. Rev.*, **60**, pp. 548–558.
101. Zhao, M., Weissleder, R. (2004). Intracellular cargo delivery using tat peptide and derivatives, *Med. Res. Rev.*, **24**, pp. 1–12.
102. Pasqualini, R., Koivunen, E., Ruoslahti, E. (1995). A peptide isolated from phage display libraries is a structural and functional mimic of an RGD-binding site on integrins, *J. Cell Biol.*, **130**, pp. 1189–1196.

103. Dijkgraaf, I., Kruijtzer, J.A., Liu, S., Soede, A.C., Oyen, W.J., Corstens, F.H., Liskamp, R.M., Boerman, O.C. (2007). Improved targeting of the $\alpha v\beta 3$ integrin by multimerisation of RGD peptides, *Eur. J. Nucl. Med. Mol. I.*, **34**, pp. 267–273.

104. Suh, W., Han, S.-O., Yu, L., Kim, S.W. (2002). An angiogenic, endothelial-cell-targeted polymeric gene carrier, *Mol. Ther.*, **6**, pp. 664–672.

105. Mitra, A., Mulholland, J., Nan, A., Mcneill, E., Ghandehari, H., Line, B.R. (2005). Targeting tumor angiogenic vasculature using polymer–RGD conjugates, *J. Control. Release*, **102**, pp. 191–201

106. Mena, I., Cotzias, G.C. (1975). Protein intake and treatment of Parkinson's disease with levodopa, *New Engl. J. Med.*, **292**, pp. 181–184.

107. Pardridge, W.M. (2002). Why is the global CNS pharmaceutical market so under-penetrated? *Drug Discovery Today*, **7**, pp. 5–7.

108. Pardridge, W.M. (2003). Blood–brain barrier drug targeting: The future of brain drug development, *Mol. Interven.*, **3**, pp. 90.

109. Muzykantov, V.R. (2013). Targeted drug delivery to endothelial adhesion molecules, *ISRN Vasc. Med.*, http://dx.doi.org/10.1155/2013/916254.

110. Simone, E., Ding, B.-S., Muzykantov, V. (2009). Targeted delivery of therapeutics to endothelium, *Cell Tissue Res.*, **335**, pp. 283–300.

Chapter 9

Nanoparticle-Based Drug Delivery to Solid Tumors

Sara A. Abouelmagd and Yoon Yeo

Department of Industrial and Physical Pharmacy, College of Pharmacy, Purdue University, West Lafayette, IN 47907, USA

sabouelm@purdue.edu

Nanoparticles have been explored as a promising drug carrier for chemotherapy, with several systems in various stages of commercial development. This chapter describes physicochemical features of nanoparticles known to critically affect their therapeutic effects. It also highlights several issues remaining in nanoparticle-based drug delivery to tumors, including biological complexity of tumors and formulation challenges.

9.1 Introduction

Chemotherapeutic agents are typically administered as intravenous injection and spread in the whole body with relatively large volumes of distribution. Therefore, the dose administered to achieve effective concentrations in tumors also affects healthy tissues and organs, resulting in adverse side effects [1]. For example, taxanes (paclitaxel and docetaxel) are associated with bone marrow suppression and

Drug Delivery Across Physiological Barriers
Edited by Silvia Muro

ISBN 978-981-4669-40-5 (Hardcover), 978-981-4669-41-2 (eBook)
www.panstanford.com

hypersensitivity [2], and cisplatin is implicated in nephrotoxicity and neurotoxicity [3]. In addition, many anticancer drugs have poor water solubility, necessitating the use of toxic solubilizers. Paclitaxel is formulated in a vehicle composed of ethanol and ethoxylated castor oil (Cremophor EL) [4], where the latter causes severe side effects such as hypersensitivity and neuropathy [5]. Similarly, docetaxel is formulated with Polysorbate 80 as a solubilizer, which has similar side effects as Cremophor EL [4]. The toxicity of active and inactive ingredients has significant impact on the clinical outcomes, including poor patient quality of life and early dropouts from treatment in extreme cases [6].

In order to maximize therapeutic outcomes of chemotherapy, it is critical to deliver anticancer agents using nontoxic carriers and concentrate the drug effect to tumors. To achieve this goal, nanoparticles (NPs) have long been pursued as a drug carrier. In 1995, the US Food and Drug Administration (FDA) approved *Doxil*®, doxorubicin sulfate loaded in long-circulating liposomes, for treatment of AIDS-associated Kaposi sarcoma and later for metastatic ovarian cancer and breast cancer [7]. The liposomal formulation contributed to reducing cardiotoxicity of doxorubicin [7]. In 2005, albumin-bound paclitaxel (*nab*-paclitaxel or *Abraxane*®), a solvent-free NP formulation, was approved by the FDA, eliminating the side effects caused by Cremophor EL [8].

An important role of NP drug carriers is to modulate pharmacokinetics and biodistribution of a drug [9]. Figure 9.1 describes biodistribution of intravenously injected NPs and their transport to tumors. Unlike low-molecular-weight drugs, NPs have a selective opportunity to access solid tumors due to the unique features of tumor vasculature. With the increasing nutritional demands, tumors rapidly recruit new blood vessels (angiogenesis) [10]. The formed vessels are structurally and functionally abnormal, with hyperpermeable endothelial lining [11], which leads to selective extravasation of blood-borne macromolecules in tumors [12]. In addition, impaired lymphatic drainage near tumors allows for longer retention of the extravasated NPs in the interstitial space of tumors. This phenomenon, collectively called enhanced permeability and retention (EPR) effect [13], has been the most pronounced principle exploited in NP-based drug delivery to tumors.

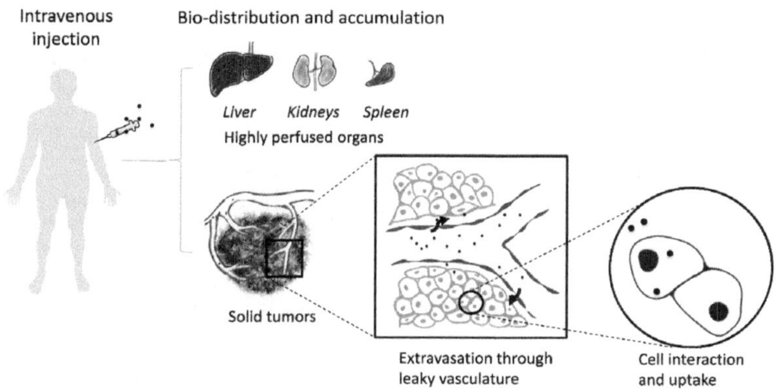

Figure 9.1 Biodistribution of intravenously injected NPs and their transport pathway to solid tumors

To take advantage of the EPR effect, particle size and surface chemistry are modified to ensure long-term circulation in the blood. To further enhance drug delivery to tumor cells and intracellular targets, NPs can be modified with surface ligands that interact with membrane proteins with a high degree of specificity. Several other variables also influence the fate and drug delivery efficiency of NPs in the body. This chapter discusses properties of NPs that influence their effectiveness as a drug carrier and remaining challenges in NP development.

9.2 Factors Affecting Effectiveness of NP Drug Delivery Systems

A systemically administered NP drug carrier needs to combat biological challenges during circulation and provide target-specific drug release. Figure 9.2 summarizes features typically considered in designing NP drug carriers.

9.2.1 *Biocompatibility*

An inactive ingredient of an NP and its degradation products are expected to have no biological effects to influence therapy. NPs are built with biocompatible biomaterials of natural and synthetic

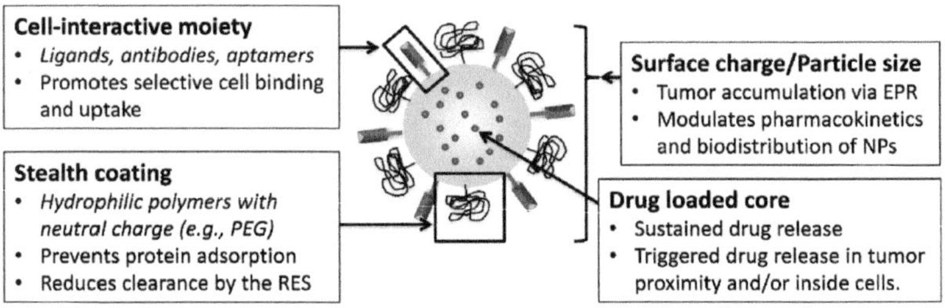

Figure 9.2 A typical design of NPs as drug carriers for systemic applications.

origin [14]. Natural materials include lipids, polysaccharides, and proteins. Lipid-based NPs include liposomes and solid lipid NPs. Polysaccharides such as chitosan, alginate, and hyaluronic acid are used in preparation of NPs [15]. NPs of synthetic origin include polymeric solid NPs, polymeric micelles, and dendrimers [16]. Poly(lactic-co-glycolic acid) (PLGA) and poly(lactic acid) (PLA) have commonly been used due to the biodegradability and excellent track records in FDA-approved products [17]. Amine-terminated dendrimers are found to have a hemolytic effect [18]; however, the terminal groups can be altered with anionic functional groups to lower the toxicity [19].

9.2.2 *Particle Size*

Ideally, NPs should be larger than 10 nm to avoid renal filtration during circulation [20] but smaller than the cutoff size of the pores in tumor vessels, which ranges from 100 to 600 nm, to extravasate and reach tumors [21]. It is generally believed that particles larger than 10 nm and smaller than 100 nm will be optimal for minimizing renal clearance and achieving effective extravasation at tumors, respectively, although NPs of this size range will still enter the liver through the sinusoid [22]. Particle size is also an important determinant of NP penetration into the tumor interstitium. Reports show that smaller particles (20–40 nm) can penetrate deeper into the tumor matrix, whereas larger particles (60–100 nm) tend to stay close to blood vessels [23, 24].

9.2.3 *Particle Shape*

Studies have shown that NP shape has strong influence on the circulation, accumulation, and cellular interaction of NPs [25, 26]. Gold nanorods, 45 nm long and 10 nm thick, were shown to have a longer circulation time, higher tumor accumulation, and lower liver accumulation than gold nanospheres with an average diameter of 50 nm [27]. In another study, filamentous micelles ("filomicelles"), 18 μm long and 20–60 nm in diameter, were injected into rodents and found circulating over 1 week post-injection, much longer than 2 days observed with ~100 nm spherical polymersomes [28, 29]. High aspect ratio of NPs also contributes to tumor penetration. Quantum-dot-based nanorods and nanospheres with an equal hydrodynamic size (33–35 nm) and similar surface were compared with respect to the depth of penetration into tumor–mimetic collagen gels and transvascular transport into tumors [30]. Nanorods consistently showed higher transport rates across porous membranes and collagen gels and faster penetration into and higher accumulation in tumors than nanospheres, attributable to the flexibility [30]. The effect of NP shape is also evident on the cellular level. Mesoporous silica NPs with different aspect ratios showed different cellular effects such as NP uptake, cytotoxicity, and cellular adhesion, with the higher-aspect-ratio NPs showing faster uptake and higher net internalization by melanoma cells [31]. NP shape also influences the extent of phagocytosis, a key clearance mechanism of NPs. It is reported that 120 nm hexagonal NPs with eight plate sides were less internalized by macrophages than spherical NPs with the same volume [32]. This result indicates that the local shape of NPs at the initial cell attachment determines the degree of phagocytic uptake. The particle replication in non-wetting templates (PRINT) technology, based on top-down lithographic fabrication method, enabled production of NPs with well-controlled sizes and shapes and facilitated systematic investigation of their effects on biodistribution [33]. Docetaxel loaded PLGA NPs were fabricated into two different aspect ratios (200 × 200 nm or 80 × 320 nm) using PRINT technology to compare biodistribution. Both 200 × 200 nm and 80 × 320 nm NPs showed greater tumor exposure than free docetaxel, with 80 × 320 nm NPs showing lower exposure to kidneys, liver, and lungs than 200 × 200 nm NPs [34].

9.2.4 *Particle Surface Charge*

Colloidal stability of NPs is largely dependent on their surface charge. Highly charged NPs repel each other in solution, avoiding aggregation during storage [35]. The cell–NP interaction is also influenced by the NP surface charge [36]. Positively charged NPs bind via electrostatic interactions with the cell membrane, and negatively charged NPs bind due to lipid components. In addition, NP surface charge influences the degree of protein adsorption and the type of the adsorbed proteins [37, 38].

NP surface charge can dictate the cellular internalization pathways of NPs. Mesoporous silica NPs with low or no quaternary ammonium groups (0–0.98 mmol/g) were taken up by human mesenchymal stem cells via clathrin- and actin-dependent endocytosis [39]. On the other hand, silica NPs with higher quaternary ammonium content (1.7 mmol/g) were taken up efficiently, not influenced by inhibitors of different cellular uptake pathways [39]. Highly charged micelles, either positive or negative, showed high liver uptake due to phagocytosis by Kupffer cells [40]. In contrast, less charged micelles of the same particle size showed relatively low liver uptake due to the reduced protein binding [40]. Similar findings were reported with polymeric NPs [41].

9.2.5 *Circulation Half-Life*

As NPs enter the circulation, plasma proteins rapidly adsorb to the surface, a process called "opsonization," followed by macrophage uptake. Therefore, NPs readily opsonized in blood end up in the organs of the reticuloendothelial system before they reach the target tissues [42]. For NPs to avoid premature clearance, they must hide surface properties prone to opsonization, such as high charge density and hydrophobicity. For this purpose, NP surface is modified with hydrophilic and electrically neutral polymers such as polyethylene glycol (PEG). Surface modification with PEG stabilizes the NPs sterically and reduces adsorption of plasma proteins, thereby extending the circulation half-life of the NPs [43]. Other hydrophilic polymers are used for similar purposes [14, 15, 44].

For optimization of the stealth effect, molecular weight, grafting density, and conformation of the surface polymers need to be carefully chosen [45]. In general, the circulation time of NPs increased with the molecular weight of PEG when all other factors are comparable [46]. In Doxil® liposomes, ~5 mol% PEG was found optimal for the stealth effect [47]. Depending on the surface density, PEG can acquire two conformations, a mushroom-like or a denser brush conformation [42]. The brush conformation, obtained at >8 mol%, is believed to coat the NP surface completely, providing the best protection and enhancement of circulation time, while in the mushroom-like conformation, a PEG layer does not extend far from the NPs surface [48]. On the other hand, excessive PEG content can compromise the integrity of membrane by the detergent effect of PEG-lipid conjugates [47] or the lateral repulsion of PEG chains [49]. At comparable surface density of PEG, the stealth effect is influenced by the size of particles covered with PEG. Chen et al. compared serum protein adsorption to gold NPs of different sizes and variable PEG densities. At an equal PEG density, relatively small NPs showed the greater protein adsorption, because the large curvature of small NPs gave a greater steric freedom to each PEG molecule [50].

9.2.6 *Tumor Retention and Cellular Uptake*

Once NPs arrive at tumors after long-term circulation and extravasation via the leaky vasculature, their interactions with target cells are important in exerting therapeutic effect on the cells. PEGylation interferes with NP–cell interactions [51]; thus, it is often necessary to functionalize NP surfaces to improve their tumor localization and retention. As a result of the small particle size, NPs have a relatively high surface area per volume, which serves as a substrate for conjugating different ligands. Monoclonal antibodies can be used as a ligand to decorate NPs, which reacts with specific antigens located on tumor cells [52]. Many other receptors overexpressed on tumor cells compared to normal ones, such as transferrin [53], folate [54], and biotin [55] receptors, have been exploited to enhance NP–cell interactions and retention in tumors. For more details, readers are referred to review articles focusing on this topic [56–58].

9.2.7 *Drug Release*

Ideal NPs should release the drug in a controlled manner at tumor tissues rather than in blood. However, due to the large surface area per volume, the loaded drug tends to leach out by diffusion, leading to premature drug release during circulation [59]. Given that it typically takes 24–48 hours for circulating NPs to achieve tumor accumulation via the EPR effect [60], it is critical to minimize drug release from NPs in blood during this period. On the other hand, if the drug is retained in NPs too well, its therapeutic effect is also diminished [61]. Therefore, an ideal NP should have a built-in mechanism to trigger drug release in a timely manner at the intended targets, whether they are extracellular matrix or intracellular organelles. Controlling drug release from NP systems is discussed in Section 9.3.2.3.

9.3 Remaining Challenges in NP Development

After decades of research on nanomedicines for chemotherapy, several nanomedicines have been approved by various regulatory bodies, and dozens of clinical trials are currently ongoing (Table 9.1) [62]. However, NPs have limitations in overcoming biological challenges imposed by the complexity of tumor physiology. Several challenges also remain in formulation development.

9.3.1 *Tumor Physiology*

A growing tumor goes through a dormant primary phase followed by a rapid growth phase in both size and nutritional demand. One consequence of rapid tumor growth is hypoxia, which leads to upregulation of hypoxia-induced factor 1, responsible for alteration of metabolism and extracellular microenvironment of tumors [75]. The microenvironmental changes influence different biochemical processes in cells, making them resistant to chemotherapy and radiation therapy [76, 77]. For example, the increasing acidity of tumor interstitium can change cell membrane fluidity and reduce active drug uptake [77]. Another consequence of the increased

Table 9.1 Some nanomedicinal formulations approved or undergoing clinical trials

Commercial name	Manufacturer	NP type	Drug	Status	Ref.
Abraxane	Celgene	Albumin NPs (*nab* technology)	Paclitaxel	Approved	[8]
Doxil/Caelyx	Ortho Biotech Schering-Plough	PEGylated liposomes	Doxorubicin	Approved	[7]
DepoCyte	Skye Pharma	Liposomes	Cytarabine	Approved	[63]
Marqibo	Talon Therapeutics, Inc.	Liposomes	Vincristine	Approved	[64]
DaunoXome	Galen Ltd.	Liposomes	Daunorubicin	Approved	[65]
Myocet	Cephalon	Liposomes	Doxorubicin	Approved	[62]
ThermoDox	Celsion	Heat activated liposomes	Doxorubicin	Phase III	[66]
Lipoplatin	Regulon Inc.	PEGylated Liposomes	Cisplatin	Phase III	[67]
OSI-211	OSI Pharmaceuticals	Liposomes	Lurtotecan	Phase II	[68]
Lipo-Dox	TTY Biopharm Co. Ltd.	PEGylated liposomes	Doxorubicin	Phase II	[69]
NK-105	Nippon Kayaku Co. Ltd	Polymeric NPs	Paclitaxel	Phase II	[62, 70]
CRLX101	Cerulean Pharma	Cyclodextrin polymeric NPs	Camptothecin	Phase II	[62, 71]
BIND-014	BIND Bioscience	Polymeric NPs	Docetaxel	Phase II	[72]
MBP-426	Mebiopharm	Liposomes	Oxaliplatin	Phase I	[73]
CPX-1	Celator Pharmaceuticals	Liposomes	Irinotecan and floxuridine	Phase I	[74]

nutritional demands is the formation of new blood vessels, mediated by the release of different angiogenic factors from tumor cells such as vascular endothelial growth factor and platelet-derived growth factor [10]. The formed blood vessels are hyperpermeable to circulating macromolecules, which results in increases in osmotic pressure as well as influx of interstitial fluid and, thus, interstitial fluid pressure (IFP) [78]. High IFP represents a physical barrier to mass transport into the tumors, resulting in poor and heterogeneous distribution of drugs and radiation [79, 80] as well as intratumoral delivery of NPs [81].

The difficulty in cancer therapy is aggravated by the heterogeneity of tumors. Intertumoral heterogeneity refers to antigenic, immunogenic, and metabolic properties of different tumors, which account for differences in drug resistance, growth rate, and metastatic capabilities. Such variability also exists at the level of a single tumor mass, causing intratumoral heterogeneity in cell morphology, phenotype, and metabolism [82–84]. Tumor heterogeneity is a significant challenge to chemotherapy as it can lead to selection of drug-resistant cell types and relapse of drug-resistant tumors [85]. The diversity and genetic instability of tumors account for, at least partly, the difficulty in advancing NPs with a single type of ligand to clinical practice.

Consequently, many have hypothesized that drug delivery to tumors can be improved by alleviating these physiological barriers. Auxiliary agents have been used to increase tumoral blood flow, normalize blood vessels, or reduce stromal barriers, thereby enhancing drug or NP penetration into tumors [81, 86, 87]. Specifically, Jain et al. demonstrated that co-administration of losartan, an angiotensin II receptor antagonist with antifibrotic activity, could enhance the penetration and efficacy of nanomedicine [88]. Au et al. reported that "priming" solid tumors with an apoptosis inducer like paclitaxel increased intratumoral transport of NPs by expanding interstitial space and facilitating blood perfusion [89, 90]. Radiation can also help increase NPs tumor accumulation by decreasing IFP and increasing vascular permeability [91]. A single dose of 15 Gy radiation decreased IFP by $\sim$40% over a period of 5 days and increased vascular permeability by 60% [91]. When PEGylated iron oxide NPs were injected 3 days post-irradiation, their accumulation increased by 2.5 folds [91].

9.3.2 *Formulation Challenges*

9.3.2.1 PEG dilemma

PEGylation of NP surface increases blood circulation time of NPs by helping NPs avoid nonspecific interactions with immune cells during circulation. However, upon the arrival of NPs at tumors, the same effect interferes with NP interaction with target cells and subsequent cellular internalization. Moreover, PEG interferes with the endosomal escape of NPs, leading to lysosomal degradation of the therapeutic payload [92]. This phenomenon, known as "PEG dilemma," has prompted alternative approaches [15, 92]. Immune responses to PEG are also reported. PEGylated liposomes showed a decrease in circulation half-life upon subsequent administration [93]. This was explained by the formation of anti-PEG-IgM at the first introduction of PEGylated liposomes, followed by the activation of the complement system, ending in accelerated blood clearance of the subsequent doses [94].

To achieve the stealth effect of PEGylation without compromising proper interaction between target cells and NPs, removable PEG coatings have been proposed. Here, PEG chains are conjugated to the NP surface via stimuli-sensitive linkers that can be cleaved by chemical or enzymatic conditions specific to the tumor stroma or the intracellular environment [92]. For example, acid-labile linkers, such as hydrazone, diorthoester, and vinylether bonds [95], were used to take advantage of slightly acidic pH of tumor stroma or lysosomes. Difference in redox potential between the extracellular and intracellular matrices is another stimulus employed to remove the PEG layer. Glutathione is present in millimolar range in the cytosol, while its extracellular concentration is as low as $\sim$10 μM [96]. Therefore, reduction-sensitive disulfide bond can remain stable in blood and be reduced at cell surface and in the cells. Another reduction-sensitive linker is dithiobenzyl carbamate, sensitive to milder thiolytic conditions [97]. This linker was used to conjugate PEG to distearoylphosphatidylethanolamine, forming reduction-sensitive liposomes [97].

Several hydrophilic polymers, including polysaccharides [14, 15] and poly(amino acids) [98], have been pursued as alternative stealth coating materials. Examples of poly(amino acids) are poly(hydroxyethyl L-asparagine) and poly(hydroxyethyl L-glutamine) [99]. Poly(amino acid)-coated liposomes showed a long circulation time comparable to that of PEG [98]. Mucin is another stealth coating inspired by cell surface glycoproteins [100].

9.3.2.2 Maintaining particle size

While nanometric particle size is one of the most fundamental properties that defines NPs, they do not necessarily maintain the particle size as the "factory setting" during circulation. Abraxane®, albumin-bound paclitaxel NPs with an average diameter of 130 nm, breaks down to paclitaxel–albumin molecules during circulation [101], which do not fully leverage the benefits of the EPR effect [102]. Conversely, liposomal systems are thermodynamically unstable and prone to aggregation and/or fusion [103], especially in plasma [104]. Liposomal membrane stability can be improved by including cholesterol and phospholipids with high phase transition temperature, which increase rigidity of the bilayer [105, 106]. Additionally, liposomes can be stabilized with hydrophilic polymers. PEG is most widely used [104], but other polymers are also reported. For example, dipalmitoyl phosphatidylcholine (DPPC) liposomes coated with chitosan showed greater resistance to detergent treatment [107]. Liposomes stabilized with alkylated polyvinyl alcohol were resistant to serum-induced aggregation and/or fusion, as compared to uncoated liposomes [108]. Recent studies propose to decorate liposomal surfaces with small, charged NPs for stabilization of the lipid bilayer [109, 110]. Liposomes based on DPPC, ~100 nm in diameter, were coated with negatively charged ~8 nm silica NPs [110]. The silica NPs provided a concentration-dependent stabilizing effect and helped inhibit liposomal aggregation in water for more than 800 h [110]. In the absence of silica NPs, the liposome size increased dramatically over 400 h. A potential advantage of this method is that the stabilizing component occupies relatively small

NP surface area, leaving a substantial portion of surface available for functionalization [109].

9.3.2.3 Controlling drug release

A number of NP designs are predicated on the assumption that an NP will retain a drug during circulation until it reaches the target cells. Otherwise, the therapeutic effect of NPs may not be different from that of free drug. However, initial burst release occurs in almost all types of delivery systems, especially in NPs with relatively large surface areas per volume ratios. Therefore, various efforts are made to ensure stable drug retention in NPs in blood. At the same time, stimuli-responsive systems are pursued in parallel to make the carried drugs available at target locations in a timely manner.

Incorporating a diffusion barrier on NP surface. To prevent the initial burst release, a self-assembled layer of water-insoluble material is added as a diffusion barrier. For example, a lecithin layer was used to coat a PLGA core encapsulating deocetaxel, resulting in a hybrid polymer–lipid NP with an attenuated docetaxel release [111]. Similarly, the release of doxorubicin entrapped in PLGA core was suppressed with an external layer of diethylenetriaminepentaacetic acid–gadolinium lipid, which also served as a paramagnetic image contrast agent [112].

Crosslinking. Polymeric micelles are prone to disassembly and dissociation upon introduction into the blood stream due to interactions with blood components and dilution below their critical micelle concentrations. Burt et al. studied biodistribution of polymeric micelles based on a block copolymer of D,L-lactic acid and methoxypolyethylene glycol (PDLLA-MePEG) encapsulating paclitaxel [113]. They reported that the drug and polymer showed distinct biodistribution profiles, indicating premature drug release and micelle dissociation [113]. The stability of polymeric micelles can be improved by crosslinking of the core, shell, or the interface. One approach is to crosslink the micelle core made of anionic polymethacrylate with Ca^{2+} [114]. In another example, the PLA core of a polymeric micelle was stabilized by introducing methacrylol end groups at the terminus of the PLA block, which could be covalently crosslinked after micelle formation [115]. Instead of NP core, the interface between hydrophilic shell and hydrophobic core was stabilized via UV-mediated crosslinking [116]. For this purpose, photo-crosslinkable poly(acryloyl carbonate) was introduced as the center block of a block copolymer. The PEG-poly(acryloyl carbonate)-polycaprolactone (PEG-PAC-PCL) formed more stable micelles in both size and drug retention as compared to those made with non-crosslinkable counterpart (PEG-PCL) [116]. Due to the improved stability, paclitaxel loaded in the photo-crosslinked PEG-PAC-PCL micelles showed greater *in vivo* tumor inhibition activity than paclitaxel in PEG-PCL micelles [116].

Stimuli-responsive systems. To confine drug release specifically to target tissues or organelles, stimuli-responsive NPs are developed. NPs are designed to change their physicochemical properties in response to intrinsic conditions of target locations or external stimuli applied on the targets [117]. For example, a drug was encapsulated in mesoporous silica NPs, capped with large molecules, such as cadmium sulfide nanocrystals [118] or PAMAM dendrimers [119] via disulfide bond. These capping materials prevented drug release but were removed in a reductive intracellular environment. In another example, reduction-sensitive NPs encapsulating doxorubicin were prepared with dextran-lipoic acid derivative crosslinked via disulfide bonds [120]. This system showed minimal drug release under extracellular condition and fast release in reductive environments [120]. Cathepsin B is used for intracellular or tumoral drug release. As a lysosomal cysteine proteinase, cathepsin B could trigger intracellular drug release from NPs containing responsive peptides [121]. Cathepsin B is also overexpressed in some tumors: it is present in the extracellular environment and on the cell surface of tumors [122]. Therefore, a peptide linker degraded by cathepsin B was incorporated in a PEG NP system for tumor-selective drug release [123]. Matrix metalloproteinases (MMPs), another type of enzymes overexpressed in tumor microenvironment, are used for a similar purpose [124]. Doxorubicin was loaded in mesoporous silica NPs coated with conjugates of PEG diacrylate and MMP-degradable peptides [125]. Drug release and cytotoxicity of doxorubicin-loaded NPs were proportional to the degradability of the peptide linkers, and the MMP-sensitive NPs resulted in greater cell death in MMP-overexpressing tumors as compared to MMP-insensitive PEG-coated NPs [125]. Acidic pH of endosomes/lysosomes is frequently used to achieve intracellular drug release. Mesoporous silica NPs were capped with calcium phosphate, soluble in pH 4–5 but insoluble in pH 7.4. To limit the calcium phosphate deposition on the surface, the silica NPs were conjugated with urease, which hydrolyzed urea and created a high pH zone around the NPs forming a layer of calcium phosphate precipitates [126]. The coated NPs significantly attenuated drug release at pH 7.4 but released drug intracellularly, indicated by sustained nuclear localization of doxorubicin [126].

Although stimuli-responsive systems offer a high level of drug release control, nonspecific triggering remains a concern [127]. For example, MMPs can be upregulated not only in tumors but also in noncancerous tissues [128]. In addition, slight acidity can develop in inflamed and infected tissues, causing off-target activation of the pH-sensitive systems [129]. Conversely, some targets may not offer sufficient stimuli. For example, folate receptor-containing endosomes are not sufficiently acidic [130], which implies that pH-sensitive NPs designed to respond to typical lysosomal pH may not show the intended intracellular drug release, if the NPs enter cells via folate receptors.

Multiple-stimuli-responsive systems. To increase the selective reactivity of the NPs, two or more mechanisms are utilized either simultaneously or sequentially. NPs are engineered to respond to dual stimuli, such as pH and temperature, pH and reductive potential, or temperature and enzyme [117, 131, 132]. For example, polymeric NPs with a dual pH sensitivity were formulated using a random copolymer that degraded by both bulk dissolution and surface degradation at weakly acidic pH [131]. The advantage of this system is the good stability at physiological pH and quick onset of degradation, which are often conflicting with each other [131]. In another study, polymeric micelles dual-responsive to acidic endosomal pH and intracellular reductive potential were prepared using a triblock copolymer made of a pH-sensitive hydrophobic block, disulfide-crosslinkable middle block, and PEG [132]. This micelle system showed minimal drug release at pH

7.4 and increasing drug release in response to dithiothreitol (10 mM) and/or pH 5 [132]. Taking this approach a step further, triple-stimuli-responsive micellar system was developed using a block copolymer with a pH-sensitive hydrophobic block and a temperature-sensitive hydrophilic block connected via a reduction-sensitive disulfide linker [133]. Notably, individual stimulus caused slow or incomplete dye release, but combined stimuli resulted in a significantly faster and greater drug release [133].

Covalent conjugation of a drug to carrier. For stable drug encapsulation, a drug may be conjugated to an NP matrix via a cleavable linker. pH-activatable NPs were made with paclitaxel conjugated to PEG-poly(acrylic acid) via an acid-labile acetal linker [134]. The NPs showed >80% drug release in pH 5 in 2 days, while they released only 29% of the total drug at pH 7.4 [134]. In another study, drug–polylactide conjugates were synthesized using a drug (paclitaxel, docetaxel, and camptothecin) as an initiator of polymerization [135]. The conjugates formed NPs with high drug contents (5–36 wt%), which showed minimal initial burst release followed by gradual drug release over a week *in vitro* [135]. Similarly, docetaxel was conjugated to PEGylated carboxymethyl cellulose (CMC) to assemble into 120 nm NPs [136]. The CMC-based NPs had an antistromal effect, increasing tumoral perfusion and lowering the IFP, with a greater antimetastatic effect than Abraxane [137]. In another study, camptothecin was conjugated to a β-sheet-forming peptide to make drug amphiphiles, which assembled into well-defined supramolecular structures [138]. Serving as a part of the carrier building block, the drug molecules constituted up to 38% of the nanotubes [138]. An important consideration in the design of drug–polymer conjugates is that cleavage of the conjugate should restore a pharmacologically active drug [139]. While it is an efficient way of controlling drug release, drug–polymer conjugate is considered a new chemical entity, which needs a new FDA approval for clinical use [140].

9.4 Conclusions and Perspectives

NP drug carriers have overcome several challenges in the administration of anticancer drugs, such as poor water solubility and side effects associated with toxic solubilizers. They also hold a great promise to potentiate the effectiveness of chemotherapy by modulating biodistribution of drugs and concentrating their effects on tumor tissues. To translate the idealized concepts into clinical practice, it is important to design NPs to satisfy multiple (often conflicting) requirements, such as stable drug retention during circulation, site-specific drug release, evasion of the immune surveillance, and preferential interaction with target cells. Moreover, the NP design must consider physiological challenges of tumors such as heterogeneity, genetic instability, and increasing resistance to drug transport on tissue and cellular levels, which counteract chemical and structural features of the NPs. Decades of research effort has brought substantial technological advancement in individual aspect. The remaining task is to integrate these technologies in a simple and scalable manner, which will justify capital investment in the translation of bench-top inventions to clinical products.

References

1. Banerjee, D., Sengupta, S. (2011). Nanoparticles in cancer chemotherapy, *Prog. Mol. Biol. Transl. Sci.*, **104**, pp. 489–507.
2. Markman, M. (2003). Managing taxane toxicities, *Support. Care Cancer*, **11**, pp. 144–147.
3. Cerri, S., Piccolini, V. M., Santin, G., Bottone, M. G., de Pascali, S. A., Migoni, D., et al. (2011). The developmental neurotoxicity study of platinum compounds. Effects of cisplatin versus a novel Pt(II) complex on rat cerebellum, *Neurotoxicol. Teratol.*, **33**, pp. 273–281.
4. Hawkins, M. J., Soon-Shiong, P., Desai, N. (2008). Protein nanoparticles as drug carriers in clinical medicine, *Adv. Drug Deliv. Rev.*, **60**, pp. 876–885.
5. ten Tije, A. J., Verweij, J., Loos, W. J., Sparreboom, A. (2003). Pharmacological effects of formulation vehicles: Implications for cancer chemotherapy, *Clin. Pharmacokinet.*, **42**, pp. 665–685.
6. Stam, H. J., Challis, G. B. (1989). Ratings of cancer chemotherapy toxicity by oncologists, nurses, and pharmacists, *J. Pain Symptom Manage.*, **4**, pp. 7–13.

7. Barenholz, Y. (2012). Doxil (R)—The first FDA-approved nano-drug: Lessons learned, *J. Control. Release*, **160**, pp. 117–134.
8. Green, M. R., Manikhas, G. M., Orlov, S., Afanasyev, B., Makhson, A. M., Bhar, P., et al. (2006). Abraxane, a novel Cremophor-free, albumin-bound particle form of paclitaxel for the treatment of advanced non-small-cell lung cancer, *Ann. Oncol.*, **17**, pp. 1263–1268.
9. Blanco, E., Hsiao, A., Mann, A. P., Landry, M. G., Meric-Bernstam, F., Ferrari, M. (2011). Nanomedicine in cancer therapy: Innovative trends and prospects, *Cancer Sci.*, **102**, pp. 1247–1252.
10. Harper, J., Moses, M. (2006). Molecular regulation of tumor angiogenesis: Mechanisms and therapeutic implications, *Cancer: Cell Structures*, **69**, pp. 223–268
11. Baish, J. W., Jain, R. K. (2000). Fractals and cancer, *Cancer Res.*, **60**, pp. 3683–3688.
12. Matsumura, Y., Maeda, H. (1986). A new concept for macromolecular therapeutics in cancer chemotherapy: Mechanism of tumoritropic accumulation of proteins and the antitumor agent smancs, *Cancer Res.*, **46**, pp. 6387–6392.
13. Torchilin, V. (2011). Tumor delivery of macromolecular drugs based on the EPR effect, *Adv. Drug Deliv. Rev.*, **63**, pp. 131–135.
14. Amoozgar, Z., Park, J. Y., Lin, Q. N., Yeo, Y. (2012). Low molecular-weight chitosan as a pH-sensitive stealth coating for tumor-specific drug delivery, *Mol. Pharm.*, **9**, pp. 1262–1270.
15. Doh, K. O., Yeo, Y. (2012). Application of polysaccharides for surface modification of nanomedicines, *Ther. Deliv.*, **3**, pp. 1447–1456.
16. Cho, K. J., Wang, X., Nie, S. M., Chen, Z., Shin, D. M. (2008). Therapeutic nanoparticles for drug delivery in cancer, *Clin. Cancer Res.*, **14**, pp. 1310–1316.
17. Jain, R. A. (2000). The manufacturing techniques of various drug loaded biodegradable poly(lactide-co-glycolide) (PLGA) devices, *Biomaterials*, **21**, pp. 2475–2490.
18. Domanski, D. M., Klajnert, B., Bryszewska, M. (2004). Influence of PAMAM dendrimers on human red blood cells, *Bioelectrochemistry*, **63**, pp. 189–191.
19. Malik, N., Wiwattanapatapee, R., Klopsch, R., Lorenz, K., Frey, H., Weener, J. W., et al. (2000). Dendrimers: Relationship between structure and biocompatibility *in vitro*, and preliminary studies on the biodistribution of I-125-labelled polyamidoamine dendrimers *in vivo* (vol 65, pg 133, 2000), *J. Control. Release*, **68**, pp. 299–302.

20. Venturoli, D., Rippe, B. (2005). Ficoll and dextran vs. globular proteins as probes for testing glomerular permselectivity: Effects of molecular size, shape, charge, and deformability, *Am. J. Physiol. Renal Physiol.*, **288**, pp. F605–F613.

21. Yuan, F., Dellian, M., Fukumura, D., Leunig, M., Berk, D. A., Torchilin, V. P., et al. (1995). Vascular permeability in a human tumor xenograft: Molecular size dependence and cutoff size, *Cancer Res.*, **55**, pp. 3752–3756.

22. Davis, M. E., Chen, Z., Shin, D. M. (2008). Nanoparticle therapeutics: An emerging treatment modality for cancer, *Nat. Rev. Drug Discov.*, **7**, pp. 771–782.

23. Perrault, S. D., Walkey, C., Jennings, T., Fischer, H. C., Chan, W. C. W. (2009). Mediating tumor targeting efficiency of nanoparticles through design, *Nano Lett.*, **9**, pp. 1909–1915.

24. Lee, H., Fonge, H., Hoang, B., Reilly, R. M., Allen, C. (2010). The effects of particle size and molecular targeting on the intratumoral and subcellular distribution of polymeric nanoparticles, *Mol. Pharm.*, **7**, pp. 1195–1208.

25. Petros, R. A., DeSimone, J. M. (2010). Strategies in the design of nanoparticles for therapeutic applications, *Nat. Rev. Drug Discov.*, **9**, pp. 615–627.

26. Champion, J. A., Katare, Y. K., Mitragotri, S. (2007). Particle shape: A new design parameter for micro- and nanoscale drug delivery carriers, *J. Control. Release*, **121**, pp. 3–9.

27. Arnida, Janat-Amsbury, M. M., Ray, A., Peterson, C. M., Ghandehari, H. (2011). Geometry and surface characteristics of gold nanoparticles influence their biodistribution and uptake by macrophages, *Eur. J. Pharm. Biopharm.*, **77**, pp. 417–423.

28. Geng, Y., Dalhaimer, P., Cai, S. S., Tsai, R., Tewari, M., Minko, T., et al. (2007). Shape effects of filaments versus spherical particles in flow and drug delivery, *Nat. Nanotechnol.*, **2**, pp. 249–255.

29. Photos, P. J., Bacakova, L., Discher, B., Bates, F. S., Discher, D. E. (2003). Polymer vesicles *in vivo*: Correlations with PEG molecular weight, *J. Control. Release*, **90**, pp. 323–334.

30. Chauhan, V. P., Popovic, Z., Chen, O., Cui, J., Fukumura, D., Bawendi, M. G., et al. (2011). Fluorescent nanorods and nanospheres for real-time *in vivo* probing of nanoparticle shape-dependent tumor penetration, *Angew. Chem. Int. Ed. Engl.*, **50**, pp. 11417–11420.

31. Huang, X. L., Teng, X., Chen, D., Tang, F. Q., He, J. Q. (2010). The effect of the shape of mesoporous silica nanoparticles on cellular uptake and cell function, *Biomaterials*, **31**, pp. 438–448.
32. Lin, S. Y., Hsu, W. H., Lo, J. M., Tsai, H. C., Hsiue, G. H. (2011). Novel geometry type of nanocarriers mitigated the phagocytosis for drug delivery, *J. Control. Release*, **154**, pp. 84–92.
33. Gratton, S. E. A., Ropp, P. A., Pohlhaus, P. D., Luft, J. C., Madden, V. J., Napier, M. E., et al. (2008). The effect of particle design on cellular internalization pathways, *Proc. Natl. Acad. Sci., U.S.A.*, **105**, pp. 11613–11618.
34. Chu, K. S., Hasan, W., Rawal, S., Walsh, M. D., Enlow, E. M., Luft, J. C., et al. (2013). Plasma, tumor and tissue pharmacokinetics of docetaxel delivered via nanoparticles of different sizes and shapes in mice bearing SKOV-3 human ovarian carcinoma xenograft, *Nanomed. Nanotechnol.*, **9**, pp. 686–693.
35. Heurtault, B., Saulnier, P., Pech, B., Proust, J. E., Benoit, J. P. (2003). Physico-chemical stability of colloidal lipid particles, *Biomaterials*, **24**, pp. 4283–4300.
36. Nel, A. E., Madler, L., Velegol, D., Xia, T., Hoek, E. M. V., Somasundaran, P., et al. (2009). Understanding biophysicochemical interactions at the nano-bio interface, *Nat. Mater.*, **8**, pp. 543–557.
37. Patil, S., Sandberg, A., Heckert, E., Self, W., Seal, S. (2007). Protein adsorption and cellular uptake of cerium oxide nanoparticles as a function of zeta potential, *Biomaterials*, **28**, pp. 4600–4607.
38. Hirsch, V., Kinnear, C., Moniatte, M., Rothen-Rutishauser, B., Clift, M. J. D., Fink, A. (2013). Surface charge of polymer coated SPIONs influences the serum protein adsorption, colloidal stability and subsequent cell interaction *in vitro*, *Nanoscale*, **5**, pp. 3723–3732.
39. Chung, T. H., Wu, S. H., Yao, M., Lu, C. W., Lin, Y. S., Hung, Y., et al. (2007). The effect of surface charge on the uptake and biological function of mesoporous silica nanoparticles 3T3-L1 cells and human mesenchymal stem cells, *Biomaterials*, **28**, pp. 2959–2966.
40. Xiao, K., Li, Y. P., Luo, J. T., Lee, J. S., Xiao, W. W., Gonik, A. M., et al. (2011). The effect of surface charge on *in vivo* biodistribution of PEG-oligocholic acid based micellar nanoparticles, *Biomaterials*, **32**, pp. 3435–3446.
41. He, C. B., Hu, Y. P., Yin, L. C., Tang, C., Yin, C. H. (2010). Effects of particle size and surface charge on cellular uptake and biodistribution of polymeric nanoparticles, *Biomaterials*, **31**, pp. 3657–3666.

42. Jokerst, J. V., Lobovkina, T., Zare, R. N., Gambhir, S. S. (2011). Nanoparticle PEGylation for imaging and therapy, *Nanomedicine (Lond.)*, **6**, pp. 715–728.

43. Torchilin, V. P., Omelyanenko, V. G., Papisov, M. I., Bogdanov, A. A., Trubetskoy, V. S., Herron, J. N., et al. (1994). Poly(ethylene glycol) on the liposome surface: On the mechanism of polymer-coated liposome longevity, *Biochim. Biophys. Acta*, **1195**, pp. 11–20.

44. Tan, J. S., Butterfield, D. E., Voycheck, C. L., Caldwell, K. D., Li, J. T. (1993). Surface modification of nanoparticles by PEO PPO block-copolymers to minimize interactions with blood components and prolong blood-circulation in rats, *Biomaterials*, **14**, pp. 823–833.

45. Gref, R., Luck, M., Quellec, P., Marchand, M., Dellacherie, E., Harnisch, S., et al. (2000). 'Stealth' corona-core nanoparticles surface modified by polyethylene glycol (PEG): Influences of the corona (PEG chain length and surface density) and of the core composition on phagocytic uptake and plasma protein adsorption, *Colloids Surf. B Biointerfaces*, **18**, pp. 301–313.

46. Gref, R., Minamitake, Y., Peracchia, M. T., Trubetskoy, V., Torchilin, V., Langer, R. (1994). Biodegradable long-circulating polymeric nanospheres, *Science*, **263**, pp. 1600–1603.

47. Li, S. D., Huang, L. (2010). Stealth nanoparticles: High density but sheddable PEG is a key for tumor targeting, *J. Control. Release*, **145**, pp. 178–181.

48. Perry, J. L., Reuter, K. G., Kai, M. P., Herlihy, K. P., Jones, S. W., Luft, J. C., et al. (2012). PEGylated PRINT nanoparticles: The impact of PEG density on protein binding, macrophage association, biodistribution, and pharmacokinetics, *Nano Lett.*, **12**, pp. 5304–5310.

49. Tirosh, O., Barenholz, Y., Katzhendler, J., Priev, A. (1998). Hydration of polyethylene glycol-grafted liposomes, *Biophys. J.*, **74**, pp. 1371–1379.

50. Walkey, C. D., Olsen, J. B., Guo, H., Emili, A., Chan, W. C. (2012). Nanoparticle size and surface chemistry determine serum protein adsorption and macrophage uptake, *J. Am. Chem. Soc.*, **134**, pp. 2139–2147.

51. Hatakeyama, H., Akita, H., Harashima, H. (2011). A multifunctional envelope type nano device (MEND) for gene delivery to tumours based on the EPR effect: A strategy for overcoming the PEG dilemma, *Adv. Drug Deliver. Rev.*, **63**, pp. 152–160.

52. Kohler, G., Milstein, C. (1975). Continuous cultures of fused cells secreting antibody of predefined specificity, *Nature*, **256**, pp. 495–497.

53. Hatakeyama, H., Akita, H., Maruyama, K., Suhara, T., Harashima, H. (2004). Factors governing the *in vivo* tissue uptake of transferrin-coupled polyethylene glycol liposomes *in vivo*, *Int. J. Pharm.*, **281**, pp. 25–33.
54. Lee, R. J., Low, P. S. (1994). Delivery of liposomes into cultured Kb cells via folate receptor-mediated endocytosis, *J. Biol. Chem.*, **269**, pp. 3198–3204.
55. Na, K., Bum Lee, T., Park, K. H., Shin, E. K., Lee, Y. B., Choi, H. K. (2003). Self-assembled nanoparticles of hydrophobically-modified polysaccharide bearing vitamin H as a targeted anti-cancer drug delivery system, *Eur. J. Pharm. Sci.*, **18**, pp. 165–173.
56. Bertrand, N., Wu, J., Xu, X., Kamaly, N., Farokhzad, O. C. (2013). Cancer nanotechnology: The impact of passive and active targeting in the era of modern cancer biology, *Adv. Drug Deliv. Rev.*, **66**, pp. 2–25.
57. Allen, T. M. (2002). Ligand-targeted therapeutics in anticancer therapy, *Nat. Rev. Cancer*, **2**, pp. 750–763.
58. Steichen, S. D., Caldorera-Moore, M., Peppas, N. A. (2013). A review of current nanoparticle and targeting moieties for the delivery of cancer therapeutics, *Eur. J. Pharm. Sci.*, **48**, pp. 416–427.
59. Chen, H., Kim, S., He, W., Wang, H., Low, P. S., Park, K., et al. (2008). Fast release of lipophilic agents from circulating PEG-PDLLA micelles revealed by *in vivo* roŕster resonance energy transfer imaging, *Langmuir*, **24**, pp. 5213–5217.
60. Iyer, A. K., Khaled, G., Fang, J., Maeda, H. (2006). Exploiting the enhanced permeability and retention effect for tumor targeting, *Drug Discov. Today*, **11**, pp. 812–818.
61. Liu, J., Qiu, Z. Y., Wang, S. Q., Zhou, L., Zhang, S. M. (2010). A modified double-emulsion method for the preparation of daunorubicin-loaded polymeric nanoparticle with enhanced *in vitro* anti-tumor activity, *Biomed. Mater.*, **5**. 065002, doi: 10.1088/1748-6041/5/6/065002.
62. Wang, R. B., Billone, P. S., Mullett, W. M. (2013). Nanomedicine in action: An overview of cancer nanomedicine on the market and in clinical trials, *J.Nanomater.* Article ID 629681, http://dx.doi.org/10.1155/2013/629681.
63. Slingerland, M., Guchelaar, H. J., Gelderblom, H. (2012). Liposomal drug formulations in cancer therapy: 15 years along the road, *Drug Discov. Today*, **17**, pp. 160–166.
64. Silverman, J. A., Deitcher, S. R. (2013). Marqibo (vincristine sulfate liposome injection) improves the pharmacokinetics and pharmacodynamics of vincristine, *Cancer Chemoth. Pharm.*, **71**, pp. 555–564.

65. Forssen, E. A. (1997). The design and development of DaunoXome for solid tumor targeting *in vivo*, *Adv. Drug Deliv. Rev.*, **24**, pp. 133–150.

66. Landon, C. D., Park, J. Y., Needham, D., Dewhirst, M. W. (2011). Nanoscale drug delivery and hyperthermia: The materials design and preclinical and clinical testing of low temperature-sensitive liposomes used in combination with mild hyperthermia in the treatment of local cancer, *Open Nanomed. J.*, **3**, pp. 38–64.

67. Stathopoulos, G. P., Antoniou, D., Dimitroulis, J., Stathopoulos, J., Marosis, K., Michalopoulou, P. (2011). Comparison of liposomal cisplatin versus cisplatin in non-squamous cell non-small-cell lung cancer, *Cancer Chemoth. Pharm.*, **68**, pp. 945–950.

68. Duffaud, F., Borner, M., Chollet, P., Vermorken, J. B., Bloch, J., Degardin, M., et al. (2004). Phase II study of OSI-211 (liposomal lurtotecan) in patients with metastatic or loco-regional recurrent squamous cell carcinoma of the head and neck: An EORTC New Drug Development Group Study, *Eur. J. Cancer*, **40**, pp. 2748–2752.

69. Chou, H. H., Wang, K. L., Chen, C. A., Wei, L. H., Lai, C. H., Hsieh, C. Y., et al. (2006). Pegylated liposomal doxorubicin (Lipo-Dox) for platinum-resistant or refractory epithelial ovarian carcinoma: A Taiwanese gynecologic oncology group study with long-term follow-up, *Gynecol. Oncol.*, **101**, pp. 423–428.

70. Kato, K., Chin, K., Yoshikawa, T., Yamaguchi, K., Tsuji, Y., Esaki, T., et al. (2012). Phase II study of NK105, a paclitaxel-incorporating micellar nanoparticle, for previously treated advanced or recurrent gastric cancer, *Invest. New Drug*, **30**, pp. 1621–1627.

71. Ng, T. S. C., Wert, D., Sohi, H., Procissi, D., Colcher, D., Raubitschek, A. A., et al. (2013). Serial diffusion MRI to monitor and model treatment response of the targeted nanotherapy CRLX101, *Clin. Cancer Res.*, **19**, pp. 2518–2527.

72. Hrkach, J., Von Hoff, D., Ali, M. M., Andrianova, E., Auer, J., Campbell, T., et al. (2012). Preclinical development and clinical translation of a PSMA-targeted docetaxel nanoparticle with a differentiated pharmacological profile, *Sci. Transl. Med.*, **4**, 128ra39.

73. Sankhala, K. K., Mita, A. C., Adinin, R., Wood, L., Beeram, M., Bullock, S., et al. (2009). A phase I pharmacokinetic (PK) study of MBP-426, a novel liposome encapsulated oxaliplatin, *J. Clin. Oncol.*, **27**, abstract 2535.

74. Batist, G., Gelmon, K. A., Chi, K. N., Miller, W. H., Chia, S. K. L., Mayer, L. D., et al. (2009). Safety, pharmacokinetics, and efficacy of CPX-1 liposome

injection in patients with advanced solid tumors, *Clin. Cancer Res.*, **15**, pp. 692–700.

75. Tian, L., Bae, Y. H. (2012). Cancer nanomedicines targeting tumor extracellular pH, *Colloids Surf. B Biointerfaces*, **99**, pp. 116–126.
76. Tannock, I., Rotin, D. (1989). Acid pH in tumors and its potential for therapeutic exploitation, *Cancer Res.*, **49**, pp. 4373–4384.
77. Wike-Hooley, J., Haveman, J., Reinhold, H. (1984). The relevance of tumour pH to the treatment of malignant disease, *Radiother. Oncol.*, **2**, pp. 343–366.
78. Heldin, C.-H., Rubin, K., Pietras, K., Ostman, A. (2004). High interstitial fluid pressure: An obstacle in cancer therapy, *Nat. Rev. Cancer*, **4**, pp. 806–813.
79. Curti, B. D. (1993). Physical barriers to drug delivery in tumors, *Crit. Rev. Oncol. Hematol.*, **14**, pp. 29–39.
80. Jain, R. K. (1998). The next frontier of molecular medicine: Delivery of therapeutics, *Nat. Med.*, **4**, pp. 655–657.
81. Holback, H., Yeo, Y. (2011). Intratumoral drug delivery with nanoparticulate carriers, *Pharm. Res.*, **28**, pp. 1819–1830.
82. Fidler, I. J., Hart, I. R. (1982). Biological diversity in metastatic neoplasms: Origins and implications, *Science*, **217**, pp. 998–1003.
83. Shibata, D. (2012). Cancer. Heterogeneity and tumor history, *Science*, **336**, pp. 304–305.
84. Gerlinger, M., Rowan, A., Horswell, S., Larkin, J., Endesfelder, D., Gronroos, E., et al. (2012). Intratumor heterogeneity and branched evolution revealed by multiregion sequencing, *N. Engl. J. Med.*, **366**, pp. 883–892.
85. Marusyk, A., Polyak, K. (2010). Tumor heterogeneity: Causes and consequences, *Biochim. Biophys. Acta*, **1805**, pp. 105–117.
86. Marcucci, F., Corti, A. (2012). Improving drug penetration to curb tumor drug resistance, *Drug Discov. Today*, **17**, pp. 1139–1146.
87. Jain, R. K., Stylianopoulos, T. (2010). Delivering nanomedicine to solid tumors, *Nat. Rev. Clin. Oncol.*, **7**, pp. 653–664.
88. Diop-Frimpong, B., Chauhan, V. P., Krane, S., Boucher, Y., Jain, R. K. (2011). Losartan inhibits collagen I synthesis and improves the distribution and efficacy of nanotherapeutics in tumors, *Proc. Natl. Acad. Sci., U.S.A.*, **108**, pp. 2909–2914.
89. Wang, J., Lu, Z., Gao, Y., Wientjes, M. G., Au, J. L. S. (2011). Improving delivery and efficacy of nanomedicines in solid tumors: Role of tumor priming, *Nanomedicine (Lond)*, **6**, pp. 1605–1620.

90. Lu, Z., Tsai, M., Lu, D., Wang, J., Wientjes, M. G., Au, J. L. S. (2008). Tumor-penetrating microparticles for intraperitoneal therapy of ovarian cancer, *J. Pharmacol. Exp. Ther.*, **327**, pp. 673–682.
91. Giustini, A. J., Petryk, A. A., Hoopes, P. J. (2012). Ionizing radiation increases systemic nanoparticle tumor accumulation, *Nanomedicine*, **8**, pp. 818–821.
92. Amoozgar, Z., Yeo, Y. (2012). Recent advances in stealth coating of nanoparticle drug delivery systems, *Wiley Interdiscip. Rev. Nanomed. Nanobiotechnol.*, **4**, pp. 219–233.
93. Dams, E. T. M., Laverman, P., Oyen, W. J. G., Storm, G., Scherphof, G. L., van der Meer, J. W. M., et al. (2000). Accelerated blood clearance and altered biodistribution of repeated injections of sterically stabilized liposomes, *J. Pharm. Exp. Ther.*, **292**, pp. 1071–1079.
94. Wang, X. Y., Ishida, T., Kiwada, H. (2007). Anti-PEG IgM elicited by injection of liposomes is involved in the enhanced blood clearance of a subsequent dose of PEGylated liposomes, *J. Control. Release*, **119**, pp. 236–244.
95. Romberg, B., Hennink, W. E., Storm, G. (2008). Sheddable coatings for long-circulating nanoparticles, *Pharm. Res.*, **25**, pp. 55–71.
96. Saito, G., Swanson, J. A., Lee, K. D. (2003). Drug delivery strategy utilizing conjugation via reversible disulfide linkages: Role and site of cellular reducing activities, *Adv. Drug Deliv. Rev.*, **55**, pp. 199–215.
97. Zalipsky, S., Qazen, M., Walker, J. A., 2nd, Mullah, N., Quinn, Y. P., Huang, S. K. (1999). New detachable poly(ethylene glycol) conjugates: Cysteine-cleavable lipopolymers regenerating natural phospholipid, diacyl phosphatidylethanolamine, *Bioconjug. Chem.*, **10**, pp. 703–707.
98. Romberg, B., Metselaar, J. M., Baranyi, L., Snel, C. J., Bunger, R., Hennink, W. E., et al. (2007). Poly(amino acid)s: Promising enzymatically degradable stealth coatings for liposomes, *Int. J. Pharm.*, **331**, pp. 186–189.
99. Metselaar, J. M., Bruin, P., de Boer, L. W., de Vringer, T., Snel, C., Oussoren, C., et al. (2003). A novel family of L-amino acid-based biodegradable polymer-lipid conjugates for the development of long-circulating liposomes with effective drug-targeting capacity, *Bioconjug. Chem.*, **14**, pp. 1156–1164.
100. Thasneem, Y. M., Rekha, M. R., Sajeesh, S., Sharma, C. P. (2013). Biomimetic mucin modified PLGA nanoparticles for enhanced blood compatibility, *J. Colloid Interface Sci.*, **409**, pp. 237–244.

101. Desai, N. (2008). Nab technology: A drug delivery platform utilising endothelial gp60 receptor-based transport and tumour-derived SPARC for targeting, *Drug Deliv. Report*, **Winter 2007/2008**, pp. 37–41.

102. Ernsting, M. J., Murakami, M., Undzys, E., Aman, A., Press, B., Li, S. D. (2012). A docetaxel-carboxymethylcellulose nanoparticle outperforms the approved taxane nanoformulation, Abraxane, in mouse tumor models with significant control of metastases, *J. Control. Release*, **162**, pp. 575–581.

103. Evans, E., Metcalfe, M. (1984). Free energy potential for aggregation of giant, neutral lipid bilayer vesicles by van der Waals attraction, *Biophys. J.*, **46**, pp. 423–426.

104. Yoshioka, H. (1991). Surface modification of haemoglobin-containing liposomes with polyethylene glycol prevents liposome aggregation in blood plasma, *Biomaterials*, **12**, pp. 861–864.

105. Gabizon, A., Papahadjopoulos, D. (1988). Liposome formulations with prolonged circulation time in blood and enhanced uptake by tumors, *Proc. Natl. Acad. Sci., U.S.A.*, **85**, pp. 6949–6953.

106. Semple, S. C., Chonn, A., Cullis, P. R. (1996). Influence of cholesterol on the association of plasma proteins with liposomes, *Biochemistry-Us*, **35**, pp. 2521–2525.

107. Mady, M. M., Darwish, M. M., Khalil, S., Khalil, W. M. (2009). Biophysical studies on chitosan-coated liposomes, *Eur. Biophys. J. Biophy.*, **38**, pp. 1127–1133.

108. Takeuchi, H., Yamamoto, H., Toyoda, T., Toyobuku, H., Hino, T., Kawashima, Y. (1998). Physical stability of size controlled small unilameller liposomes coated with a modified polyvinyl alcohol, *Int. J. Pharm.*, **164**, pp. 103–111.

109. Zhang, L. F., Granick, S. (2006). How to stabilize phospholipid liposomes (using nanoparticles), *Nano Lett.*, **6**, pp. 694–698.

110. Michel, R., Plostica, T., Abezgauz, L., Danino, D., Gradzielski, M. (2013). Control of the stability and structure of liposomes by means of nanoparticles, *Soft Matter.*, **9**, pp. 4167–4177.

111. Zhang, L. F., Chan, J. M., Gu, F. X., Rhee, J. W., Wang, A. Z., Radovic-Moreno, A. F., et al. (2008). Self-assembled lipid-polymer hybrid nanoparticles: A robust drug delivery platform, *ACS Nano*, **2**, pp. 1696–1702.

112. Liao, Z. Y., Wang, H. J., Wang, X. D., Zhao, P. Q., Wang, S., Su, W. Y., et al. (2011). Multifunctional nanoparticles composed of a poly(D,L-lactide-coglycolide) core and a paramagnetic liposome shell for simultaneous

magnetic resonance imaging and targeted therapeutics, *Adv. Funct. Mater.*, **21**, pp. 1179–1186.

113. Burt, H. M., Zhang, X., Toleikis, P., Embree, L., Hunter, W. L. (1999). Development of copolymers of poly(D,L-lactide) and methoxypoly-ethylene glycol as micellar carriers of paclitaxel, *Colloids Surf. B Biointerfaces*, **16**, pp. 161–171.
114. Bronich, T. K., Keifer, P. A., Shlyakhtenko, L. S., Kabanov, A. V. (2005). Polymer micelle with cross-linked ionic core, *J. Am. Chem. Soc.*, **127**, pp. 8236–8237.
115. Iijima, M., Nagasaki, Y., Okada, T., Kato, M., Kataoka, K. (1999). Core-polymerized reactive micelles from heterotelechelic amphiphilic block copolymers, *Macromolecules*, **32**, pp. 1140–1146.
116. Yang, R., Meng, F. H., Ma, S. B., Huang, F. S., Liu, H. Y., Zhong, Z. Y. (2011). Galactose-decorated cross-linked biodegradable poly(ethylene glycol)-b-poly(epsilon-caprolactone) block copolymer micelles for enhanced hepatoma-targeting delivery of paclitaxel, *Biomacromolecules*, **12**, pp. 3047–3055.
117. Cheng, R., Meng, F. H., Deng, C., Klok, H. A., Zhong, Z. Y. (2013). Dual and multi-stimuli responsive polymeric nanoparticles for programmed site-specific drug delivery, *Biomaterials*, **34**, pp. 3647–3657.
118. Lai, C. Y., Trewyn, B. G., Jeftinija, D. M., Jeftinija, K., Xu, S., Jeftinija, S., et al. (2003). A mesoporous silica nanosphere-based carrier system with chemically removable CdS nanoparticle caps for stimuli-responsive controlled release of neurotransmitters and drug molecules, *J. Am. Chem. Soc.*, **125**, pp. 4451–4459.
119. Gruenhagen, J. A., Lai, C. Y., Radu, D. R., Lin, V. S., Yeung, E. S. (2005). Real-time imaging of tunable adenosine 5-triphosphate release from an MCM-41-type mesoporous silica nanosphere-based delivery system, *Appl. Spectrosc.*, **59**, pp. 424–431.
120. Li, Y. L., Zhu, L., Liu, Z. Z., Cheng, R., Meng, F. H., Cui, J. H., et al. (2009). Reversibly stabilized multifunctional dextran nanoparticles efficiently deliver doxorubicin into the nuclei of cancer cells, *Angew. Chem. Int. Edit.*, **48**, pp. 9914–9918.
121. Meng, F. H., Cheng, R., Deng, C., Zhong, Z. Y. (2012). Intracellular drug release nanosystems, *Mater. Today*, **15**, pp. 436–442.
122. Campo, E., Munoz, J., Miquel, R., Palacin, A., Cardesa, A., Sloane, B. F., et al. (1994). Cathepsin B expression in colorectal carcinomas correlates with tumor progression and shortened patient survival, *Am. J. Pathol.*, **145**, pp. 301–309.

123. Glangchai, L. C., Caldorera-Moore, M., Shi, L., Roy, K. (2008). Nanoimprint lithography based fabrication of shape-specific, enzymatically-triggered smart nanoparticles, *J. Control. Release*, **125**, pp. 263–272.

124. Danhier, F., Feron, O., Preat, V. (2010). To exploit the tumor microenvironment: Passive and active tumor targeting of nanocarriers for anti-cancer drug delivery, *J. Control. Release*, **148**, pp. 135–146.

125. Singh, N., Karambelkar, A., Gu, L., Lin, K., Miller, J. S., Chen, C. S., et al. (2011). Bioresponsive mesoporous silica nanoparticles for triggered drug release, *J. Am. Chem. Soc.*, **133**, pp. 19582–19585.

126. Rim, H. P., Min, K. H., Lee, H. J., Jeong, S. Y., Lee, S. C. (2011). pH-Tunable calcium phosphate covered mesoporous silica nanocontainers for intracellular controlled release of guest drugs, *Angew. Chem. Int. Ed. Engl.*, **50**, pp. 8853–8857.

127. Zhu, L., Torchilin, V. P. (2013). Stimulus-responsive nanopreparations for tumor targeting, *Integr. Biol-Uk*, **5**, pp. 96–107.

128. Spallarossa, P., Altieri, P., Garibaldi, S., Ghigliotti, G., Barisione, C., Manca, V., et al. (2006). Matrix metalloproteinase-2 and -9 are induced differently by doxorubicin in H9c2 cells: The role of MAP kinases and NAD(P)H oxidase, *Cardiovasc. Res.*, **69**, pp. 736–745.

129. Torchilin, V. P., Zhou, F., Huang, L. (1993). pH-sensitive liposomes, *J. Liposome Res.*, **3**, pp. 201–255.

130. Yang, J., Chen, H. T., Vlahov, I. R., Cheng, J. X., Low, P. S. (2007). Characterization of the pH of folate receptor-containing endosomes and the rate of hydrolysis of internalized acid-labile folate-drug conjugates, *J. Pharmacol. Exp. Ther.*, **321**, pp. 462–468.

131. Sankaranarayanan, J., Mahmoud, E. A., Kim, G., Morachis, J. M., Almutairi, A. (2010). Multiresponse strategies to modulate burst degradation and release from nanoparticles, *ACS Nano*, **4**, pp. 5930–5936.

132. Dai, J., Lin, S. D., Cheng, D., Zou, S. Y., Shuai, X. T. (2011). Interlayer-crosslinked micelle with partially hydrated core showing reduction and pH dual sensitivity for pinpointed intracellular drug release, *Angew. Chem. Int. Ed. Engl.*, **50**, pp. 9404–9408.

133. Klaikherd, A., Nagamani, C., Thayumanavan, S. (2009). Multi-stimuli sensitive amphiphilic block copolymer assemblies, *J. Am. Chem. Soc.*, **131**, pp. 4830–4838.

134. Gu, Y. D., Zhong, Y. N., Meng, F. H., Cheng, R., Deng, C., Zhong, Z. Y. (2013). Acetal-linked paclitaxel prodrug micellar nanoparticles as a versatile

and potent platform for cancer therapy, *Biomacromolecules*, **14**, pp. 2772–2780.

135. Tong, R., Cheng, J. (2008). Paclitaxel-initiated, controlled polymerization of lactide for the formulation of polymeric nanoparticulate delivery vehicles, *Angew. Chem. Int. Ed. Engl.*, **47**, pp. 4830–4834.
136. Ernsting, M. J., Tang, W. L., MacCallum, N., Li, S. D. (2011). Synthetic modification of carboxymethylcellulose and use there of to prepare a nanoparticle forming conjugate of docetaxel for enhanced cytotoxicity against cancer cells, *Bioconjug. Chem.* **22**, pp. 2474–2486.
137. Murakami, M., Ernsting, M. J., Undzys, E., Holwell, N., Foltz, W. D., Li, S. D. (2013). Docetaxel conjugate nanoparticles that target alpha-smooth muscle actin-expressing stromal cells suppress breast cancer metastasis, *Cancer Res.*, **73**, pp. 4862–4871.
138. Cheetham, A. G., Zhang, P., Lin, Y. A., Lock, L. L., Cui, H. (2013). Supramolecular nanostructures formed by anticancer drug assembly, *J. Am. Chem. Soc.*, **135**, pp. 2907–2910.
139. Stella, V. J., Nti-Addae, K. W. (2007). Prodrug strategies to overcome poor water solubility, *Adv. Drug Deliv. Rev.*, **59**, pp. 677–694.
140. Kim, S., Kim, J. H., Jeon, O., Kwon, I. C., Park, K. (2009). Engineered polymers for advanced drug delivery, *Eur. J. Pharm. Biopharm.*, **71**, pp. 420–430.

Section C

Drug Transport into Cells and Subsequent Intracellular Trafficking

Chapter 10

Membrane Lipids and Drug Transport

Vinod Labhasetwar
Department of Biomedical Engineering, Lerner Research Institute, Cleveland Clinic, Cleveland, OH 44195, USA
labhasv@ccf.org

Cell membrane lipids are involved in a variety of cellular activities. However, the role of membrane lipids in drug discovery and drug delivery has not been fully explored. In this chapter, we review biophysical interactions of drugs and nanoparticles with membrane lipids and discuss how these interactions influence drug delivery. Biophysical interaction studies can improve our understanding of the role of membrane lipids in the transport of drugs/nanocarriers across biological barriers and can provide a simple yet effective mechanistic and rational approach for designing efficient drugs/nanocarrier systems for targeting.

10.1 Introduction

Lipids are involved in a broad spectrum of cellular activities such as membrane trafficking, signal transduction [1], regulating the activities of membrane proteins (e.g., P-glycoprotein) [2], and recruitment of peripheral (extrinsic) proteins (e.g., G protein). The significance of

Drug Delivery Across Physiological Barriers
Edited by Silvia Muro

ISBN 978-981-4669-40-5 (Hardcover), 978-981-4669-41-2 (eBook)
www.panstanford.com

lipids in various disease conditions prompted the NIH-funded multi-institutional initiative, LIPID MAPS, to identify new paradigms for diseases such as cancer, diabetes, and atherosclerosis and to develop novel therapies targeted at lipids, known as *membrane-lipid therapy* [3].

Lipid alterations have been reported in cancer [4, 5], diabetes [1], and cardiovascular disorders such as hypertension, atherosclerosis, and coronary heart disease [6–8]. In cancer, the onset of disease may be genetic, but significant evidence shows that changes occur in the lipid profile and biophysical properties of membrane lipids during malignant transformation. For example, anionic phospholipids, which are largely absent from the external leaflet of plasma membrane of mammalian cells under normal conditions, are also exposed during malignant transformation and stress conditions of the tumor microenvironment [9]. These changes in tumor tissue lipids have been investigated to understand the mechanism of cancer progression and metastasis, for prognostic evaluation of the disease state [10] and to monitor response to drug/radiation therapies [11]. There is also significant evidence suggesting the role of faulty lipid metabolism in neurodegenerative diseases [12].

In this chapter, we highlight the significance of biophysics of drug/nanoparticle (NP)–lipid interactions in drug delivery. These interactions could depend on biophysical characteristics of cell membrane as well as on physical/interfacial properties of NPs. Biophysical interaction studies can improve our basic understanding of the role of membrane lipids in drug/nanocarrier transport across biological barriers and could provide a new approach for designing efficient and target-specific drugs/nanocarrier systems.

10.2 Membrane Lipids

Biological membranes contain lipid bilayers as their basic structural unit. Lipid bilayers are sheet-like assemblies of thousands of amphiphilic lipid molecules held together by hydrophobic interactions between their acyl chains. Such bilayers form the boundaries between intracellular cytoplasm and the cell's outside environment, as well as between the interior of many of the cell's organelles and their cytoplasm. Although this lipid bilayer structure was first recognized as the basis for cell membrane architecture in 1925 [13], detailed study of the lipid structure of cell membranes and their dynamic and functional nature began very recently. Singer and Nicholson [14] in 1972 first proposed a fluid mosaic model, which suggested that lipids and proteins diffuse freely within the plane of the cell membrane. Discovery of large membrane domains (e.g., basal, lateral, and apical membrane regions of glandular, endothelial, and epithelial cells) and lateral microdomain structures (e.g., lipid rafts, caveolae, and coated pits) revealed the complex nature of the cell membrane structure [15, 16]. In the last decade or so, it has been realized that cell membrane lipids are not simply the physical scaffold for supporting membrane proteins or act as a barrier that isolate and define cells and organelles but, rather, are dynamic and actively involved in a broad spectrum of cellular functions, including the activity of membrane proteins [1]. In recent years, there has been significant interest in understanding cancer progression and other diseases as well as biological processes on the basis of physical and biophysical principles (*Nature Reprint Collections, Physical Sciences in Oncology, Dec 2011*).

10.3 Models Used for Biophysical Interaction Studies

The dynamic nature and complexity of the processes involved in drug/nanomaterial uptake by cells and the intricacy of the cell membranes make these biophysical interactions more difficult to study and understand in real time. Therefore, biomimetic model cell membranes are used to study such interactions. Among the various model cell membranes, the most common ones are supported lipid bilayers, liposome membranes, and the lipid monolayers [17, 18]. Suitability of the model to use also depends on the purpose of investigation. A liposome-based model is most commonly used to study interaction with drugs. In Langmuir lipid monolayer model, where lipids are compressed to biological surface pressure (SP = 30), lipid molecule composition, subphase composition, and temperature can be set to imitate biological conditions so that the data obtained from this model membrane are relevant and can be

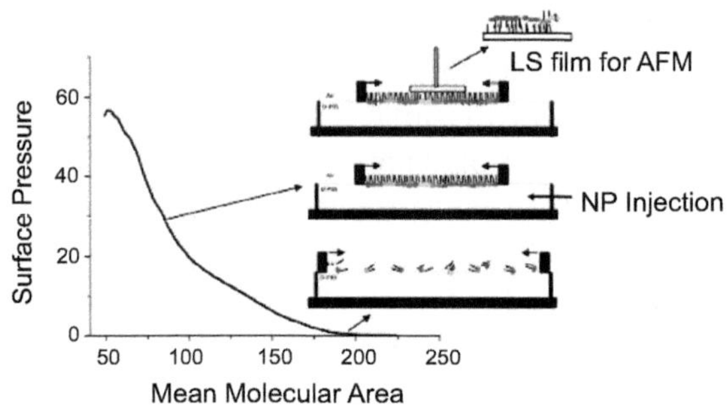

Figure 10.1 Schematic of the Langmuir model to study the biophysical interactions of NPs with model membrane. Interactions can be monitored by change in isotherm and from the Langmuir–Schaefer film transferred on to a substrate by atomic force microscopy (AFM). Reprinted with permission from Ref. [19]. Copyright 2009, American Chemical Society.

useful in predicting interactions with cell membranes, both *in vitro* and *in vivo*. Further, lipid monolayer in this model can be transferred to a substrate before and after interaction to understand the nature of lipid–drug/NP interactions using atomic force microscopy (AFM) (Fig. 10.1) [19]. In addition, one can study the lipid domain structures of membrane lipids isolated from different cells such as cancer and normal cells, which can help in understanding the biophysical characteristics of cell membrane. In our study, we have used this model to study interaction of drugs/NPs of different physical characteristics with model membranes [20] as well as with the lipids extracted from cancer (drug sensitive and resistant) and normal cells to determine whether NPs of specific characteristics demonstrate selective interactions with lipids of cancer cells [21]. Such interaction studies could provide a new strategy for developing target-specific NPs.

10.4 Drug–Lipid Interactions

Drug–lipid interactions have been investigated in drug development and discovery to understand the mechanism of drug action and to predict pharmacokinetic properties of drugs, such as their transport, biodistribution, accumulation, and hence efficacy [22]. Drugs with high affinity to the lipids of target tissue are being developed to avoid nonspecific toxicity [11]. A variety of drugs—antibiotics [23–25], antihypertensive drugs [26, 27], antifungal drugs [28], antipsychotic drugs [29, 30], and anticancer agents [31–33]—have been investigated for their interaction with lipids using different biophysical techniques in drug discovery efforts or to synthesize more effective derivatives. This model of drug–lipid interaction is considered to provide more relevant information than octanol/water partition coefficient that is conventionally used to understand the structure–activity correlation. Many drug molecules have intracellular targets, and hence they must pass across one or more phospholipid bilayers to reach the intracellular targets and elicit a biological response; hence, drug–lipid interactions are inevitable.

Lipid composition differences with different cells and tissue or normal versus diseased tissue, as well as their effect on the biophysical interactions between drugs, show the importance of such studies. Developing drugs that specifically bind to particular lipids or lipid domain structures in diseased cells or tissue is a new therapeutic approach. This also signifies the importance of choosing model membranes that closely resemble the tissue or cells under consideration to ensure that the behavior seen with lipid model membrane reflects the behavior of cells *in vitro* and/or *in vivo*. Conventionally, biophysical investigations of drug with lipid model membranes are retrospective since most studies are performed after the drug behavior *in vitro* or *in vivo* is known in order to understand the mechanism of drug efficacy. However, with more data and a good correlation between drug–lipid interactions, biophysical interaction studies can become a powerful, rapid, and cost-effective predictive technique for drug discovery and drug development.

In our study, we isolated lipids from doxorubicin-sensitive (MCF-7) and- resistant (MCF-7/ADR) breast cancer cells to characterize the changes in biophysical properties of membrane lipids (particularly lipid packing and membrane fluidity) and to understand the role of the interaction of cell membrane lipids with drug on drug transport and efficacy. Resistant cell membrane lipids showed significantly different composition, were high in cholesterol and sphingomyelin, and formed more condensed, less fluid monolayers than did lipids from sensitive cells. Doxorubicin, a most commonly used drug for the treatment of breast cancer, showed a strong hydrophobic interaction with resistant cell membrane lipids but significantly less interaction, as well as a different pattern of interaction (i.e., ionic), with sensitive cell membrane lipids. In addition to the biophysical characteristics of resistant cell membrane lipids, lipid–doxorubicin interactions appeared to decrease intracellular drug transport via diffusion as the drug is trapped in the lipid bilayer [34]. This may have happened due to the hydrophobic nature of lipids of resistant cell membrane. The drug transport in resistant cells occurred only at high concentration of the drug, perhaps only after the membrane is saturated with the drug. Although there are different mechanisms of drug resistance, based on our results in these breast cancer cells, we suggested that cancer cells acquire a drug-resistant phenotype through the alteration of lipid synthesis to inhibit intracellular drug transport to protect from cytotoxic effect of the drug [35].

In certain cancers, epigenetic changes (e.g., DNA hypermethylation) are essential to maintain drug-resistant phenotype. Therefore, altered lipid synthesis in this cell line may be linked to epigenetic mechanisms of drug resistance. We demonstrated that treating MCF-7/ADR cells with the epigenetic drug 5-aza-2′-deoxycytidine (decitabine) significantly altered cell lipid composition and biophysical properties, causing the resistant cells to acquire biophysical characteristics similar to those of sensitive cell lipids. Following decitabine treatment, resistant cells demonstrated increased sphingomyelinase activity, resulting in a decreased sphingomyelin level that influenced lipid domain structures, increased membrane fluidity, and reduced P-glycoprotein expression. Changes in the biophysical characteristics of resistant cell lipids facilitated doxorubicin transport, enhancing the drug efficacy. Further, resistant cells regained endocytic function following treatment with epigenetic drug, which is impaired due to rigid nature of resistant cell membrane [36] (Fig. 10.2). Based on our results, we proposed a new mechanism of efficacy of an epigenetic drug that is mediated through changes in lipid metabolism, changing membrane compo-

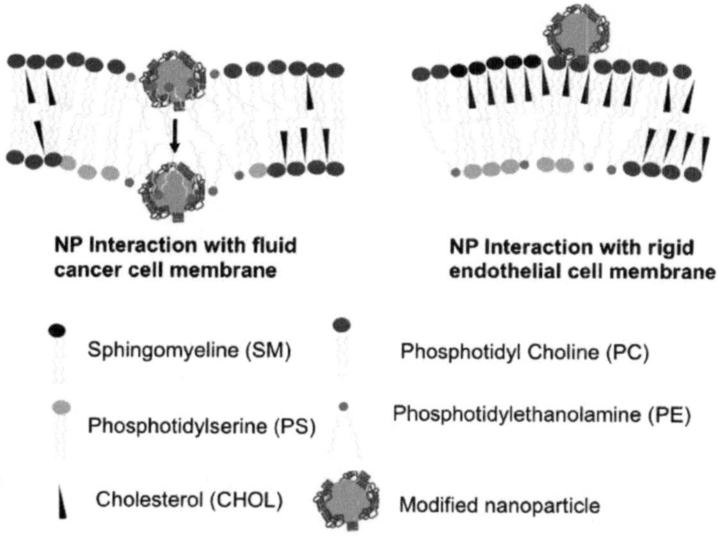

Figure 10.2 Schematic depicting how membrane rigidity influences nanoparticle internalization process. Treating breast cancer resistant cells with epigenetic drug altered lipid synthesis that increased membrane fluidity, thus regaining the endocytic function.

sition and hence biophysical properties in reversing cancer drug resistance.

10.5 Significance of Biophysical Interactions with Nanomaterials

Nanoscale carrier systems, such as NPs, block copolymer micelles, and liposomes, with different surface composition and characteristics are being investigated for their various biomedical applications, including drug delivery. Interfacial properties of NPs—such as their surface charge [37–39], presence of surface functional groups [40, 41], particle size [42, 43], and surface hydrophilicity/hydrophobicity [44, 45]—are known to influence the efficiency of NPs in transporting biotherapeutic agents to target tissue by affecting their interactions with biological environment.

10.5.1 *Effect of Size and Charge of Nanoparticles on Biophysical Interactions*

We developed an endothelial model cell membrane (EMM) using a mixture of lipids that is typically present in normal endothelial cells and Langmuir balance to study its interaction with NPs. For our initial studies, we used polystyrene NPs of different surface chemistry and sizes as a model nanomaterial, and changes in the membrane's SP were used as a parameter to monitor its interactions with NPs. Our results demonstrated that aminated NPs (60 nm) increased SP, plain NPs reduced it, and carboxylated NPs of the same size had no effect. However, smaller NPs (20 nm) increased SP irrespective of surface chemistry and whether the interaction studies were carried out in the presence or absence of fetal bovine serum (FBS), whereas the presence of FBS masked the effect of interaction of larger ($\geq$60 nm) plain and carboxylated, but not of aminated, NPs. The particular effect of NP characteristics on SP, determined using AFM and π-A (surface pressure-area) isotherm, explained whether the interaction results in condensation of phospholipids (increase in SP) or their displacement from the interface into the subphase (decrease in SP), causing destabilization of the membrane. Displacement of lipids could be the cause of cytotoxicity. Based on the change in SP, the interaction of NPs with the EMM significantly depends on their surface characteristics and sizes. In general, small aminated NPs and plain NPs have greater interactions with the EMM than do carboxylated and large plain NPs. Further, FBS does not seem to influence the interaction of small NPs with the membrane [46]. These results thus suggested the role of protein corona that is formed around NPs following interactions with biological fluids on biophysical interactions with membrane lipids, but the effect depends on physical characteristics of NPs. In our ongoing studies, we are identifying the composition of protein carona formed around NPs of different surface characteristics to determine what role they might play in interactions with cells, biodistribution, and targeting. However, this is a complex process, not only because of the number of proteins involved but also due to the dynamic nature of protein–NP interactions [47].

10.5.2 *Effect of Molecular Structure of Cationic Surfactant on Biophysical Interactions of Surface-Modified Nanoparticles*

NPs are surface modified with different polymers and peptides, including cationic surfactants to increase their interactions with anionic cell membrane. In this study, we demonstrated that the molecular structure of cationic surfactants at the NP interface influences the biophysical interactions of NPs with a model membrane and cellular uptake of NPs. Polystyrene NPs (surfactant free, 130 nm) were modified with cationic surfactants. These surfactants were of either di-chained [didodecyldimethyl-ammonium bromide (DMAB)] or single-chained [cetyltrimethylammonium bromide (CTAB) and dodecyltrimethyl-ammonium bromide (DTAB)] forms, the latter two with different hydrophobic chain lengths. Changes in the SP of EMM as a function of time following interaction with NPs and in the compression isotherm (π-A) of the lipid mixture of EMM in the presence of NPs were analyzed. Langmuir–Schaeffer (LS) films, which are EMMs that have been transferred onto a suitable substrate, were imaged by AFM, and the images were analyzed to determine the mechanisms of the NP–EMM interaction. DMAB-modified NPs showed a greater increase in SP and a shift toward higher mean molecular area than CTAB- and DTAB-modified NPs, indicating stronger interactions of DMAB-modified NPs with the EMM. However, analysis of the AFM phase and height images of the LS films revealed that both DMAB- and CTAB-modified NPs interacted with the EMM but via different mechanisms: DMAB-modified NPs penetrated the EMM, thus explaining the increase in SP, whereas CTAB-modified NPs anchored onto the EMM's condensed lipid domains and hence did not cause any significant change in SP. Human umbilical vein endothelial cells (HUVECs) showed greater uptake of DMAB- and CTAB-modified NPs than of DTAB-modified or unmodified NPs. We are currently investigating the biomechanics and thermodynamics of interactions of NPs with model membrane and its role in endocytic process of cellular uptake of DMAB-modified NPs as compared to CTAB-modified and unmodified NPs. From this study, we concluded that (1) the di-chained and single-chained cationic surfactants on NPs have different mechanisms of interaction with the model membrane than single-chain and unmodified NPs and (2) NPs that demonstrate greater biophysical interactions with the membrane also show greater cellular uptake.

Surface charge of NPs is conventionally used as a predictor of NP-cell interactions. Here we demonstrated that it is not the charge since all the surfactants used here imparted cationic charge to NPs, but the molecular structure of surfactant at the NP interface determines the extent and mechanism of interaction with lipid membrane [19]. This has to do with how surfactant molecules arrange at the interface and how that arrangement influences biophysical interactions (Fig. 10.3). Based on the nature of interactions, cellular uptake and intracellular trafficking of NPs could also vary, which could influence the efficacy of the encapsulated therapeutics. The results of these studies illustrate that biophysical interactions of NPs with a model membrane could thus be effectively used for developing nanocarriers with optimized surface properties for drug delivery applications.

10.5.3 *Peptide-Modified Nanoparticles and Biophysical Interactions with Model Membrane*

Functionalization of nanocarriers with cell-penetrating peptides (CPPs) is one of the successful strategies that has been developed to overcome the low cellular permeability of the encapsulated agents. Antennapedia [48], transportan [49, 50], and HIV-1 TAT (trans-activating transcriptor) peptide [51–53] are examples of different CPPs with membrane-translocating properties that are being used to transport different cargoes. CPP-mediated cellular delivery has been demonstrated to occur rapidly across various cell types, and specific integral protein transduction domains of the CPPs are responsible for their cellular uptake. Among the different CPPs with membrane-translocating properties, the HIV-1 TAT peptide and its derivatives are highly utilized for cellular delivery [54]. In our study, we tested the effects of the TAT peptide sequence and the amount of peptide conjugated to NPs on biophysical interactions with EMM, and the HUVECs were used to determine the uptake of the encapsulated therapeutic. Ritonavir was chosen as a model drug since it possesses

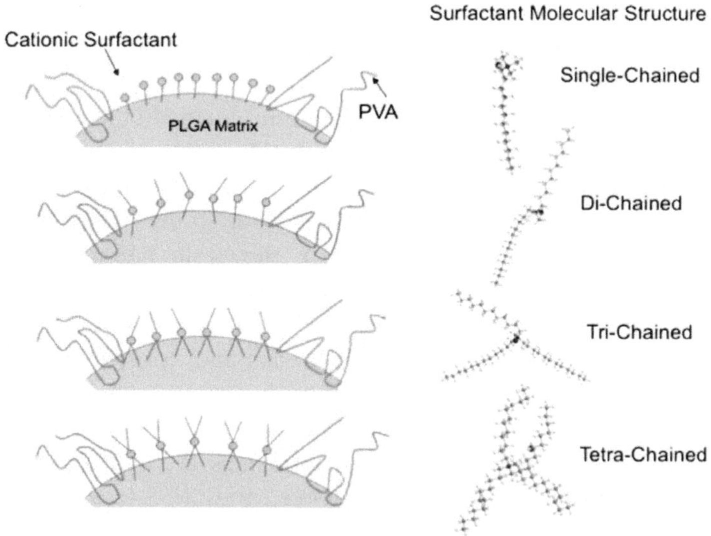

Figure 10.3 Schematic representation of how cationic surfactants depending on their structure could assemble at the nanoparticle interface, which could influence biophysical interaction of surface-modified NPs with membrane lipids.

limited cellular permeability and transport, attributed mainly to its P-gp-mediated efflux [55]. Our results demonstrate that the TAT peptide sequence and the amount of TAT conjugated to NPs significantly affect the biophysical interactions of NPs with the EMM, and these interactions correlated with the cellular delivery of the encapsulated drug [56]. The broader significance of the study lies in optimizing the peptide sequence and the amount to be conjugated to nanocarriers using biophysical interaction studies.

The central nervous system (CNS) serves as one of the principal anatomical reservoirs for the replicating HIV-1 virus [57]. The virus enters the CNS at an early stage of infection and re-infects the peripheral tissues resulting in reactivation of the infection [58]. A spectrum of CNS complications ensue due to the residence of HIV within the CNS, including progressive dementia, memory loss, HIV-mediated encephalopathy, and cerebrovascular complications [59]. These neurological dysfunctions contribute toward significant morbidity and mortality among HIV-infected individuals [60]. We have shown that TAT-peptide-conjugated NPs, as characterized above, bypassed the efflux action of P-glycoprotein and increased the transport of the encapsulated ritonavir across the blood–brain barrier (BBB) to the CNS [61]. We found a steady increase in the drug parenchyma/capillary ratio with time without disrupting the BBB integrity, which suggests that TAT-conjugated NPs are first immobilized in the brain vasculature prior to their transport into parenchyma. Localization of NPs in the brain parenchyma was further confirmed with histological analysis of the brain sections. The brain drug level with conjugated NPs was 800-fold higher than that with drug in solution at two weeks and drug clearance was seen within four weeks. This data thus demonstrate the significance of biophysical characterization of peptide-conjugated NPs to their *in vivo* efficacy in transporting drug across the BBB to the CNS. TAT-conjugated NPs not only enhanced the CNS bioavailability of the encapsulated protease inhibitor but also maintained therapeutic drug level in the brain for a sustained period. These NPs could very well be effective in reducing the viral load in the CNS, which acts as a reservoir for replicating HIV-1 virus.

10.6 Specificity of Biophysical Interactions with Cancer Cell Membrane Lipids versus Normal Cell Lipids

The success of NP-based delivery systems for cancer therapy relies on effective localization and uptake of NPs within tumors. Conventional strategies for tumor targeting of NPs typically exploit the proteins/receptors overexpressed on tumor cells that distinguish them from normal cells. Ligands, such as antibodies, peptides, or aptamers, may be conjugated to the NP surface to actively bind to specific molecules on the surface of cancer cells or the tumor vasculature [62]. This approach has met with some success compared with similar nontargeted/passively targeted drug delivery systems [63]; however, several limitations exist: (1) the number of targeting sites available or accessible may be limited and insufficient for achieving a therapeutic dose of NPs in tumor tissue; (2) ligands with low affinity binding for their receptor can limit targeting efficiency; and (3) slow recycling of receptors may limit the dose of NPs that can be delivered to the tumor site at one time.

We investigated a targeting strategy that takes advantage of the overall differences in the biophysical properties of cell membranes of malignant versus normal cells, rather than aiming at specific molecules. Cell membrane properties are determined, to a large extent, by their lipid composition. Lipid metabolism is altered during the malignant transformation of cells; in this context, tumor cells undergo increased metabolic demands, adapt to hypoxia, and respond to specific oncogenic pathways [64]. Consequently, the lipid composition and biophysical characteristics of the cell membranes of cancerous versus normal cells differ substantially. Targeting gene- or drug-loaded NPs to tumors and ensuring their intratumoral retention after systemic administration remain key challenges to improving the efficacy of NP-based therapeutics. We exploited changes in lipid metabolism and cell membrane biophysics that occur during malignancy for developing tumor-targeted NPs. We have demonstrated that different surfactants, incorporated onto the NPs' surface, affect the biophysical interactions of NPs with the lipids of cancer cells and normal HUVECs. NPs surface modified with DMAB, a dichain cationic surfactant, demonstrated greater interaction with cancer cell lipids, which was 6.7-fold greater than with unmodified NPs and 5.5-fold greater than with endothelial cell lipids. This correlated with increased uptake of DMAB-modified NPs with incubation time by cancer cells (PC-3) compared to other formulations of NPs and to uptake by endothelial cells. Upon systemic injection, DMAB-NPs demonstrated a 4.6-fold increase in tumor accumulation compared to unmodified NPs, which also correlated to improved efficacy of p53 gene therapy [21]. The results of this study thus suggest that characterization of the biophysical interactions between NPs and lipid membranes of tumors or other diseased tissues/organs may hold promise for engineering targeted delivery of therapeutics.

10.7 Conclusions and Perspectives

There is growing body of evidence supporting the major role of membrane lipids in various cellular functions. Changes in lipid biosynthesis in various disease conditions could significantly influence biophysical properties of the cell membrane that could impact drug transport and endocytic functions. Understanding such changes in membrane lipids and their effects on drug delivery could be of importance in drug design and development and for developing effective nanocarrier systems. With additional data, it may be feasible to develop a computer model that could correlate biophysical properties of cell membrane, physical/interfacial characteristics of NPs to cellular uptake, toxicity, biodistribution, and targeting of NPs. Such a model could be a powerful tool in designing NPs for specific characteristics for drug delivery applications.

Acknowledgment

The work described here from authors' laboratory is funded by grant R01CA149359 (to V.L.) from the National Cancer Institute of the National Institutes of Health.

References

1. Vigh, L., Escriba, P. V., Sonnleitner, A., Sonnleitner, M., Piotto, S., Maresca, B., Horvath, I., and Harwood, J. L. (2005). The significance of lipid composition for membrane activity: New concepts and ways of assessing function, *Prog. Lipid Res.*, **44**, pp. 303–344.
2. Romsicki, Y., and Sharom, F. J. (1999). The membrane lipid environment modulates drug interactions with the P-glycoprotein multidrug transporter, *Biochemistry*, **38**, pp. 6887–6896.
3. van Meer, G., Voelker, D. R., and Feigenson, G. W. (2008). Membrane lipids: Where they are and how they behave, *Nat. Rev. Mol. Cell Biol.*, **9**, pp. 112–124.
4. Patel, P. S., Shah, M. H., Jha, F. P., Raval, G. N., Rawal, R. M., Patel, M. M., Patel, J. B., and Patel, D. D. (2004). Alterations in plasma lipid profile patterns in head and neck cancer and oral precancerous conditions, *Indian J. Cancer*, **41**, pp. 25–31.
5. Shah, F. D., Shukla, S. N., Shah, P. M., Patel, H. R., and Patel, P. S. (2008). Significance of alterations in plasma lipid profile levels in breast cancer, *Integr. Cancer Ther.*, **7**, pp. 33–41.

6. Escriba, P. V., Sanchez-Dominguez, J. M., Alemany, R., Perona, J. S., and Ruiz-Gutierrez, V. (2003). Alteration of lipids, G proteins, and PKC in cell membranes of elderly hypertensives, *Hypertension,* **41**, pp. 176–182.
7. Ma, Q. F., and Yang, F. Y. (1997). Lipids may not be involved in the recognition of apocytochrome C during its transportation, *Biochem. Mol. Biol. Int.,* **41**, pp. 487–496.
8. Morrill, G. A., Gupta, R. K., Kostellow, A. B., Ma, G. Y., Zhang, A., Altura, B. T., and Altura, B. M. (1997). Mg^{2+} modulates membrane lipids in vascular smooth muscle: A link to atherogenesis, *FEBS Lett.,* **408**, pp. 191–194.
9. Ran, S., Downes, A., and Thorpe, P. E. (2002). Increased exposure of anionic phospholipids on the surface of tumor blood vessels, *Cancer Res.,* **62**, pp. 6132–6140.
10. Preetha, A., Banerjee, R., and Huilgol, N. (2005). Surface activity, lipid profiles and their implications in cervical cancer, *J. Cancer Res. Ther.,* **1**, pp. 180–186.
11. Preetha, A., Banerjee, R., and Huilgol, N. (2005). Dynamic surface tensiometry of tissues using Langmuir films, *Colloids Surf. B. Biointerfaces,* **40**, pp. 35–43.
12. Yadav, R. S., and Tiwari, N. K. (2014). Lipid integration in neurodegeneration: An overview of Alzheimer's disease, *Mol. Neurobiol.,* PMID: 24590317.
13. Gorter, E., and Grendel, F. (1925). On bimolecular layers of lipoids on the chromocytes of the blood, *J. Expt. Med.,* **41**, pp. 439–443.
14. Singer, S. J., and Nicolson, G. L. (1972). The fluid mosaic model of the structure of cell membranes, *Science,* **175**, pp. 720–731.
15. Escriba, P. V. (2006). Membrane-lipid therapy: A new approach in molecular medicine, *Trends Mol. Med.,* **12**, pp. 34–43.
16. Vereb, G., Szollosi, J., Matko, J., Nagy, P., Farkas, T., Vigh, L., Matyus, L., Waldmann, T. A., and Damjanovich, S. (2003). Dynamic, yet structured: The cell membrane three decades after the Singer–Nicolson model, *Proc. Natl. Acad. Sci., U.S.A.,* **100**, pp. 8053–8058.
17. Brezesinski, G., and Mohwald, H. (2003). Langmuir monolayers to study interactions at model membrane surfaces, *Adv. Colloid Interface Sci.,* **100-102**, pp. 563–584.
18. Marsh, D. (1996). Intrinsic curvature in normal and inverted lipid structures and in membranes, *Biophys. J.,* **70**, pp. 2248–2255.
19. Peetla, C., and Labhasetwar, V. (2009). Effect of molecular structure of cationic surfactants on biophysical interactions of surfactant-modified

nanoparticles with a model membrane and cellular uptake, *Langmuir,* **25**, pp. 2369–2377.

20. Peetla, C., and Labhasetwar, V. (2008). Biophysical characterization of nanoparticle-endothelial model cell membrane interactions, *Mol. Pharmaceutics,* **5**, pp. 418–429.
21. Sharma, B., Peetla, C., Adjei, I. M., and Labhasetwar, V. (2013). Selective biophysical interactions of surface modified nanoparticles with cancer cell lipids improve tumor targeting and gene therapy, *Cancer Lett.,* **334**, pp. 228–236.
22. Peetla, C., Stine, A., and Labhasetwar, V. (2009). Biophysical interactions with model lipid membranes: Applications in drug discovery and drug delivery, *Mol. Pharm.,* **6**, pp. 1264–1276.
23. Grancelli, A., Morros, A., Cabanas, M. E., Domenech, O., Merino, S., Vazquez, J. L., Montero, M. T., Vinas, M., and Hernandez-Borrell, J. (2002). Interaction of 6-fluoroquinolones with dipalmitoylphosphatidylcholine monolayers and liposomes, *Langmuir,* **18**, pp. 9177–9182.
24. Berquand, A., Fa, N., Dufrene, Y. F., and Mingeot-Leclercq, M. P. (2005). Interaction of the macrolide antibiotic azithromycin with lipid bilayers: Effect on membrane organization, fluidity, and permeability, *Pharm. Res.,* **22**, pp. 465–475.
25. Fa, N., Schanck, A., and Mingeot-Leclecq, M. P. (2005). Interaction of the macrolide antibiotic azithromycin with model of membranes, *Biophys. J.,* **88**, pp. 241a–241a.
26. Barcelo, F., Prades, J., Funari, S. S., Frau, J., Alemany, R., and Escriba, P. V. (2004). The hypotensive drug 2-hydroxyoleic acid modifies the structural properties of model membranes, *Mol. Membr. Biol.,* **21**, pp. 261–268.
27. Cheng, H. Y., Randall, C. S., Holl, W. W., Constantinides, P. P., Yue, T. L., and Feuerstein, G. Z. (1996). Carvedilol-liposome interaction: Evidence for strong association with the hydrophobic region of the lipid bilayers, *Biochim. Biophys. Acta. Biomembranes,* **1284**, pp. 20–28.
28. Corvis, Y., Barzyk, W., Brezesinski, G., Mrabet, N., Badis, M., Hecht, S., and Rogalska, E. (2006). Interactions of a fungistatic antibiotic, griseofulvin, with phospholipid monolayers used as models of biological membranes, *Langmuir,* **22**, pp. 7701–7711.
29. Agasosler, A. V., Tungodden, L. M., Cejka, D., Bakstad, E., Sydnes, L. K., and Holmsen, H. (2001). Chorpromazine-induced increase in dipalmitoylphosphatidylserine surface area in monolayers at room temperature, *Biochem. Pharmacol.,* **61**, pp. 817–825.

30. Hidalgo, A. A., Caetano, W., Tabak, M., and Oliveira, O. N. (2004). Interaction of two phenothiazine derivatives with phospholipid monolayers, *Biophys. Chem.*, **109**, pp. 85–104.
31. Feng, S. S., Gong, K., and Chew, J. (2002). Molecular interactions between a lipid and an antineoplastic drug paclitaxel (taxol) within the lipid monolayer at the air/water interface, *Langmuir*, **18**, pp. 4061–4070.
32. Preetha, A., Huilgol, N., and Banerjee, R. (2006). Comparison of paclitaxel penetration in normal and cancerous cervical model monolayer membranes, *Colloids Surf. B. Biointerfaces*, **53**, pp. 179–186.
33. Preetha, A., Huilgol, N., and Banerjee, R. (2007). Effect of fluidizing agents on paclitaxel penetration in cervical cancerous monolayer membranes, *J. Memb. Biol.*, **219**, pp. 83–91.
34. Peetla, C., Bhave, R., Vijayaraghavalu, S., Stine, A., Kooijman, E., and Labhasetwar, V. (2010). Drug resistance in breast cancer cells: Biophysical characterization of and doxorubicin interactions with membrane lipids, *Mol. Pharmaceutics*, **7**, pp. 2334–2348.
35. Peetla, C., Vijayaraghavalu, S., and Labhasetwar, V. (2013). Biophysics of cell membrane lipids in cancer drug resistance: Implications for drug transport and drug delivery with nanoparticles, *Adv. Drug Deliv. Rev.*, **65**, pp. 1686–1698.
36. Vijayaraghavalu, S., Peetla, C., Lu, S., and Labhasetwar, V. (2012). Epigenetic modulation of the biophysical properties of drug-resistant cell lipids to restore drug transport and endocytic functions, *Mol. Pharmaceutics*, **9**, pp. 2730–2742.
37. Chung, T. H., Wu, S. H., Yao, M., Lu, C. W., Lin, Y. S., Hung, Y., Mou, C. Y., Chen, Y. C., and Huang, D. M. (2007). The effect of surface charge on the uptake and biological function of mesoporous silica nanoparticles in 3T3-L1 cells and human mesenchymal stem cells, *Biomaterials*, **28**, pp. 2959–2966.
38. Kumar, M. N., Mohapatra, S. S., Kong, X., Jena, P. K., Bakowsky, U., and Lehr, C. M. (2004). Cationic poly(lactide-co-glycolide) nanoparticles as efficient *in vivo* gene transfection agents, *J. Nanosci. Nanotechnol.*, **4**, pp. 990–994.
39. Lockman, P. R., Koziara, J. M., Mumper, R. J., and Allen, D. D. (2004). Nanoparticle surface charges alter blood-brain barrier integrity and permeability, *J. Drug Target*, **12**, pp. 635–641.
40. Hauck, T. S., Ghazani, A. A., and Chan, W. C. (2008). Assessing the effect of surface chemistry on gold nanorod uptake, toxicity, and gene expression in mammalian cells, *Small*, **4**, pp. 153–159.

41. Pan, J., and Feng, S. S. (2008). Targeted delivery of paclitaxel using folate-decorated poly(lactide)-vitamin E TPGS nanoparticles, *Biomaterials,* **29**, pp. 2663–2672.
42. Balogh, L., Nigavekar, S. S., Nair, B. M., Lesniak, W., Zhang, C., Sung, L. Y., Kariapper, M. S., El-Jawahri, A., Llanes, M., Bolton, B., Mamou, F., Tan, W., Hutson, A., Minc, L., and Khan, M. K. (2007). Significant effect of size on the *in vivo* biodistribution of gold composite nanodevices in mouse tumor models, *Nanomedicine,* **3**, pp. 281–296.
43. Tabata, Y., and Ikada, Y. (1988). Effect of the size and surface charge of polymer microspheres on their phagocytosis by macrophage, *Biomaterials,* **9**, pp. 356–362.
44. Sahoo, S. K., Panyam, J., Prabha, S., and Labhasetwar, V. (2002). Residual polyvinyl alcohol associated with poly(D,L-lactide-co-glycolide) nanoparticles affects their physical properties and cellular uptake, *J. Control Release,* **82**, pp. 105–114.
45. Woodle, M. C. (1998). Controlling liposome blood clearance by surface-grafted polymers, *Adv. Drug Del. Rev.,* **32**, pp. 139–152.
46. Peetla, C., and Labhasetwar, V. (2008). Biophysical characterization of nanoparticle-endothelial model cell membrane interactions, *Mol. Pharmaceutics,* **5**, pp. 418–429.
47. Tenzer, S., Docter, D., Kuharev, J., Musyanovych, A., Fetz, V., Hecht, R., Schlenk, F., Fischer, D., Kiouptsi, K., Reinhardt, C., Landfester, K., Schild, H., Maskos, M., Knauer, S. K., and Stauber, R. H. (2013). Rapid formation of plasma protein corona critically affects nanoparticle pathophysiology, *Nat. Nanotechnol.,* **8**, pp. 772–781.
48. Derossi, D., Calvet, S., Trembleau, A., Brunissen, A., Chassaing, G., and Prochiantz, A. (1996). Cell internalization of the third helix of the antennapedia homeodomain is receptor-independent, *J. Biol. Chem.,* **271**, pp. 18188–18193.
49. Soomets, U., Lindgren, M., Gallet, X., Hallbrink, M., Elmquist, A., Balaspiri, L., Zorko, M., Pooga, M., Brasseur, R., and Langel, U. (2000). Deletion analogues of transportan, *Biochim. Biophys. Acta. Biomembranes,* **1467**, pp. 165–176.
50. Yandek, L. E., Pokorny, A., Floren, A., Knoelke, K., Langel, U., and Almeida, P. F. F. (2007). Mechanism of the cell-penetrating peptide transportan 10 permeation of lipid bilayers, *Biophys. J.,* **92**, pp. 2434–2444.
51. Ziegler, A., Blatter, X. L., Seelig, A., and Seelig, J. (2003). Protein transduction domains of HIV-1 and SIV TAT interact with charged lipid vesicles. Binding mechanism and thermodynamic analysis, *Biochemistry,* **42**, pp. 9185–9194.

52. Eum, W. S., Jang, S. H., Kim, D. W., Choi, H. S., Choi, S. H., Kim, S. Y., An, J. J., Lee, S. H., Han, K., Kang, J. H., Kang, T. C., Won, M. H., Cho, Y. J., Choi, J. H., Kim, T. Y., Park, J., and Choi, S. Y. (2005). Enhanced transduction of Cu,Zn-superoxide dismutase with HIV-1 Tat protein transduction domains at both termini, *Mol. Cells,* **19**, pp. 191–197.

53. Al-Taei, S., Penning, N. A., Simpson, J. C., Futaki, S., Takeuchi, T., Nakase, I., and Jones, A. T. (2006). Intracellular traffic and fate of protein transduction domains HIV-1 TAT peptide and octaarginine. Implications for their utilization as drug delivery vectors, *Bioconjugate Chem.,* **17**, pp. 90–100.

54. Chen, B., Liu, Q., Zhang, Y., Xu, L., and Fang, X. (2008). Transmembrane delivery of the cell-penetrating peptide conjugated semiconductor quantum dots, *Langmuir,* **24**, pp. 11866–11871.

55. Alsenz, J., Steffen, H., and Alex, R. (1998). Active apical secretory efflux of the HIV protease inhibitors saquinavir and ritonavir in Caco-2 cell monolayers, *Pharma. Res.,* **15**, pp. 423–428.

56. Borgmann, K., Rao, K. S., Labhasetwar, V., and Ghorpade, A. (2011). Efficacy of Tat-conjugated ritonavir-loaded nanoparticles in reducing HIV-1 replication in monocyte-derived macrophages and cytocompatibility with macrophages and human neurons, *AIDS Res. Hum. Retroviruses,* **27**, pp. 853–862.

57. Schrager, L. K., and D'Souza, M. P. (1998). Cellular and anatomical reservoirs of HIV-1 in patients receiving potent antiretroviral combination therapy, *JAMA,* **280**, pp. 67–71.

58. Banks, W. A., Freed, E. O., Wolf, K. M., Robinson, S. M., Franko, M., and Kumar, V. B. (2001). Transport of human immunodeficiency virus type 1 pseudoviruses across the blood–brain barrier: Role of envelope proteins and adsorptive endocytosis, *J. Virol.,* **75**, pp. 4681–4691.

59. Letendre, S., Ances, B., Gibson, S., and Ellis, R. J. (2007). Neurologic complications of HIV disease and their treatment, *Top. HIV Med.,* **15**, pp. 32–39.

60. Patsalides, A. D., Wood, L. V., Atac, G. K., Sandifer, E., Butman, J. A., and Patronas, N. J. (2002). Cerebrovascular disease in HIV-infected pediatric patients: Neuroimaging findings, *AJR Am. J. Roentgenol.,* **179**, pp. 999–1003.

61. Rao, K. S., Reddy, M. K., Horning, J. L., and Labhasetwar, V. (2008). TAT-conjugated nanoparticles for the CNS delivery of anti-HIV drugs, *Biomaterials,* **29**, pp. 4429–4438.

62. Ruoslahti, E., Bhatia, S. N., and Sailor, M. J. (2010). Targeting of drugs and nanoparticles to tumors, *J. Cell Biol.,* **188**, pp. 759–768.

63. Cheng, Z., Al Zaki, A., Hui, J. Z., Muzykantov, V. R., and Tsourkas, A. (2012). Multifunctional nanoparticles: Cost versus benefit of adding targeting and imaging capabilities, *Science*, **338**, pp. 903–910.

64. Santos, C. R., and Schulze, A. (2012). Lipid metabolism in cancer, *FEBS J.*, **279**, pp. 2610–2623.

Chapter 11

Drug Delivery Systems that Fuse with Plasmalemma

Muriel Blanzat,[a] **Marie-Pierre Rols,**[b,c] **and Justin Teissié**[b,c]

[a]*Laboratoire IMRCP, UMR 5623, Université Paul Sabatier, 118 Route de Narbonne, F-31062 Toulouse cedex 9, France*

[b]*CNRS, IPBS (Institut de Pharmacologie et de Biologie Structurale), 205 route de Narbonne BP64182, F-31077 Toulouse, France*

[c]*Université de Toulouse, UPS, IPBS, F-31077 Toulouse, France*

blanzat@chimie.ups-tlse.fr

Membrane fusion is a crucial event occurring between lipid membranes in living organisms. However, so far very few drug delivery systems have been identified to spontaneously fuse with the plasmalemma of cells. The fusion process is often associated with endocytic mechanisms. After a physical description of membrane fusion, this chapter reviews all the natural or synthetic vectors that fuse with membranes, spontaneously or with physical assistance.

11.1 Introduction

The cell membrane, also named plasmalemma, is a selective barrier that regulates the transfer of hydrophilic drugs into the cell. During the last 40 years, a number of drug delivery systems have been

Drug Delivery Across Physiological Barriers
Edited by Silvia Muro

ISBN 978-981-4669-40-5 (Hardcover), 978-981-4669-41-2 (eBook)
www.panstanford.com

developed to control drug release, absorption, and distribution in an effort to improve efficacy and safety. The direct release of active molecules inside the cytoplasm is often considered the most efficient and safest mechanism for drug delivery. This promising strategy, which can proceed through membrane fusion between suitable carriers and the cell membrane, is not straightforward and has resulted in extensive research and development on the design of synthetic delivery systems [1]. Important knowledge has consequently been generated regarding the pivotal membrane fusion step, which is a key event occurring in various intracellular and intercellular processes. This diversity of cellular fusion processes points to the variation of mechanisms that could exist in natural membrane fusion events. However, some general molecular mechanisms appear in all membrane fusion processes. As reported in the following sections, the fusion of drug delivery systems with the plasmalemma is a process characterized by major steps and is governed by energy barriers that have to be overcome to undergo fusion.

11.2 Physical Description of Bilayer Fusion

To increase the knowledge of the molecular mechanisms of membrane fusion processes, a number of biophysical studies have been carried out on simplified membrane models, such as pure lipidic systems [2–5].

It has been found that the fusion of lipid bilayers in an aqueous environment occurs in two main events. First, the two membranes are brought very close. However, the inability to counteract the electrostatic repulsive forces prevents an intimate interaction of the lipids of the proximal leaflets [6]. Second, the boundary between the hydrophilic and hydrophobic portions of each monolayer (i.e., leaflet) must be destabilized. It is assumed that nonbilayer transition states may appear, resulting in local merging of the two bilayers in contact, forming a so-called stalk [7]. This step is then followed either by a very fast formation of an aqueous fusion pore or by the metastable hemifusion intermediate. All transitions between the different putative states are governed by forces that

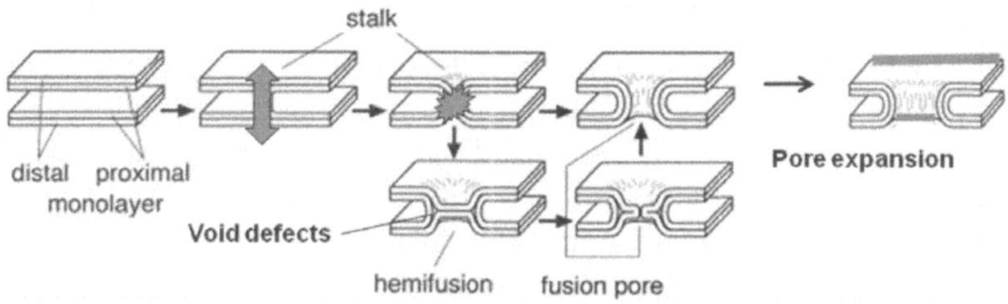

Figure 11.1 Transition states in membrane fusion. The monolayers are smooth and bendable sheets as described by the stalk hypothesis. Reprinted from Ref. 2, Copyright 2003, with permission from Elsevier.

minimize the exposure of nonpolar surfaces to water. According to the stalk-to-fusion pore transition hypothesis (accepted by the scientific community), fusion proceeds in the order of structural transitions that include the merging of the proximal monolayers, stalk formation, and fusion pore opening (Fig. 11.1) [8]. The generation of hemifusion intermediates remains problematic.

Due to the convenience of experimental assays, the early stages of phosphatidylserine (PS) liposome fusion induced by Ca^{2+} were the most extensively characterized fusion model [3]. Although they must be considered primitive approaches to the understanding of biological fusion, they nevertheless led to the identification of three important physical steps in fusion:

1. Stable aggregation of lipid vesicles up to a separation of about 1 nm
2. Close approach, under the control of external forces or perturbations, leading to molecular contact between the headgroups of the partner bilayers and loss of interfacial water molecule (both first steps in the vesicle contact)
3. Structural changes involving lipids of the apposed bilayers resulting in lipid mixing (membrane fusion is present) and finally, contents mixing (i.e., delivery).

11.2.1 *Contact of Membranes*

Aggregation of liposomes is mainly governed by (attractive) long-range van der Waals forces and short-range (double layer) electrostatic interactions between membranes (repulsive for negatively charged liposomes), as described by the Derjaguin–Landau–Verwey–Overbeek theory of the interaction between colloids [6]. In a high ionic content buffer and in the absence of divalent ions, liposomes made of pure PS will not aggregate due to classical electrostatic repulsion. Addition of millimolar concentrations of divalent cations into such systems leads to screening of the charged residues on the headgroups of the phospholipids and direct binding of the cations. As a result, the profile of interaction energy between the bilayers is completely changed. A secondary minimum at a distance of about 3 to 10 nm and a primary minimum at about 1 nm are present with a modulation depending on the lipid composition [6].

For fusion, stable aggregation in the primary minimum is required, but aggregation by itself does not lead to fusion. If membranes are forced to approach closer than this minimum, a repulsive force becomes apparent. At shorter distances, physical techniques (osmotic stress and surface force apparatus) revealed an exponentially increasing repulsive force between the membranes, preventing a closer approach [6, 9]. The nature of this force, which forms the main barrier to fusion, is complex. A classical interpretation is that it is needed to remove polar water molecules bound to and oriented by the polar lipid headgroups (and forming a hydrogen bonded network). The role of dipole potential inside the lipid membrane interfacial layer remains an open question as its definition is poorly understood. Besides the aforementioned forces, an additional long-range repulsive force, extending several nanometers from the membrane, was observed. It has been shown that this force is due to the confinement of the thermal undulatory motion and peristaltic waves of the lipid membrane, of protrusion forces (overlap of molecular scale hair brush or mushroom protrusions) if two membranes are forced to approach each other [6].

11.2.2 *Merging of Membranes*

With the headgroups of phospholipids in contact, membrane merging (fusion) must necessarily start with the formation of nonbilayer metastable lipid structures. Because divalent cations induce structural changes in negatively charged phospholipid bilayers, it has been suggested that related changes would be involved in fusion. Cation-induced isothermal phase transitions in lipids were thought to create local domain boundaries (interfacial defects) between lipids in the gel and liquid crystalline phases, resulting in "focal" points for fusion. However, other less classical phase states should be considered more relevant. Inverted micellar intermediates (IMI), involved in the Lα to hexagonal (HII) phase transition were proposed to be involved in fusion-related short-lived structures. Although no evidence was found for a role of IMIs in PS/Ca^{2+} fusion, it is thought to be involved in Ca^{2+}-induced fusion of other negatively charged liposomes containing cardiolipin.

In the PS/Ca^{2+} system, a number of studies have emphasized the role of dehydration of the bilayer surface and local packing defects in bilayers in initiating fusion. In contrast to Ca^{2+}, Mg^{2+} leads to stable aggregation but not to fusion of PS liposomes with a diameter of 100 nm [10]. Much less water is left between the membranes in the presence of Ca^{2+} than in the presence of Mg^{2+}. Ca^{2+} probably builds an electrostatic bridge between two PS molecules from apposed bilayers forming a "trans" complex between the bilayers, removing the interfacial water molecules, rupturing the water network and, as a result, dehydrating the interface. Mg^{2+} causes no observable fusion with 100 nm liposomes, where the complex between the PS is mediated within the same vesicle. Mg^{2+} does fuse metastable PS vesicles with a diameter of 25 nm that are stressed due to the vesicle's high curvature, making them particularly prone to undergo fusion as many packing defects are present [10]. Complete dehydration of the opposed bilayers is required for fusion, unless compensatory surface energy can be released as in the case of the strain associated to highly curved defect-rich membranes.

The stalk-to-fusion pore transition hypothesis was obtained by macroscopic models considering bilayers as homogeneous elastic surfaces and not as an assembly of individual molecules interacting by a network of weak forces [11]. Therefore, the existence of nonbilayer transition states during fusion is a problem. In the elastic models, the high curvature of the intermediates and the lack of compressibility of the lipid molecule result in the presence of void spaces. This results in the prediction of unrealistically high activation energies if a metastable state (with a long lifetime) is assumed to be present (the so-called hemifusion) (Fig. 11.1) [12]. It was assumed that lipid tilting would reduce these energies [8], but such a movement is linked to unphysical discontinuities. Fusogenic molecules may help by increasing membrane curvature due to their insertion between the polar headgroups [13]. But more presumably, they act by inducing local perturbations of phospholipid packing as described for the PS/Ca^{2+} model. Stalk formation results from local fluctuations and thermal movements of lipid molecules. The transition states appear more disordered than assumed by the elastic models.

Fusion does not result from one single crater-like stalk formation; rather, numerous intramembrane invaginations cause small intervesicular connections to form. A collective reorganization of the contact area follows, resulting in the formation of fusion pores. Fusion pores build an aqueous connection across the fusing bilayers (Fig. 11.1). As shown by electrophysiological approaches in exocytosis in cells [14], fusion pores open within a few microseconds with a diameter in the nanometer range. In the next 10–20 ms, irregular and rapid pore openings and closures are observed (flickering) during a few milliseconds up to many seconds [15]. Fusion pores can then increase their conductance giving rather stable conducting pathways with variable conductance states [16]. The reaction can then reverse, pore expansion does not proceed further, and the pore closes again (the so-called kiss-and-run phenomenon). In most events, a gradual irreversible expansion of the fusion pore occurs resulting in vesicle fusion. Lack of homogeneity in the initial conductance of fusion pores and their flickering upon fusion of protein-free liposomes [15] support the notion that fusion reactions involve an out-of-equilibrium transient disorganization of the two bilayers, resulting in that spontaneous resealing into one single bilayer. This is possible by a transient rupture collective organization of each single lipid bilayer. Similar events in conductance are observed when the fusion is mediated by proteins. This supports the model where, in biological fusion events, fusion pores are mainly lipidic.

Fusion associates with the dynamics of lipid bilayers or domains in biological membranes. It appears as a very fast process, where an out-of-equilibrium configuration of the collective assembly is the driving force. Fluctuations in packing assemblies, undulations, and interfacial protrusions are events responsible for the initiation of membrane merging. Hence, the nature of the lipids (shape, ionization of the polar heads, gel-to-fluid state phase transition) and their interactions with ions present in the solution (Ca^{2+}) are responsible for controlling fusion.

11.3 Biological Vectors that Fuse with Plasmalemma

Membrane fusion is one of the key events occurring in life. It is an extremely controlled process present during viral infection, fertilization, tissue formation, exocytosis, and intracellular trafficking [2, 17, 18]. Fusion occurs when two separate membranes merge into a single one. Membrane fusion can be described as a multistep process where the proximal leaflets of the two membranes merge first and the distal membrane leaflets remain separate until the opening of a fusion pore [19]. Cell fusion is catalyzed by proteins. These proteins are involved in the initial recognition of the membranes that are destined to fuse and pull the membranes close together to destabilize the lipid/water interface. A single fusion protein may be sufficient, but protein complexes may be required for intracellular fusion reactions to ensure control in space and time. Cellular fusion machines are, therefore, adapted but operate by similar principles [2].

11.3.1 *Fusion Occurring during Viral Infection*

Viral entry is the first stage of infection. Viruses that have an envelope, such as HIV and herpes simplex virus, fuse their membrane with the host cell membrane, thereby releasing their genome into the cytoplasm and initiating the viral replication cycle. For that, viruses have to first come into contact with the host cell. In all cases, virus–cell fusion is implemented by one or several viral surface glycoproteins, including one that is generally designated as the fusion protein. Viral fusion proteins can vary in structure and in their modes of triggering cell fusion. Three classes of proteins, based on their structure, have been identified, and at least four distinct mechanisms, by which they are triggered to undergo fusion-inducing conformational changes, have been described [20]. They differ in sequence and architecture, in the way they are triggered for fusion, in their need for accessory proteins, and in the type of fusion peptide, but common principles apply. Membrane fusion requires bringing the two separate membrane bilayers—one of the virus and the other of the cell—into close contact and then merging them into one. In response to the activating trigger, such as low pH and receptor binding, the metastable fusion protein changes to a more extended conformation, which inserts into the target membrane via its fusion peptide. A subsequent conformational change causes the fusion protein to fold back upon itself, thereby bringing its fusion peptide and its transmembrane domain (and their attached target and viral membranes) into intimate contact. The key step is called hemifusion. During hemifusion, small regions of the outer contacting monolayers merge, while the inner monolayers remain intact. That leads to the merger of the inner monolayers and the creation of a small fusion pore, which further expands. In each case, one or a small set of viral surface transmembrane glycoproteins mediates fusion. Highly diverse fusion proteins mediate a common pathway of membrane fusion. This process of hemifusion and fusion pore formation is also present during cellular fusion reactions [21, 22].

11.3.2 *Fusion Occurring during Intracellular Events*

The other most known and studied example of natural fusion is the one occurring in intracellular membrane fusions, which are highly precise and controlled processes mediated by SNARE (soluble NSF attachment protein receptor) proteins. SNARE proteins are small, abundant plasma-membrane-bound proteins that represent a large protein superfamily [23, 24]. Although they vary considerably in structure and size, they all share a segment in their cytosolic domain called a SNARE motif. This motif consists of 60–70 amino acids able of reversible assembly into tight, 4-α-helix bundles called "trans"-SNARE complexes. The primary role of SNARE proteins is to mediate vesicle fusion, occurring during exocytosis. SNAREs are divided into two categories: (1) vesicle or v-SNAREs, which are incorporated into the membranes of transport vesicles during budding, and (2) target or t-SNAREs, which are located in the membranes of target compartments. During membrane fusion, the SNARE proteins combine to form a SNARE complex, that is, a 4-α-helix bundle [23].

The most widely studied SNARE proteins are involved in cell-to-cell communication in the nervous system. They mediate the docking of synaptic vesicles with the presynaptic membrane [25]. The fusion process can be described as a three-step process: collision, docking, and fusion. First, synaptic vesicles approach the neuronal membrane. Then, they dock to the target membrane thanks to a coiled-coil bundle. *In vitro* experiments using reconstituted SNARE proteins have shown that the formation of the coiled-coil bundle docking the vesicles to the target membrane is sufficient to locally disrupt the lipid monolayers and cause the membranes to merge [26, 27]. Finally, opening of a fusion pore allows the transfer of the neurotransmitters out of the neuron.

11.3.3 *Application to Controlled Delivery*

Understanding such membrane fusion processes paves the way for controlled nanoreactor mixing and controlled delivery of cargo to cells [28]. In that context, Marsden et al. have used a simple bottom-up approach in which synthetic fusogens replicate the essential features of SNARE proteins [28, 29]. This reduced SNARE model meets all the characteristics of native membrane fusion and makes the system an accurate model for SNARE-mediated membrane fusion. They indeed replaced the coiled-coil molecular recognition motif of SNARE proteins by a peptide complex anchored in liposome membranes. By this way, depending on the experimental conditions and on the fusion regime, two (or thousands) of small liposomes can fuse into a single (giant) one. In another study, sets of complementary lipidated coiled-coil forming peptides have been designed. The influence of the coiled-coil motif on the rate of liposome fusion was studied showing that an increased coiled-coil stability of complementary peptides translates into increased rates of membrane fusion [30]. The most highly anticipated application of controlled membrane fusion is the targeted transport of cargo such as drugs or gene therapies into cells, which could reduce side effects in patients and increase the efficacy of treatments.

11.4 Membrane Fusion Induced by Physical Methods

Physical methods should affect the bilayer organization to induce a destabilization of the interface to facilitate stalk formation. This can result either from a direct action on the interface (electrically charged with dipoles) or an action on the core of the bilayer. This second approach is supported by compensating membrane reorganization that should minimize the internal lateral pressure [31–33]. The membrane fusion step requires the intravesicular nucleation of defects, that is, a method to bring the partners in contact. Physical methods can induce the formation of defects in a semi-controlled way and will be shortly described.

11.4.1 *Ultrasound*

Acoustic cavitation is known to increase cell permeability. When a mixture of cells and liposomes is ultrasonic pulsed, the content of the liposomes was observed to be internalized. This is the result of a direct fusion process at the stressed membrane level. Cells mixed with liposomes were exposed to a focused 1 MHz ultrasound beam at its focal region for 40 s with a spatial peak acoustic pressure amplitude of 0.25 MPa [34]. Acoustic radiation force pushes liposomes to collide with cells effectively and facilitate the cell-fusion delivery process. Fusogenic defects are mechanically induced by the kinetic energy provide by the acoustic cavitation.

11.4.2 *Osmotic Shock*

Fusion of phospholipid vesicles with planar bilayer membranes is triggered by an osmotically mediated swelling of the vesicles after the replacement in their medium of an impermeant solute by a permeant one [35]. Osmotic swelling generates lipid bilayer tension, which acts as a driving force for vesicle-planar membrane fusion.

11.4.3 *Laser Light Processes*

Local illumination of giant liposomes is known to induce giant (several microns large) pores that spontaneously reseal [36]. They are due to the stretching of the collective assembly. Such stable lipidic pores appear similar to exocytotic fusion pores. These properties suggested that lipidic pores can exhibit the properties of exocytotic fusion pores [37]. Visible light stretches the vesicles, increasing the membrane tension until the membrane responds by a sudden transient rupture of a large pore. Indeed, a cascade of successive transient lipidic (fusogenic) pores appears, up to 30-40 in succession [38]. A similar bursting process is observed on UV-sensitive polymersomes made of block copolymer amphiphiles. A spontaneous curvature of the membrane forms, highly suggesting its potential for stalk formation if polymersomes are in contact [39]. Obviously, to obtain fusion, the illumination spot must be focused on the contact zone between the two partners.

11.4.4 *Electrical Methods*

The membrane autoassembly in cells can be destabilized when short and intense electric pulses are applied to cells. This new organization of the membrane supports a spontaneous fusion process when two cells are in contact [40]. This was first described in the early 1980s and is called the contact-first protocol (cells are previously brought in contact and then the electric pulse train is delivered) [41–43]. But in fact the fusion is obtained even if the contact is obtained between cells already electropermeabilized [44, 45].

The differences between the approaches are in the protocols used to bring the partners in contact, not in biophysical processes supporting membrane mixing. This can be a natural biological contact, the addition of chemicals bringing aggregation, a biochemical manipulation of the surfaces of the two partners (to improve their specific recognition) [46], or physical methods such as dielectrophoresis [42], sucking on filters [47], attachment on dishes [48], ultrasonic standing wave [49], or mild centrifugation [45].

As a final conclusion, one should mention that only the electrical method is currently used routinely at the lab bench to transfer drugs or nucleic acids.

11.5 Synthetic Vectors that Fuse with Plasmalemma

The fusion of drug delivery systems with the plasmalemma of cells is a process characterized by the merging of lipid membranes and mixing of inner contents. As we previously discussed, these major steps of the fusion process are governed by energy barriers that have to be overcome to undergo fusion. In order to fuse, the two membranes need to be brought together. After this close contact, membrane merging requires a local disorder in the lipid bilayers. The spontaneous curvature of the external monolayer of both membranes should allow for merging of the lipid membranes.

11.5.1 *Membrane Fusion Induced by Fusogenic Agents*

This section presents the major fusogenic agents that are used with drug delivery systems to promote membrane fusion. Most of the research to validate the input of these agents was monitored using pure lipidic systems. The models consist of various types of liposomes associated with one or more fusogens.

DOPE (dioleylphosphatidylethanolamine) is commonly used for the preparation of synthetic drug delivery systems that can fuse with membranes [50]. Because of its inverted cone molecular geometry, DOPE supports the formation of inverted hexagonal phase that is favorable for the merging of lipids [51]. In this respect, DOPE appears as a leader in the field of fusogenic lipids [52].

Another factor that facilitates the fusion of membranes is the accumulation of defects in the contact zone. Defects can be introduced by the incorporation of fatty acids or surfactants [53] in the vesicle bilayer. Addition of fatty acids as oleic acid [54] or the six-fold unsaturated docosahexaenoic acid (DHA) [55] is proposed to promote lipid merging. The high degree of unsaturation or the ionizability is proposed to be responsible for their fusogenic properties due to the modifications on bilayer's phase transitions.

In general, small molecules get partitioned in the membrane and give rise to destabilization of the lipid bilayer. This destabilization can increase the permeability of the membrane or may induce defects in the membrane.

In the case of ion-induced fusion, ions bind to membranes and favor the screening of electric charges on the membrane surface. This phenomenon is then favored with cations that can decrease the surface polarity [56]. Among all the positive ions that have some potential to induce fusion, Ca^{2+} is the principle protagonist. Ca^{2+} is able to promote fusion between negatively charged vesicles of PS, as described by Papahadjopoulos et al. about 40 years ago [57]. More recently, Tanaka et al. have evidenced fusion of neutral giant mixed (DOPC/DOPE) vesicles induced by the trivalent cation La^{3+} [58]. The presence of DOPE is known to favor the formation of inverted hexagonal phase [59] that is stabilized in the present case by La^{3+}. This spontaneous negative curvature is essential for the mixing of lipids and favors fusion [52].

Another method to allow a close approach of the membranes may be obtained with the addition of hydrophilic molecules. The effects of water-soluble molecules may be explained through surface potential effects, as proposed by Maggio et al. with dimethylsulfoxide or polyethylene glycol (PEG) on phospholipid monolayers [60]. It is also proposed that hydrophilic polymers compete with the vesicles for water molecules. Thus, modification of the phospholipid polar groups' hydration in the cell membrane promotes the fusion of membranes.

11.5.2 *Drug Delivery Systems that Undergo Spontaneous Membrane Fusion*

As we just discussed, membrane fusion is a complex mechanism that can be promoted using various fusogenic agents. However, most of the systems studied focus on simplified lipidic systems. The plasmalemma is a much more intricate system that undergoes fusion in very specific situations. As detailed below, a few examples of spontaneous fusion between synthetic vectors free from natural fusogen proteins and the plasmalemma have thus far been identified.

In the early 1970s, Papahadjopoulos et al. proposed that liposomes composed of pure PS could fuse with plasma membranes and release their contents into the cytoplasm [61]. More recently, Szoka et al. found that fluorescent lipid probes were not free to diffuse in the plasma membrane, meaning there is no fusion [62]. Supported by these fluorescence recovery after photobleaching experiments, the authors proposed a different mechanism: adsorption on cell surface and endocytosis. This example demonstrates that the cellular internalization pathway of synthetic vectors is sometimes an area of controversy.

11.5.2.1 pH-sensitive vesicles

pH-sensitive liposomes are usually prepared from phospholipid mixtures containing DOPE, which adopts an inverted hexagonal phase favorable for membrane fusion [51] and a fatty acid [54]. At pH values above the pK_a of the fatty acid, the negative charge is used to stabilize the DOPE in the bilayer organization of the liposomes. When the pH decreases under the pK_a, the protonation of the fatty acid diminishes its capacity to stabilize DOPE in the bilayer organization, which is favorable for membrane fusion [63]. pH-sensitive liposomes enable cytosolic delivery of their contents in response to acidic pH within endosomes, while remaining stable in the plasmalemma. The membrane fusion occurs with the endosome's membrane after the endocytosis process. However, Huth et al. proposed one example of pH-sensitive liposomes, composed of DOPE and cholesteryl hemisuccinate (CHEMS), that directly fuse with the plasmalemma of HUVEC and COS-7 cells [64]. Using endocytosis inhibitors coupled to flow cytometry and fluorescent-labeled vesicles, the authors verified that these pH-sensitive liposomes are mostly internalized via endocytotic pathways, but also via membrane fusion. This example confirms the capacity of synthetic delivery systems to spontaneously fuse with plasmalemma.

11.5.2.2 Cationic/catanionic vesicles

Cationic liposomes are prone to strongly interact with negatively charged molecules, such as most of phospholipids, proteins, peptides, or DNA. A great number of cationic liposomes have been designed and synthesized for the transport and delivery of nucleic acids into cells [65]. Moreover, electrostatic interactions between the positively charged cationic lipid complexes and the negatively charged membrane surface drastically improve their internalization by cells. Even if an endocytosis internalization process is proposed as the main pathway, confocal fluorescence microscopy and fluorescence resonance energy transfer experiments have emphasized the capacity of cationic liposomes to fuse with the plasmalemma [66]. Several groups have also studied the DNA-induced fusion of vesicles, taking advantage of the high selectivity of the complementary bases [67]. DNA strands can be chemically modified to ensure spontaneous incorporation into the lipid bilayers.

Liposomal systems can also be generated from mixtures of cationic and anionic lipids. Variation on the stoichiometry of both lipids allows adaptation of the lipid bilayer structure and to favor the inverted hexagonal phase to promote vesicles' fusion [68]. A particular vesicular system has been described by Kaler et al. with mixtures of cationic and anionic surfactants [69]. These so-called "catanionic vesicles" have shown to be exceptionally stable in aqueous solution because of the reduction in the headgroup area when the ion pairs are formed, leading to transition from micellar to lamellar structures. Recently, Boudier et al. have studied the cell internalization pathways of sugar-derived catanionic vesicles, confirming the contribution of endocytotic and fusion processes with the plasma membranes of cells [70]. The authors confirmed the ability of sugar-derived catanionic vesicles to fuse with lipidic membranes using pure lipidic systems [71]. The synergy between the heterogeneity of the lipid bilayer and the endogenous interfacial electric field is explained to be responsible for the spontaneous membrane fusion process.

11.5.2.3 Programmable fusogenic vesicles

Programmable fusogenic vesicles (PFVs) are lipid-based drug delivery systems that exhibit time-dependent destabilization [72]. The first examples of these systems were proposed with nonbilayer-forming lipid DOPE liposomes, stabilized by PEG–lipid conjugates

[73]. The PEG coating ensures the stability of the system in the circulation, as well as its protection from rapid clearance in the plasma. Depending on its lipid chain hydrophobicity, the partition of the PEG–lipid conjugates between the bilayer and the aqueous phase is modified and the destabilization of the vesicles is controlled. Fusion between the PFV and the plasmalemma is thereby promoted, allowing cytosolic delivery of drugs [72].

The concept of PFVs can be expanded to vesicular systems that can fuse with membranes after a programmed event or stimulus. As an example, spontaneous membrane fusion has been promoted by Kunishima et al. using morphological changes induced in the vesicle's membrane after endogenous synthesis of ceramides [74]. The authors proposed local modification of the curvature [75] after a cone-shape to inverted-cone-shape transition of lipids in the outer monolayer of liposomes.

11.6 Conclusions and Perspectives

As we have discussed in this chapter, very few drug delivery systems have been identified that spontaneously fuse with plasmalemma. The fusion process is often associated with endocytotic mechanisms that use endolysosomal compartments to deliver drugs into cells. Indeed, the fusion of drug delivery systems with the plasmalemma is far from being straightforward as different energy barriers have to be overcome to allow membrane fusion. First, the membranes have to get in close contact. This can be provided by increasing the attraction forces between the membranes or decreasing the hydration repulsion forces. Then, an endogenous or exogenous programmed destabilization of the bilayers, in the contact region, is necessary to allow membrane fusion. This perturbation can be provoked by fusogenic agents or physical methods. However, these perturbation-induced mechanisms are different from the ones of membrane fusion occurring in natural events mediated by proteins or peptides, which are allowed via structural reorganization of macromolecules. One of the key issues for the design of fusion-based drug delivery systems is to overcome this discrepancy. In this regard, the examples illustrated earlier show that recent research on membrane fusion processes has allowed us to refine the design of systems that can be programmed to fuse with the plasmalemma and to provide subsequent cytosolic drug delivery.

References

1. Soussan, E., Cassel, S., Blanzat, M., and Rico-Lattes, I. (2009). Drug delivery by soft matter: Matrix and vesicular carriers, *Angew. Chem. Int. Ed.,* **48**, pp. 274–288.
2. Jahn, R., Lang, T., and Sudhof, T. C. (2003). Membrane fusion, *Cell,* **112**, pp. 519–533.
3. Ohki, S., Doyle, D., Flanagan, T. D., Hui, S. W., and Mayhew, E. (1987). *Molecular Mechanisms of Membrane Fusion* (Plenum Press, New York).
4. Sowers, A. E. (1988). *Cell Fusion* (Plenum Press, New York).
5. Wilschut, J., and Hoeckstra, D. (1990). *Membrane Fusion* (Marcel Dekker, New York).
6. Israelachvili, J. (1992). *Intermolecular and Surface Forces* (Academic Press, San Diego).
7. Markin, V. S., Kozlov, M. M., and Borovjagin, V. L. (1984). On the theory of membrane fusion: The stalk mechanism, *Gen. Physiol. Biophys.,* **5**, pp. 361–377.
8. Kozlovsky, Y., and Kozlov, M. M. (2002). Stalk model of membrane fusion: Solution of energy crisis, *Biophys. J.,* **82**, pp. 882–895.
9. Rand, R. P., and Parsegian, V. A. (1989). Hydration forces between phospholipid bilayers, *BBA-Rev. Biomembranes,* **988**, pp. 351–376.
10. Lentz, B. R., Carpenter, T. J., and Alford, D. R. (1987). Spontaneous fusion of phosphatidylcholine small unilamellar vesicles in the fluid phase, *Biochemistry,* **26**, pp. 5389–5397.
11. Jahn, R., and Grubmuller, H. (2002). Membrane fusion, *Curr. Opin. Cell Biol.,* **14**, pp. 488–495.
12. Siegel, D. P. (1993). Energetics of intermediates in membrane fusion: Comparison of stalk and inverted micellar intermediate mechanisms, *Biophys. J.,* **65**, pp. 2124–2140.
13. Epand, R. M., and Epand, R. F. (2000). Modulation of membrane curvature by peptides, *Biopolymers,* **55**, pp. 358–363.
14. Jahn, R., and Sudhof, T. C. (1999). Membrane fusion and exocytosis, *Annu. Rev. Biochem.,* **68**, pp. 863–911.

15. Chanturiya, A., Chernomordik, L. V., and Zimmerberg, J. (1997). Flickering fusion pores comparable with initial exocytotic pores occur in protein-free phospholipid bilayers, *Proc. Natl. Acad. Sci. U.S.A.*, **94**, pp. 14423–14428.

16. Breckenridge, L. J., and Almers, W. (1987). Currents through the fusion pore that forms during exocytosis of a secretory vesicle, *Nature*, **328**, pp. 814–817.

17. Earp, L. J., Delos, S. E., Park, H. E., and White, J. M. (2005). The many mechanisms of viral membrane fusion proteins, *Curr. Top. Microbiol. Immunol.*, **285**, pp. 25–66.

18. Mohler, W. A., Shemer, G., del Campo, J. J., Valansi, C., Opoku-Serebuoh, E., Scranton, V., Assaf, N., White, J. G., and Podbilewicz, B. (2002). The type I membrane protein EFF-1 is essential for developmental cell fusion, *Dev. Cell*, **2**, pp. 355–362.

19. Chernomordik, L. V., and Kozlov, M. M. (2005). Membrane hemifusion: Crossing a chasm in two leaps, *Cell*, **123**, pp. 375–382.

20. White, J. M., Delos, S. E., Brecher, M., and Schornberg, K. (2008). Structures and mechanisms of viral membrane fusion proteins: Multiple variations on a common theme, *Crit. Rev. Biochem. Mol. Biol.*, **43**, pp. 189–219.

21. Chernomordik, L. V., and Kozlov, M. M. (2008). Mechanics of membrane fusion, *Nat. Struct. Mol. Biol.*, **15**, pp. 675–683.

22. Jahn, R., and Scheller, R. H. (2006). SNAREs: Engines for membrane fusion, *Nat. Rev. Mol. Cell Biol.*, **7**, pp. 631–643.

23. Sutton, R. B., Fasshauer, D., Jahn, R., and Brunger, A. T. (1998). Crystal structure of a SNARE complex involved in synaptic exocytosis at 2.4 A resolution, *Nature*, **395**, pp. 347–353.

24. Gerald, K. (2002). *Cell and Molecular Biology*, 4th Ed. (John Wiley and Sons, Inc).

25. Takamori, S., Holt, M., Stenius, K., Lemke, E. A., Gronborg, M., Riedel, D., Urlaub, H., Schenck, S., Brugger, B., Ringler, P., Muller, S. A., Rammner, B., Grater, F., Hub, J. S., De Groot, B. L., Mieskes, G., Moriyama, Y., Klingauf, J., Grubmuller, H., Heuser, J., Wieland, F., and Jahn, R. (2006). Molecular anatomy of a trafficking organelle, *Cell*, **127**, pp. 831–846.

26. Hernandez, J. M., Stein, A., Behrmann, E., Riedel, D., Cypionka, A., Farsi, Z., Walla, P. J., Raunser, S., and Jahn, R. (2012). Membrane fusion intermediates via directional and full assembly of the SNARE complex, *Science*, **336**, pp. 1581–1584.

27. Melia, T. J., Weber, T., McNew, J. A., Fisher, L. E., Johnston, R. J., Parlati, F., Mahal, L. K., Sollner, T. H., and Rothman, J. E. (2002). Regulation of membrane fusion by the membrane-proximal coil of the t-SNARE during zippering of SNARE pins, *J. Cell Biol.,* **158**, pp. 929–940.

28. Marsden, H. R., Korobko, A. V., Zheng, T., Voskuhl, J., and Kros, A. (2013). Controlled liposome fusion mediated by SNARE protein mimics, *Biomater. Sci.,* **1**, pp. 1046–1054.

29. Marsden, H. R., Elbers, N. A., Bomans, P. H. H., Sommerdijk, N. A. J. M., and Kros, A. (2009). A reduced SNARE model for membrane fusion, *Angew. Chem. Int. Ed.,* **48**, pp. 2330–2333.

30. Zheng, T., Voskuhl, J., Versluis, F., Zope, H. R., Tomatsu, I., Marsden, H. R., and Kros, A. (2013). Controlling the rate of coiled coil driven membrane fusion, *Chem. Commun. (Camb.),* **49**, pp. 3649–3651.

31. Cantor, R. S. (1997). Lateral pressures in cell membranes: A mechanism for modulation of protein function, *J. Phys. Chem. B.,* **101**, pp. 1723–1725.

32. Cantor, R. S. (1999). Lipid composition and the lateral pressure profile in bilayers, *Biophys. J.,* **76** pp. 2625–2639.

33. van den Brink-van der Laan, E., Killian, J. A., and de Kruijff, B. (2004). Nonbilayer lipids affect peripheral and integral membrane proteins via changes in the lateral pressure profile, *Biochim. Biophys. Acta.,* **1666** pp. 275–288.

34. Xi, X., Yang, F., Chen, D., Luo, Y., Zhang, D., Gu, N., and Wu, J. (2008). A targeting drug delivery model via interactions among cells and liposomes under ultrasonic excitation, *Phys. Med. Biol.,* **53**, pp. 3251–3265.

35. Cohen, F. S., Akabas, M. H., and Finkelstein, A. (1982). Osmotic swelling of phospholipid vesicles causes them to fuse with a planar phospholipid bilayer membrane, *Science,* **217**, pp. 458–460.

36. Sandre, O., Moreaux, L., and Brochard-Wyart, F. (1999). Dynamics of transient pores in stretched vesicles, *Proc. Natl. Acad. Sci. U.S.A.,* **96**, pp. 10591–10596.

37. Nanavati, C., Markin, V. S., Oberhauser, A. F., and Fernandez, J. M. (1992). The exocytotic fusion pore modeled as a lipidic pore, *Biophys. J.,* **63**, pp. 1118–1132.

38. Karatekin, E., Sandre, O., Guitouni, H., Borghi, N., Puech, P. H., and Brochard-Wyart, F. (2003). Cascades of transient pores in giant vesicles: Line tension and transport, *Biophys. J.,* **84**, pp. 1734–1749.

39. Mabrouk, E., Cuvelier, D., Brochard-Wyart, F., Nassoy, P., and Li, M. H. (2009). Bursting of sensitive polymersomes induced by curling, *Proc. Natl. Acad. Sci. U.S.A.*, **106**, pp. 7294–7298
40. Chang, D. C., Chassy, B. M., Saunders, J. A., and Sowers, A. E. (1992). *Guide to Electroporation and Electrofusion* (Academic Press, California).
41. Neumann, E., Gerisch, G., and Opatz, K. (1980). Cell fusion induced by electric impulses applied to Dictyostelium, *Naturwissenschaften*, **67** pp. 414–415.
42. Pilwat, G., Richter, H. P., and Zimmermann, U. (1981). Giant culture cells by electric field-induced fusion, *FEBS Lett.*, **133**, pp. 169–174.
43. Senda, M., Takeda, J., Abe, S., and Nakamura, T. (1979). Induction of cell fusion of plant protoplasts by electrical stimulation, *Plant Cell Physiol.*, **20** pp. 1441–1443.
44. Sowers, A. E. (1986). A long-lived fusogenic state is induced in erythrocyte ghosts by electric pulses, *J. Cell Biol.*, **102**, pp. 1358–1362.
45. Teissié, J., and Rols, M. P. (1986). Fusion of mammalian cells in culture is obtained by creating the contact between cells after their electropermeabilization, *Biochem. Biophys. Res. Commun.*, **140**, pp. 258–266.
46. Lo, M. M., Tsong, T. Y., Conrad, M. K., Strittmatter, S. M., Hester, L. D., and Snyder, S. H. (1984). Monoclonal antibody production by receptor-mediated electrically induced cell fusion, *Nature*, **310**, pp. 792–794.
47. Ramos, C., Bonenfant, D., and Teissié, J. (2002). Cell hybridization by electrofusion on filters, *Anal. Biochem.*, **302**, pp. 213–219.
48. Teissié, J., Knutson, V. P., Tsong, T. Y., and Lane, M. D. (1982). Electric pulse-induced fusion of 3T3 cells in monolayer culture, *Science*, **216**, pp. 537–538.
49. Bardsley, D. W., Coakley, W. T., Jones, G., and Liddell, J. E. (1989). Electroacoustic fusion of millilitre volumes of cells in physiological medium, *J. Biochem. Biophys. Methods*, **19**, pp. 339–348.
50. Litzinger, D. C., and Huang, L. (1992). DOPE liposomes: Drug delivery, gene transfer and immunodiagnostic applications, *Biochim. Biophys. Acta.*, **1113**, pp. 201–227.
51. Cullis, P. R., and de Kruijff, B. (1978). The polymorphic phase behaviour of phosphatidylethanolamines of natural and synthetic origin. A 31P-NMR study, *Biochim. Biophys. Acta.*, **513**, pp. 31–42.
52. Hafez, I. M., and Cullis, P. R. (2001). Roles of lipid polymorphism in intracellular delivery, *Adv. Drug Deliv. Rev.*, **47**, pp. 139–148.

53. Madden, T. D., and Cullis, P. R. (1982). Stabilization of bilayer structure for unsaturated phosphatidylethanolamines by detergents, *Biochim. Biophys. Acta.*, **684**, pp. 149–153.
54. Fattal, E., Couvreur, P., and Dubernet, C. (2004). "Smart" delivery of antisense oligonucleotides by anionic pH-sensitive liposomes, *Adv. Drug Deliv. Rev.*, **56**, pp. 931–946.
55. Teague, W. E., Fuller, N. L., Rand, R. P., and Gawrisch, K. (2002). Polyunsaturated lipids in membrane fusion events, *Cell. Mol. Biol. Lett.*, **7**, pp. 262–264.
56. Cevc, G. (1990). Membrane electrostatics, *Biochim. Biophys. Acta.*, **1031**, pp. 311–382.
57. Papahadjopoulos, D., Poste, G., Schaeffer, B. E., and Vail, W. J. (1974). Membrane fusion and molecular segregation in phospholipid vesicles, *Biochim. Biophys. Acta.*, **352**, pp. 10–28.
58. Tanaka, T., and Yamazaki, M. (2004). Membrane fusion of giant unilamellar vesicles of neutral phospholipid membranes induced by La^{3+}, *Langmuir*, **20**, pp. 5160–5164.
59. Lewis, R. N. A. H., and McElhaney, R. N. (1993). Calorimetric andspectroscopic studies of the polymorphic phase behavior of a homologous series of n-saturated DOPE, *Biophys. J.*, **64**, pp. 1081–1096.
60. Maggio, B., and Lucy, J. A. (1978). Interactions of water-soluble fusogens with phospholipids in monolayers, *FEBS Lett.*, **94**, pp. 301–304.
61. Papahadjopoulos, D., Poste, G., and Schaeffer, B. E. (1973). Fusion of mammalian cells by unilamellar lipid vesicles: Influence of lipid surface charge, fluidity and cholesterol, *BBA-Rev. Biomembranes*, **323**, pp. 23–42.
62. Szoka, F., Jacobson, K., Derzko, Z., and Papahadjopoulos, D. (1980). Fluorescence studies on the mechanism of liposome-cell interactions *in vitro*, *BBA-Rev. Biomembranes*, **600**, pp. 1–18.
63. Lai, M. Z., Vail, W. J., and Szoka, F. C. (1985). Acid and calcium induced structural changes in phosphatidylethanolamine membranes stabilized by cholesteryl hemisuccinate, *Biochemistry*, **24**, pp. 1654–1661.
64. Huth, U. S., Schubert, R., and Peschka-Süss, R. (2006). Investigating the uptake and intracellular fate of pH-sensitive liposomes by flow cytometry and spectral bioimaging, *J. Control. Release*, **110**, pp. 490–504.
65. Wasungu, L., and Hoekstra, D. (2006). Cationic lipids, lipoplexes and intracellular delivery of genes, *J. Control. Release*, **116**, pp. 255–264.

66. Koynova, R., Tarahovsky, Y. S., Wang, L., and MacDonald, R. C. (2007). Lipoplex formulation of superior efficacy exhibits high surface activity and fusogenicity, and readily releases DNA, *Biochim. Biophys. Acta.*, **1768**, pp. 375–386.
67. Stengel, G., Zahn, R., and Hook, F. (2007). DNA-induced programmable fusion of phospholipid vesicles, *J. Am. Chem. Soc.*, **129**, pp. 9584–9585.
68. Hafez, I. M., Ansell, S., and Cullis, P. R. (2000). Tunable pH-sensitive liposomes composed of mixtures of cationic and anionic lipids, *Biophys. J.*, **79**, pp. 1438–1446.
69. Kaler, E. W., Murthy, A. K., Rodriguez, B. E., and Zasadzinski, J. A. (1989). Spontaneous vesicle formation in aqueous mixtures of single-tailed surfactants, *Science*, **245**, pp. 1371–1374.
70. Boudier, A., Castagnos, P., Soussan, E., Beaune, G., Belkhelfa, H., Ménager, C., Cabuil, V., Haddioui, L., Roques, C., Rico-Lattes, I., and Blanzat, M. (2011). Polyvalent catanionic vesicles: Exploring the drug delivery mechanisms, *Int. J. Pharma.*, **403**, pp. 230–236.
71. Castagnos, P., Mauroy, C., Blache, M.-C., Teissié, J., Rico-Lattes, I., Rols, M.-P., and Blanzat, M. (2012). Spontaneous membrane fusion allows direct delivery from catanionic vesicles into pure lipid systems, *Chem. Commun.*, **48**, pp. 6648–6650.
72. Adlakha-Hutcheon, G., Bally, M. B., Shew, C. R., and Madden, T. D. (1999). Controlled destabilization of a liposomal drug delivery system enhances mitoxantrone antitumor activity, *Nat. Biotechnol.* **17**, pp. 775–779.
73. Holland, J. W., Cullis, P. R., and Madden, T. D. (1996). Poly(ethylene glycol)-lipid conjugates promote bilayer formation in mixtures of non-bilayer-forming lipids, *Biochemistry*, **35**, pp. 2610–2617.
74. Kunishima, M., Tokaji, M., Matsuoka, K., Nishida, J., Kanamori, M., Hioki, K., and Tani, S. (2006). Spontaneous membrane fusion induced by chemical formation of ceramides in a lipid bilayer, *J. Am. Chem. Soc.*, **128**, pp. 14452–14453.
75. McMahon, H. T., and Gallop, J. L. (2005). Membrane curvature and mechanisms of dynamic cell membrane remodelling, *Nature*, **438**, pp. 590–596.

Chapter 12

Endocytosis and the Endolysosomal Route in Drug Delivery

Jeff Rappaport,[a] Iason Papademetriou,[a] and Silvia Muro[a,b]

[a]*Fischell Department of Bioengineering and*
[b]*the Institute for Bioscience and Biotechnology Research, University of Maryland-College Park 5115 Plant Sciences Building, College Park, MD 20742-4450, USA*
muro@umd.edu

Intracellular delivery of therapeutics is key for many applications. The design of drug carriers capable of entering cells and trafficking to precise intracellular destinations is an area of extensive research. Many drug carriers exploit the endocytic route to gain intracellular access, resulting in engulfment and transport within membrane-bound endosomes. The cell-surface receptor or pathway targeted often define the efficiency, kinetics, and final destination, which can be also modulated by tuning carrier valency, surface charge, geometry, composition, etc. Drug carriers contained in endosomes can traffic to recycling routes, transcytosis across the cell body or, most often, lysosomes. Lysosomal degradation of carriers allows for drug release, needed in this compartment or at other sites within the cell. Lysosomal escape may be required, for which carriers are designed to detect lysosomal cues and disrupt the compartment. The

Drug Delivery Across Physiological Barriers
Edited by Silvia Muro

ISBN 978-981-4669-40-5 (Hardcover), 978-981-4669-41-2 (eBook)
www.panstanford.com

potential side effects of these endolysosomal interactions of drug carriers are just beginning to be explored.

12.1 Introduction

Many drugs, such as small molecules, proteins, peptides, nucleic acids, etc., must interact with and enter the cell in order to reach their intended target; however, drugs generally have little intrinsic affinity for the cells of interest. In addition, the plasma membrane and membranes of intracellular compartments present barriers to accessing intracellular targets, hindering drug uptake and diminishing therapeutic effects.

To address these challenges, drugs can be designed to target the tissue, cellular, and subcellular sites required for therapeutic intervention. The simplest approach is termed passive targeting, where drug accumulation in a certain location is achieved simply based on natural biodistribution patterns in the body [1, 2]. For example, leaky tumor vasculature enables drugs to escape from the bloodstream and accumulate within tumor tissue, a paradigm known as enhanced permeation and retention (EPR) effect, commonly used in cancer therapies [1]. Accumulation in macrophages and tissues of the reticuloendothelial system (e.g., liver and spleen) is also observed due to their roles in the clearance of foreign objects from the body [3]. This can be beneficial where therapeutics need to exert action over these cells, but limits accumulation elsewhere. Alternatively, drugs may be coupled with a targeting moiety (e.g., an antibody, antibody fragment, peptide, aptamer, etc.) with high affinity for specific cell surface molecules. Termed active targeting, these modifications can improve binding to the site of interest and enable access to the cell interior [4].

Coupling the drug and targeting moiety into a single system can be accomplished by direct conjugation or by incorporation into a macromolecular assembly (Fig. 12.1). Means of drug conjugation include co-synthesis of the drug and ligand, covalent crosslinking, or non-covalent attachment (e.g., biotin–avidin pairing) [2]. Conjugated systems are typically small and capable of greater intracellular diffusion, but also experience more rapid clearance

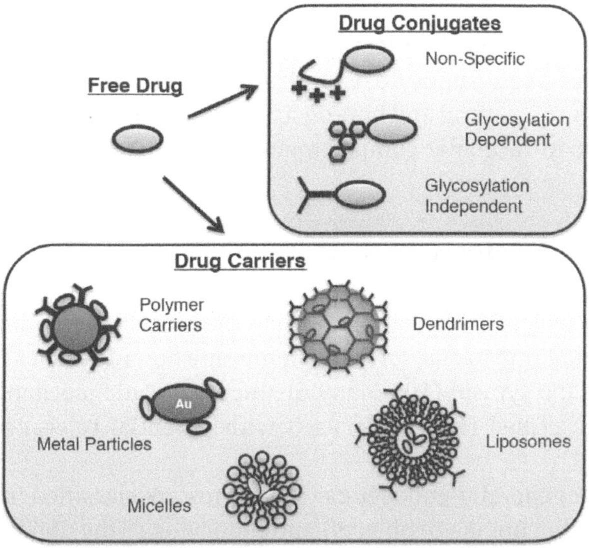

Figure 12.1 Drug Delivery Strategies. Several strategies can be employed to improve drug circulation time, diminish degradation in the bloodstream, modify release kinetics, and target cellular and subcellular features. Targeting moieties can be directly conjugated to the drugs themselves, or incorporated into larger drug-carrying assemblies with tunable features.

by the lymphatic system [2]. Alternatively, drugs can be carried by nano- and micron-scale drug carriers (e.g., liposomes, micelles, nanoparticles, dendrimers, etc.). These systems are often more complex in composition, but have greater potential to enhance drug pharmacokinetics [4]. For example, drug carriers may increase drug absorption into the bloodstream, prolong circulation, limit premature degradation or expulsion by efflux transporters, and modulate drug release [2, 4].

Drugs, drug conjugates, and drug carriers access intracellular targets by traversing the plasma membrane. This can occur by passive diffusion through the membrane or active passage via transmembrane solute transporters [5, 6]. Also, physical mechanisms (e.g., electroporation, magnetofection, ultrasound, etc.), cell penetrating peptides, or components of certain toxins enable transport via transient cell membrane disruption [7]. These approaches are most feasible for small drugs and conjugates, rather than nanocarriers, due to their comparatively smaller size [2, 7]. On the other hand, endocytosis of drugs in membrane-bound vesicles represents a natural and non-destructive means to enter the cell and traffic to intracellular compartments [8, 9].

12.2 Endocytosis

Endocytosis is the process by which the plasma membrane engulfs extracellular material, forming membrane bound vesicles that pinch off into the cytosol [10]. Many distinct endocytic mechanisms have been described (Fig. 12.2), each with potential relevance to drug delivery.

Most generally, endocytic mechanisms are classified into phagocytosis (the uptake of objects) and pinocytosis (the uptake of fluid). Phagocytosis is primarily observed in immune cells and facilitates clearance of foreign pathogens and particulate matter [11]. On their surface, these cells express transmembrane receptors that recognize sugars (mannose receptors), antibodies (Fc receptors), lipids (scavenger receptors), adhesion molecules (integrins), and other features on the target object, triggering the assembly of intracellular actin cups that extend the plasma membrane around the object and engulf it [12, 13]. These newly formed vesicles, called phagosomes, ultimately merge with lysosomes to degrade the contents [11]. Phagocytic uptake of drug carriers has several applications. For example, liposomes and polymeric nanoparticles have been used to deliver cytotoxic agents to Kupffer cells and hepatic tumor tissue (doxorubicin-carrying PACA nanospheres are under clinical trial under the name Transdrug®), and to deliver antibiotics to murine macrophages infected with *Micobacterium tuberculosis, Salmonella typhimurium, and Listeria monocytogenes* [13].

In contrast to phagocytosis, pinocytosis is defined by the uptake of extracellular fluid. Pinocytic mechanisms are sub-characterized according to the vesicle size, contents, molecular composition, etc. Many pinocytic pathways concentrate receptors in nascent vesicles that bind to specific ligands within the extracellular fluid [9]. As a result, compounds may enter these vesicles passively within the

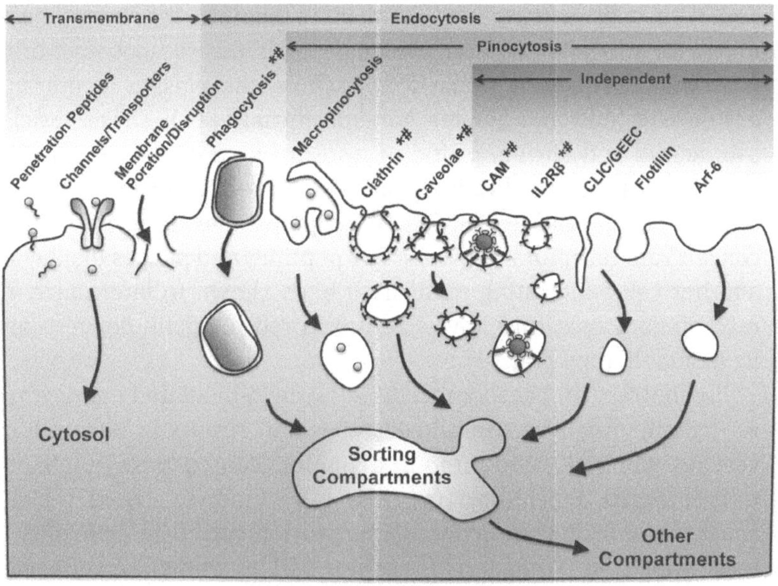

Figure 12.2 Mechanisms of Cell Entry. Drugs can traverse the plasma membrane by several means. *Transmembrane* approaches largely facilitate small molecule transport directly from the extracellular milleu to the cytosol. In contrast, *endocytosis* encompasses several mechanisms of plasma membrane restructuring to internalize objects (phagocytosis) and extracellular fluid (pinocytosis) in intracellular vesicles. Many of these vesicles concentrate receptors that mediate uptake of specific ligands via receptor-mediated endocytosis (denoted by *), while others (e.g., macropinocytosis) largely internalize fluid and solutes by non-specific means. Pathway activity varies between cell types; macropinocytosis and phagocytosis are largely restricted to immune cells and clathrin- and caveolin-mediated pathways are more ubiquitous. Additional pathways (independent of clathrin/caveolin) appear to serve specialized functions and are often categorized by dynamin dependence (denoted by #).

extracellular fluid, or actively by targeting receptors within the budding membrane [9].

Macropinocytosis differs from other pinocytic routes by forming large vesicles, termed macropinosomes, often greater than 0.5 μm in diameter [11]. Like phagocytosis, macropinocytosis is actin-driven and typically observed in immune cells (although it can be transiently induced in other cell types [11]). For example, activated dendritic cells use this mechanism to sample the extracellular milieu for antigens [11]. Unlike phagocytosis, macropinosomes arise from membrane ruffles that fold back on the plasma membrane, enveloping fluid and solutes non-specificially [11]. These vesicles can acidify and shrink, traffic to lysosomes, or recycle to the cell surface, and they tend to be leaky, which may be advantageous for cytosolic drug release [13, 14]. Macropinosomes do not appear to concentrate receptors; however TAT peptides and phages displaying another cell penetrating motif have been shown to internalize via macropinocytosis, and have been explored for gene delivery and leukemia therapy [14, 15].

In contrast to macropinocytosis, clathrin-mediated endocytosis is virtually ubiquitous in all cell types and results in intracellular vesicles of ~100–150 nm in diameter, coated with a characteristic lattice of clathrin proteins [13]. Clathrin-coated vesicles concentrate a diverse array of receptors involved in transport of macromolecules (transferrin receptor, LDL receptor), cell adhesion molecules (selectins, VCAM-1, $\alpha_v\beta_3$ integrin), and others [2, 13]. Nascent vesicles commonly transport to endosomes, where the ligand-receptor complex disassociates and contents are sorted to the lysosome and recycling compartments [13]. In polarized cells, contents can also transport across the cell to the opposite surface [13, 16], which may be a valuable means to transport drugs across cellular barriers such as the blood–brain barrier [13]. For example, PEGylated albumin nanoparticles bearing transferrin were shown to deliver the antiviral AZT into rat brain tissue [17]. Because the transferrin receptor is overexpressed in tumor tissue, the ligand has also been studied as a tool for cancer therapy [13]. Transferrin-conjugated-PLGA nanoparticles have been shown to accumulate in MCF-7 cells, and transferrin-conjugated-PACA particles have been shown to deliver paclitaxel to S-180 tumor bearing mice [13].

Independent of clathrin recruitment, ~60–80 nm flask shaped invaginations containing the transmembrane protein caveolin-1 (termed caveolae) form in lipid-raft microdomains on the plasma membrane of most cell types [10, 18]. Caveolae are abundant in adipocytes, smooth muscle, fibroblasts, and endothelial cells, but are largely absent in certain neuronal cells and lymphocytes [19]. Upon scission from the membrane, caveolar vesicles fuse with intermediate compartments known as caveosomes, and contents can traffic in some cases to the lysosome, but rather across polarized cells, or to the Golgi and rough endoplasmic reticulum [10, 18]. Uptake is largely receptor-mediated, regulated by ligand interactions with albumin-binding receptors (gp60), glycosphingolipids (GM1), enzymes involved in immune response (aminopeptidase N and P), the folate receptor in the case of crosslinking engagement (without crosslinking the mechanism is still under debate), and others [2, 11]. Folate has been widely demonstrated as a drug targeting ligand. For example, polymer, liposome, dendrimer and other nanoparticles displaying folate have been shown to improve the accumulation of chemotherapeutic and imaging agents in folate receptor expressing cancer cells and murine tumor models [20].

Several clathin- and caveolin-independent mechanisms have been described as well [10]. Pathway distinctions are generally drawn based on dynamin-dependence, a protein that mediates clathrin and caveolar vesicle scission from the membrane. Dynamin dependency has been implicated for the cell adhesion molecule (CAM)-mediated pathway [21] and the interleukin-2 receptor β (IL2Rβ)-mediated pathway [22]. Dynamin dependence has not been demonstrated (or remains unclear) for the clathrin-independent carrier/GPI-AP enriched early endosomal compartment (CLIC/GEEC) pathway [10], the flotillin-1 dependent pathway, [10] or the ADP ribosylation factor-6 (Arf6) pathway [23]. While our understanding of these alternate pathways remains limited in most cases, they may also prove advantageous for drug delivery; for example, an FDA approved therapy for patients with therapy-refractory cutaneous T-cell lymphoma combines diphtheria toxin with interleukin-2 as a targeting moiety [24].

An example of a specific clathrin- and caveolin-independent pathway with therapeutic promise is CAM-mediated endocytosis [21]. Multivalent engagement of intercellular adhesion molecule-1 (ICAM-1; a transmembrane glycoprotein involved in leukocyte adhesion and extravasation across the endothelial wall) induces enzyme-mediated remodeling of the plasma membrane, enabling engulfment of polymeric ICAM-1-targeted nanoparticles ranging from 0.2 to several μm [25]. Carriers targeted to ICAM-1 transport to lysosomes [26], and also mimic the extravasation process and traffic via transcytosis, enabling transport across cellular barriers (e.g., the gastrointestinal epithelium and the blood–brain barrier) without opening intercellular junctions or otherwise disrupting barrier integrity [27, 28]. Applications of this pathway to the treatment of lysosomal storage diseases are discussed in Section 12.5.

12.3 Endosomes and Lysosomes as Obstacles or Advantages to Drug Delivery

Following their uptake in endocytic vesicles, internalized receptors, ligands, and dissolved solutes pass through a series of sorting compartments that mediate ligand-receptor dissociation, recycling to the cell surface, and transport to downstream intracellular sites. Extracellular material commonly traffics on a degradative route, presenting both advantages and disadvantages from the perspective of drug delivery.

The major organelles in this sorting pathway traditionally consist of early endosomes, recycling endosomes, late endosomes, and lysosomes [9, 29]. Early endosomes are the first destination for most incoming vesicles (typically clathrin-coated vesicles) and are characterized by a slightly acidic pH of ~6.5, enabling the dissociation of ligands from their receptors, such as transferrin and LDL [9, 29]. The early endosomes are the primary sorting organelles, consisting of a network of tubules and vesicular structures that traffic receptors into recycling endosomes and ligands/solutes into late endosomes [29]. Recycling endosomes fuse with the plasma membrane, returning empty integral membrane receptors to the cell surface and releasing a fraction of endosomal fluids into the extracellular milieu [29]. Late endosomes are acidic compartments that initiate degradation and ultimately deliver their contents to lysosomes [29, 30]. Lysosomes can be distinguished from endosomes by their greater density, lack of mannose-6-phosphate receptors (these and other receptors are sorted out of early and late endosomes), and presence of lysosome-associated membrane proteins (LAMPs) [8, 29]. Degradation within lysosomes occurs due to the activity of enzymes known as acid hydrolases, which function at the acidic lysosomal pH of 4.5–5.0 [30]. While endolysosomal traffic ultimately ends in these terminal degradative compartments, entry into the pathway is highly dynamic and input comes from several sources. Intermediate endocytic compartments (e.g., phagosomes, macropinosomes, caveosomes, etc.), as well as intracellular compartments from other cellular processes (e.g., autophagosomes, trans Golgi vesicles, etc.) also merge with, or otherwise deliver contents to endosomes and lysosomes [2, 10, 30].

One problematic aspect of the endolysosomal route is that of lysosomes acting as a trap for drugs, restricting access to intracellular targets and/or resulting in premature degradation [2]. Several strategies that exploit lysosomal features have been employed to address this issue. Cationic polymers have been used to construct dendrimers and nanoparticles that induce osmotic swelling in endosomes [31–33]. For example, protonated polyamine DNA polyplexes were found to buffer H^+ in the endosome, inducing osmotic swelling and vesicle lysis by increasing Cl^- influx [31]. Surprisingly, some DNA-built devices have been also shown to escape endosomal compartments, enabling cytosolic and nuclear drug delivery [34–36]. Alternatively, the bacterial-derived peptide listeriolysin O (from *Listeria monocytogenes*) can induce pore formation in the membrane at endosomal pH, enabling drug escape into the cytosol [37]. pH-sensitve fusogenic peptides derived from viruses (e.g., HIV, influenza, West Nile virus, etc.) or designed synthetically (e.g., GALA, KALA) offer another means of escape by fusing with the membrane and destabilizing the lipid bilayer [38–42].

Lysosomal features may also be exploited as a cue for controlled delivery of drugs. Drug-carrier linkages comprised of pH sensitive conjugation chemistries or lysosomal enzyme cleavable components can enable controlled release in these compartments [43–47]. Alternatively, nanoparticles can be designed to expand and release the drug payload at lysosomal pH, a parameter that may be optimized to modulate release kinetics [48, 49]. These approaches can also lessen toxic effects by requiring the drug to enter the late endosome/lysosome prior to release.

It is important to note that these means of overcoming lysosomal trapping may also lead to undesirable side effects [2]. There is potential for toxicity due to the formation of carrier metabolites

[50], or the permeabilization of endosomes and lysosomes [51]. Oxidative stress and/or release of hydrolases into the cytosol can damage other intracellular compartments and lead to apoptosis [51]. Alkalinazation of lysosomes by certain nanocarrier materials (e.g., zinc oxide and gold) can result in membrane permeabilization and reduced enzyme activity, while lysosomal accumulation of non-degradable nanocarriers (e.g., gold, polystyrene, and titanium dioxide) can impair or activate autophagy [8].

In certain cases however, trapping of drugs in lysosomes represents an important advantage. This is the case for treatment of lysosomal storage disorders, where enzyme replacement therapies can gain access to the therapeutic target by trafficking via the endolysosomal route (discussed in Section 12.5).

12.4 Manipulating Endocytosis and Intracellular Trafficking

Endocytosis and intracellular trafficking are highly regulated processes, where the surface receptors responsible for uptake and the intracellular protein complexes responsible for sorting typically dictate the subcellular destination [2, 9, 52]. Many vesicles merge into the endolysosomal pathway, which may or may not represent a desirable therapeutic destination, and controlling this process can be a challenging and unpredictable effort [2]. Here we describe biological and engineered parameters of drug delivery systems that have been shown to modulate the endocytic gateway, intracellular trafficking, and final destination.

12.4.1 *The Targeting Moiety*

Internalized material can recycle to the cell surface, traffic to alternate subcellular destinations, or traverse the cell via transcytosis, potentially bypassing the degradative pathway and expanding drug delivery options [2, 9]. Several studies demonstrate manipulation of routing, where the final destination is governed by the ligand used as a targeting moiety. For example, folate (a vitamin B9 carrier) and transferrin (an iron carrier) induce compartment recycling [20, 52]. These ligands enter the cell via receptor-mediated endocytosis, traffic to endosomes, and release their vitamin/iron cargo in the acidic environment [20, 52]. The carrier-receptor complex (free of cargo) returns to the cell surface within a recycling endosome, providing a potential means for sustained extracellular drug uptake and intracellular delivery [2, 20, 52]. In contrast, pathogenic compounds, such as cholera and shiga toxins, have been shown to disrupt sorting in order to bypass lysosomal degradation [53, 54]. Following delivery to early endosomes, free and drug-conjugated toxins induce formation of compartments that traffic to the Golgi apparatus and endoplasmic reticulum [53, 54]. Finally, several surface proteins (e.g., gp60, LDL, transferrin, and insulin receptors) have been shown to participate in transcellular transport, permitting drug carriers targeting these proteins to traverse cellular barriers such as the vascular endothelium [2, 55].

Carriers bearing ligands to multiple endocytic receptors are beginning to be explored, such as the case of multiple endothelial cell adhesion molecules, transferrin receptor and integrins, folate receptor and the epidermal growth factor receptor, etc. [56, 57]. For instance, polymer nanoparticles bearing antibodies to the transferrin receptor (clathrin pathway) and ICAM-1 (CAM pathway) enhanced liver accumulation, diminished brain accumulation, and had an intermediate effect versus single targeted nanoparticles on lung accumulation by modulating tissue affinity and cellular uptake [58].

12.4.2 *Valency*

Drugs conjugated to a single targeting moiety (monovalent) have been shown to localize to different compartments than drugs conjugated to multiple copies of the same moiety (multivalent). Among others, this pattern has been observed for folate conjugates, where monovalent conjugates recycle to the plasma membrane and multivalent conjugates localize within the lysosome [20]. Both transport modes can be valuable; the recycling of monovalent conjugates permits continuous drug uptake, whereas the lysosomal delivery of multivalent conjugates may be valuable for pH sensitive compounds. Valency-dependent trafficking has also been observed for the neonatal Fc receptor, where bivalent ligands more efficiently

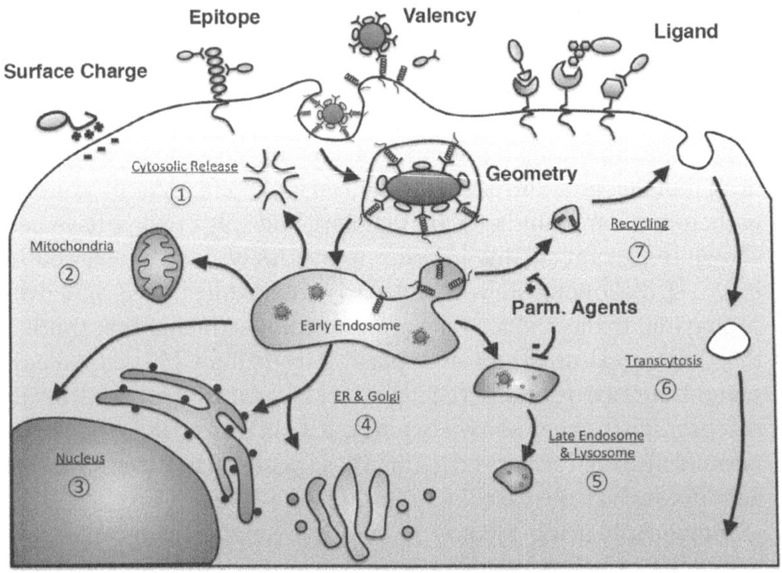

Figure 12.3 Manipulating Endocytosis and Trafficking for Drug Delivery. Several drug and carrier properties (targeting moiety, valency, epitope, carrier geometry, etc.) can be tuned to regulate internalization, subcellular destination, and the kinetics of these processes. Conjugation with affinity moieties allows targeting of cell surface receptor and also intracellular organelles, for example, (1) cytosolic delivery using GALA and KALA peptides, (2) nuclear delivery using nuclear localization sequences, (3) ER/Golgi delivery using cholera and shiga toxins, (4) mitochondrial delivery using localization peptides from cytochrome oxidase subunits, (5) lysosomal delivery by targeting cell adhesion molecules or mannose-6-phosphate, (6) transcytosis by targeting ICAM-1, aminopeptidase P, and the LDL, transferrin, and insulin receptors, and (7) recycling by targeting ICAM-1, the folate receptor, and the transferrin receptor [2].

undergo transcytosis in polarized kidney cells, while monovalent ligands are sorted to lysosomes [59]. Valency can also impact trafficking kinetics, providing a tunable parameter for controlled drug release in desirable compartments. For example, multivalent transferrin was shown to slow internalization kinetics and quadruple the endosomal retention time as compared to the monovalent ligand [60].

In certain circumstances, multivalency can trigger or enhance receptor-mediated uptake, for example, antibodies to ICAM-1 are poorly internalized by cell versus particles displaying multiple copies of anti-ICAM, which induce engulfment, uptake, and endolysosomal trafficking [61]. Ligand density has also been correlated with internalization efficiency for nanoparticles bearing transferrin or antibodies to the transferrin receptor, where increasing the valency was demonstrated to increase uptake and cytoxicity in Ramos B-cell lymphoma cells [62].

12.4.3 *Receptor Epitope*

Contributing an additional layer of complexity, transmembrane proteins can expose several binding sites to the extracellular environment, each with different steric accessibility and intracellular signaling cues. The relevance to the design of drug delivery systems has been explored for several endothelial proteins, such as angiotensin-converting enzyme, ICAM-1, PECAM-1, and VCAM [63]. For example, monoclonal antibodies to five PECAM-1 epitopes were shown to bind to this glycoprotein, whereas multivalent anti-PECAM nanocarriers could only access the four distal epitopes [64]. Only three of these epitopes permitted carrier internalization, and each with different trafficking kinetics and specificity for endosomal versus lysosomal delivery [64]. Furthermore, targeting multiple epitopes simultaneously can be advantageous; pulmonary accumulation of a fusion conjugate targeting PECAM-1 was enhanced when a second PECAM-1 epitope was concomitantly engaged by an antibody [65]. This is not a universal case however; polymer nanocarriers bearing antibodies to two different transferrin receptor epitopes accumulated to a lesser degree in the mouse brain and liver, and to an intermediate degree in the lungs, relative to the singly targeted carriers [58].

12.4.4 *Geometry and Other Physical Parameters*

The size of drugs, drug conjugates, and drug carriers often influence internalization efficacy and subcellular trafficking kinetics. Multivalent carriers tend to be larger than their monovalent counterparts, and therefore experience size-exclusion from small endocytic compartments [2]. Indeed, studies on transferrin-bearing liposomes (targeting clathrin uptake) and cholera toxin-bearing particles (targeting caveloar uptake) illustrated poor internalization of carriers larger than the compartment diameter [66, 67]. Alternate ligand-mediated pathways such as phagocytosis and CAM endocytosis are capable of internalizing relatively large objects, and often demonstrate size-dependent trafficking kinetics. For example, micron-scale opsonized carriers phagocytized by macrophages are sorted to lysosomes faster than nanoscale carriers [68]. The opposite has been observed for CAM endocytosis in endothelial cells, where larger carriers traffic more slowly to the lysosome [25], additionally demonstrating how size effects can vary between pathways and cell types.

Other relevant physical parameters include the shape, stiffness, binding force, and surface charge. For example, endothelial cells internalize spherical carriers more rapidly than disk-shaped carriers via CAM endocytosis, and the same is true of macrophages using phagocytosis [25, 69]. This depends on the carrier aspect ratio, where engulfment of non-spherical objects requires greater cytoskeletal and plasma membrane remodeling [69]. Carrier stiffness also plays a role, as macrophages preferentially engulf rigid objects over soft objects [70]. With regards to binding force, a study using atomic force microscopy measured a five-fold greater force between the cell surface and poly-L-lysine functionalized PLGA particles versus non-functionalized particles, correlating with significantly faster internalization [71]. Positive surface charge on chitosan nanoparticles has also been correlated with faster internalization in multiple cell lines, likely due to enhanced association with the negatively charged plasma membrane [72].

12.4.5 *Pharmacological Agents*

Phamacological agents, delivered along with drug delivery systems, may offer additional tools to influence endolysosomal routing. For example, monensin, an agent known to alkalize endosomes by altering Na^+/H^+ exchange, has been demonstrated to alter endosomal sorting, diminishing recycling of the beta-2-androgenic receptor,

LDL receptor, and transferrin receptor [73–75]. The agent has also been shown to switch the trafficking of ICAM-1-targeted carriers from a lysosomal to a recycling route, a potential means of avoiding drug degradation [61]. Other means of endosome/lysosome pH modulation, such as treatment with chloroquine (a weak base), have been shown to diminish ICAM-1-targeted carrier degradation by endothelial cells [61], and to enhance phagosome–lysosome fusion in aveolar macrophages [76]. Inhibition of specific proteins that regulate trafficking may also be desirable. For example, inhibition of phosphoinositide 3-kinase elicits endosomal retention of megalin and its ligands, cathepsin and albumin, diminishing recycling and lysosomal trafficking in kidney proximal tubular cells [77]. Inhibition of microtubule polymerization with nocodazole has also been shown prevent lysosomal trafficking, prolonging the presence of ICAM-1-targeted carriers in the endosome [61].

12.5 Lysosomes as Therapeutic Targets

Drug release in the lysosome can be desirable for certain pathologies such as lysosomal storage diseases (LSDs), a fatal group of ~50 hereditary conditions involving aberrant accumulation of undigested material in the lysosome [78]. LSDs are characterized by the deficiency of a lysosomal protein, commonly an enzyme, but also proteins involved in the processing and trafficking of lysosomal enzymes (e.g., N-acetylglucosamine-1-phosphotranseferase in I-Cell disease) and proteins of the lysosomal membrane (e.g., NPC1 in Niemann–Pick type C disease) (Fig. 12.4) [78]. Irrespective of the underlying cause, the storage disrupts an array of cellular functions, ranging from gene expression, to cell signaling, to vesicular trafficking, ultimately leading to widespread organ damage [78]. Less severe pathologies progress slowly into adulthood, affecting peripheral organs such as the liver and kidneys and requiring transplantation [79]. More than half of LSDs result in aggressive symptoms of the central nervous system including seizures, dementia, and dyskinesia, leading to death at an early age [79].

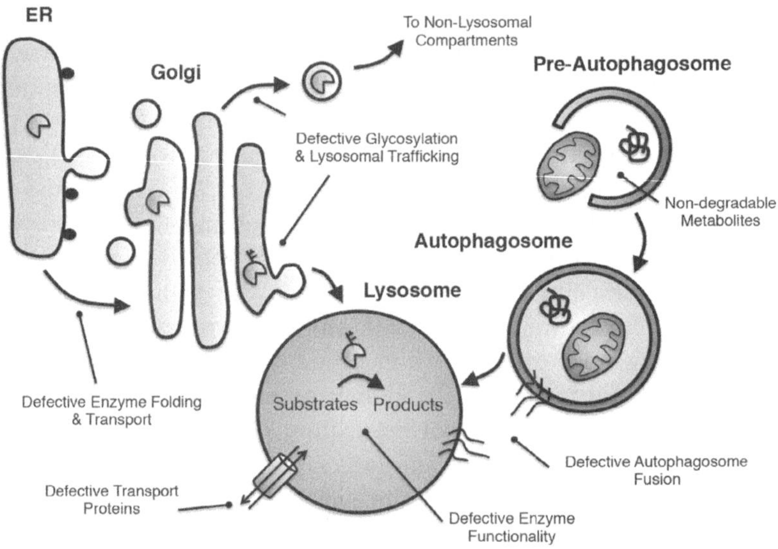

Figure 12.4 Defects Resulting in Aberrant Lysosomal and Autophagic Storage. LSDs are characterized by genetic mutations affecting lysosomal hydrolase synthesis, trafficking, and activity. For example, aberrant enzyme folding in the ER can prevent delivery to the Golgi apparatus (e.g., Gaucher disease) and improper enzyme glycosylation can misdirect enzyme trafficking to non-lysosomal sites (e.g., I Cell disease). Enzymes may reach the lysosome but remain inactive due to misfolding (e.g., Niemann–Pick A/B disease). Alternatively, faulty transport proteins in the lysosomal membrane may disrupt metabolite flux (e.g., cystinosis). Lysosomal storage can lead to secondary accumulation in upstream autophagosomes when compartment fusion is impaired (e.g., Pompe disease). Furthermore, several non-LSDs characterized by a buildup of non-degradable macromolecules lead to autophagic distress as well (e.g., Huntington's disease).

Metabolite accumulation in the lysosome leads to secondary accumulation in upstream compartments that traffic to the lysosome, such as those of the endolysosomal and autophagic pathways [8, 80]. Autophagy is a housekeeping process whereby cellular components such as mitochondria, RNA, ER fragments, and others, are removed from the cytosol and routed for degradation [8, 80]. Such objects become internalized within autophagosomes, double-membrane compartments that ultimately fuse with degradative compartments [8, 80]. Impaired fusion is observed in LSD due to abnormal SNARE protein distribution, leading to retention of "expired" materials [8, 80]. Autophagic impairment arises in non-LSDs as well, such as Huntington's, Alzheimer's, and Parkinson's disease [81]. Despite different underlying causes, the atypical storage behavior in these and other neurodegenerative diseases leads to similar CNS symptoms observed in LSDs [8].

Several small molecule therapies have been demonstrated experimentally and clinically to relieve aberrant lysosomal storage and autophagy. For example, miglustat (Zavesca®) can inhibit metabolite synthesis in Gaucher, Fabry, Tay–Sachs, and type C Niemann–Pick diseases, while Cysteamine has been used similarly to treat cystinosis [8]. Chemical chaperones that improve protein folding include *N*-nonyldeoxynojirimycin for Gaucher disease, 1-deoxygalac-tonoojirimycin for Fabry disease, and *N*-octyl-4-epi-β-valienamine for GM1 gangliosidosis, all three of which are undergoing clinical trials [8]. Substrate reduction can be achieved with compounds such as cyclodextrins, ring-like polysaccharides that remove excess cholesterol in type C Niemann–Pick disease cells, potentially restoring SNARE function and autophagosome–lysosome fusion [82]. Alternatively, δ-Tocopherol has been shown to relieve lipid accumulation from Wolmann and type C Niemann–Pick disease patient cells by inducing exocytosis of engorged compartments [83]. Compounds that induce autophagy may also play a therapeutic role, such as rapamycin, which enhances substrate degradation *in vitro* for Huntington's and Parkinson's disease [8].

Another therapeutic focus for LSD is "cross-correction," the finding that healthy cells producing functional lysosomal enzymes can relieve the symptoms of the surrounding tissue [79]. For example, transplanted bone marrow or mesenchymal stem cells have the potential to secrete functional lysosomal enzymes into the extracellular environment, which can be internalized and trafficked to the lysosomes of diseased cells [8]. Alternatively, the delivery of wild type cDNA sequences to diseased cells may restore healthy enzyme levels in a sufficient population of cells to supply whole tissues [8]. Clinical studies are underway to assess the efficacy of gene therapy for Sanfillipo and Pompe disease [84]. Only small doses (<10%–20% of the natural enzyme quantity) are needed for therapeutic effect, an important feature of cross correction strategies [85].

A more established technique, enzyme replacement therapy (ERT), involves the intravenous delivery of glycosylated wild-type enzymes that bind to the mannose-6-phosphate receptor, enter diseased cells via clathrin-mediated endocytosis, and become active in the lysosome [85]. FDA-approved ERT products currently exist for six different LSDs: Hurler disease, Hunter disease, Maroteaux–Lamy syndrome, Fabry disease, Pompe disease, and Gaucher type I and III [84, 85]. However, obstacles to the efficacy of ERT are both systemic (e.g., bloodstream clearance, proteases, immune reactions, cellular barriers), as well as subcellular (e.g., impaired endocytic activity, intracellular trafficking, and compartment fusion that characterize lysosomal storage and autophagic disorders) [2, 8, 86, 87]. Glycosylation-independent strategies offer another promising therapeutic avenue, where enzymes are fused to affinity peptides capable of targeting alternate cell surface features. For example, conjugation of α-L-iduronidase (absent in Hurler's syndrome) with an antibody to the insulin receptor enabled transport across the blood–brain barrier in rhesus monkeys, demonstrating a potential therapeutic approach for LSDs with CNS pathologies [86].

Despite these advances, injection of unprotected recombinant enzymes into the systemic circulation often remains an inefficient and prohibitively expensive therapy, with treatment costing $50,000–$165,000 per patient per year [8]. Coupling recombinant enzymes to nanoscale carriers offers an alternative solution to present obstacles, providing a means of enzyme protection, specific cell targeting, barrier transport, and tunable drug release. For example, functionalized liposomes carrying lysosomal enzymes have been explored since the 1970s (e.g., with PEG to impart stealth from the immune system or with IgG to specifically target immune cells) [8]. Strategies to avoid intravenous delivery include implantable calcium alginate microspheres, which offer a means to prolong enzyme release and access otherwise difficult-to-reach tissues, such as bone in Gaucher disease [8]. Other carriers including quantum dots, gold nanoparticles conjugated with cell penetrating peptides, and pH sensitive trimethyl chitosan complexes have also been reported to enhance lysosomal enzyme delivery [8].

In addition to these approaches, enzyme-carrying nanocarriers targeting ICAM-1 offer a number of attractive features. ICAM-1 is overexpressed at sites of inflammation, a hallmark of tissues affected by LSDs, and carriers targeting the glycoprotein traffic both to the lysosome (the target for ERT) and across the cell, enabling transport through the blood–brain barrier [28]. Importantly, unlike the carrier size restrictions posed by clathrin and caveolar endocytosis, the cell-type restrictions posed by phagocytosis, and the non-specific nature of macropinocytosis, the ICAM-1-mediated pathway permits effective internalization of targeted carriers ranging from 0.2 to several μm and has been demonstrated in all tested cell types expressing this glycoprotein [8, 25].

Given these features, ICAM-1 has been demonstrated as a promising therapeutic target for inflammatory conditions such as LSDs. Both polystyrene and biodegradable PLGA nanocarriers targeted to ICAM-1 have been shown to deliver acid sphingomyelinase, α-galactosidase, and α-glucosidase to the tissues of type A-B Niemann–Pick, Fabry, and Pompe disease models respectively [26, 88, 89]. Carriers injected in mice demonstrated reduced circulation time and increased accumulation in the liver, spleen, heart, kidney, lung, muscle, and brain compared to free-enzyme (as much as $\sim$200 fold in the lungs and $\sim$7 fold in the brain) [26, 88, 89]. In cell culture, these carriers associate with cells in proportion with ICAM-1 expression, and demonstrate markedly improved internalization, lysosomal delivery, and substrate reduction compared to free enzyme [26, 87–89]. Current studies are exploring means of increasing biocompatibility, modulating carrier properties, and enhancing delivery across barriers, offering an increasingly flexible and effective technique for ERT via the endolysosomal route.

12.6 Conclusions and Perspectives

Drug delivery to intracellular sites can be achieved via fluid phase or receptor-mediated endocytosis, and commonly results in delivery to the lysosome, presenting both obstacles and advantages that can be incorporated into the therapeutic strategy. Recent decades have therefore witnessed an explosion of interest in drugs, drug conjugates, and drug carriers capable of targeting different mechanisms of endocytosis and manipulating their intracellular fate. Insights into the role of the targeting moiety, valency, receptor epitopes, carrier shape and size, and pharmacological modulation have provided remarkable means to improve therapeutic efficacy. For example, nanocarriers modified to improve encapsulation, prolong release, enhance lysosomal targeting, and traverse cellular barriers have expanded the promise of enzyme replacement therapies for lysosomal storage diseases. Nevertheless, the side effects of new drug delivery systems remain underexplored and their complex behavior in different cell types and diseases has largely delayed clinical translation. Future progress will require further elucidation of the underlying mechanisms governing endocytosis and endolysosomal traffic, and development of rational delivery systems capable of harnessing these pathways with therapeutic effect.

References

1. Allen, T.M., Cullis, P.R. (2004). Drug delivery systems: Entering the mainstream, *Science*, **303**, pp. 1818–1822.
2. Muro, S. (2012). Challenges in design and characterization of ligand-targeted drug delivery systems, *J. Control. Release*, **164**, pp. 125–137.
3. Chrastina, A., Massey, K.A., Schnitzer, J.E. (2011). Overcoming *in vivo* barriers to targeted nanodelivery, *Wiley Interdiscip. Rev. Nanomed. Nanobiotechnol.*, **3**, pp. 421–437.
4. Torchilin, V.P. (2006). Multifunctional nanocarriers, *Adv. Drug Deliv. Rev.*, **58**, pp. 1532–1555.
5. Sai, Y., Tsuji, A. (2004). Transporter-mediated drug delivery: Recent progress and experimental approaches, *Drug Discov. Today*, **9**, pp. 712–720.
6. Masson, M., Loftsson, T., Masson, G., Stefansson, E. (1999). Cyclodextrins as permeation enhancers: Some theoretical evaluations and *in vitro* testing, *J. Control. Release*, **59**, pp. 107–118.
7. Mura, S., Nicolas, J., Couvreur, P. (2013). Stimuli-responsive nanocarriers for drug delivery, *Nat. Mater.*, **12**, pp. 991–1003.

8. Manthe, R., Muro, S. (2014). *Handbook of Nanobiomedical Research*, Torchilin, V. (Ed.) Chapter 9, "Lysosomes and therapeutics: Diseases, treatments, and side effects."
9. Duncan, R., Richardson, S.C. (2012). Endocytosis and intracellular trafficking as gateways for nanomedicine delivery: Opportunities and challenges, *Mol. Pharm.*, **9**, pp. 2380–2402.
10. Doherty, G.J., McMahon, H.T. (2009). Mechanisms of endocytosis, *Annu. Rev. Biochem.*, **78**, pp. 857–902.
11. Gong, Q., Huntsman, C., Ma, D. (2008). Clathrin-independent internalization and recycling, *J. Cell. Mol. Med.*, **12**, pp. 126–144.
12. Aderem, A., Underhill, D.M. (1999). Mechanisms of phagocytosis in macrophages, *Annu. Rev. Immunol.*, **17**, pp. 593–623.
13. Hillaireau, H., Couvreur, P. (2009). Nanocarriers' entry into the cell: Relevance to drug delivery, *Cell. Mol. Life Sci.*, **66**, pp. 2873–2896.
14. Khalil, I.A., Kogure, K., Akita, H., Harashima, H. (2006). Uptake pathways and subsequent intracellular trafficking in nonviral gene delivery, *Pharmacol. Rev.*, **58**, pp. 32–45.
15. Nishimura, S., Takahashi, S., Kamikatahira, H., Kuroki, Y., Jaalouk, D.E., O'Brien, S., Koivunen, E., Arap, W., Pasqualini, R., Nakayama, H., Kuniyasu, A. (2008). Combinatorial targeting of the macropinocytotic pathway in leukemia and lymphoma cells, *J. Biol. Chem.*, **283**, pp. 11752–11762.
16. Tuma, P., Hubbard, A.L. (2003). Transcytosis: Crossing cellular barriers, *Physiol. Rev.*, **83**, pp. 871–932.
17. Mishra, V., Mahor, S., Rawat, A., Gupta, P.N., Dubey, P., Khatri, K., Vyas, S.P. (2006). Targeted brain delivery of AZT via transferrin anchored pegylated albumin nanoparticles, *J. Drug Target*, **14**, pp. 45–53.
18. Kirkham, M., Parton, R.G. (2005). Clathrin-independent endocytosis: New insights into caveolae and non-caveolar lipid raft carriers, *Biochim. Biophys. Acta.*, **1745**, pp. 273–286.
19. Stan, R.V. (2002). Structure and function of endothelial caveolae, *Microsc. Res. Tech.*, **57**, pp. 350–364.
20. Xia, W., Low, P.S. (2010). Folate-targeted therapies for cancer, *J. Med. Chem.*, **53**, pp. 6811–6824.
21. Muro, S., Wiewrodt, R., Thomas, A., Koniaris, L., Albelda, S.M., Muzykantov, V.R., Koval, M. (2003). A novel endocytic pathway induced by clustering endothelial ICAM-1 or PECAM-1, *J. Cell. Sci.*, **116**, pp. 1599–1609.

22. Lamaze, C., Dujeancourt, A., Baba, T., Lo, C.G., Benmerah, A., Dautry-Varsat, A. (2001). Interleukin 2 receptors and detergent-resistant membrane domains define a clathrin-independent endocytic pathway, *Mol. Cell*, **7**, pp. 661–671.

23. Naslavsky, N., Weigert, R., Donaldson, J.G. (2004). Characterization of a nonclathrin endocytic pathway: Membrane cargo and lipid requirements, *Mol. Biol. Cell*, **15**, pp. 3542–3552.

24. Schrama, D., Reisfeld, R.A., Becker, J.C. (2006). Antibody targeted drugs as cancer therapeutics, *Nat. Rev. Drug Discov.*, **5**, pp. 147–159.

25. Muro, S., Garnacho, C., Champion, J.A., Leferovich, J., Gajewski, C., Schuchman, E.H., Mitragotri, S., Muzykantov, V.R. (2008). Control of endothelial targeting and intracellular delivery of therapeutic enzymes by modulating the size and shape of ICAM-1-targeted carriers, *Mol. Ther.*, **16**, pp. 1450–1458.

26. Hsu, J., Serrano, D., Bhowmick, T., Kumar, K., Shen, Y., Kuo, Y.C., Garnacho, C., Muro, S. (2011). Enhanced endothelial delivery and biochemical effects of α-galactosidase by ICAM-1-targeted nanocarriers for Fabry disease, *J. Control. Release*, **149**, pp. 323–331.

27. Ghaffarian, R., Bhowmick, T., Muro, S. (2012). Transport of nanocarriers across gastrointestinal epithelial cells by a new transcellular route induced by targeting ICAM-1, *J. Control. Release*, **163**, pp. 25–33.

28. Hsu, J., Rappaport, J., Muro, S. (2014). Specific binding, uptake, and transport of ICAM-1-targeted nanocarriers across endothelial and subendothelial cell components of the blood–brain barrier, *Pharm. Res.*, **31**, pp. 1855–1866.

29. Mellman, I. (1996). Endocytosis and molecular sorting, *Annu. Rev. Cell Dev. Biol.*, **12**, pp. 575–625.

30. Luzio, J.P., Pryor, P.R., Bright, N.A. (2007). Lysosomes: Fusion and function, *Nat. Rev. Mol. Cell Biol.*, **8**, pp. 622–632.

31. Sonawane, N.D., Szoka, F.C., Jr., Verkman, A.S. (2003). Chloride accumulation and swelling in endosomes enhances DNA transfer by polyamine-DNA polyplexes, *J. Biol. Chem.*, **278**, pp. 44826–44831.

32. Thomas, T.P., Majoros, I., Kotlyar, A., Mullen, D., Holl, M.M., Baker, J.R., Jr. (2009). Cationic poly(amidoamine) dendrimer induces lysosomal apoptotic pathway at therapeutically relevant concentrations, *Biomacromolecules*, **10**, pp. 3207–3214.

33. Arote, R.B., Jiang, H.L., Kim, Y.K., Cho, M.H., Choi, Y.J., Cho, C.S. (2011). Degradable poly(amido amine)s as gene delivery carriers, *Expert Opin. Drug Deliv.*, **8**, pp. 1237–1246.

34. Muro, S. (2014). A DNA device that mediates selective endosomal escape and intracellular delivery of drugs and biologicals, *Adv. Funct. Mater.*, **24**, pp. 2899–2906.

35. Roh, Y.H., Lee, J.B., Kiatwuthinon, P., Hartman, M.R., Cha, J.J., Um, S.H., Muller, D.A., Luo, D. (2011). DNAsomes: Multifunctional DNA-based nanocarriers, *Small*, **7**, pp. 74–78.

36. Zhang, K., Hao, L., Hurst, S.J., Mirkin, C.A. (2012). Antibody-linked spherical nucleic acids for cellular targeting, *J. Am. Chem. Soc.*, **134**, pp. 16488–16491.

37. Kullberg, M., Owens, J.L., Mann, K. (2010). Listeriolysin O enhances cytoplasmic delivery by Her-2 targeting liposomes, *J. Drug Target*, **18**, pp. 313–320.

38. Plank, C., Oberhauser, B., Mechtler, K., Koch, C., Wagner, E. (1994). The influence of endosome-disruptive peptides on gene transfer using synthetic virus-like gene transfer systems, *J. Biol. Chem.*, **269**, pp. 12918–12924.

39. Wagner, E., Plank, C., Zatloukal, K., Cotten, M., Birnstiel, M.L. (1992). Influenza virus hemagglutinin HA-2 N-terminal fusogenic peptides augment gene transfer by transferrin-polylysine-DNA complexes: Toward a synthetic virus-like gene-transfer vehicle, *Proc. Natl. Acad. Sci. U.S.A.*, **89**, pp. 7934–7938.

40. Kamper, N., Day, P.M., Nowak, T., Selinka, H.C., Florin, L., Bolscher, J., Hilbig, L., Schiller, J.T., Sapp, M. (2006). A membrane-destabilizing peptide in capsid protein L2 is required for egress of papillomavirus genomes from endosomes, *J. Virol.*, **80**, pp. 759–768.

41. Kimura, T., Ohyama, A. (1988). Association between the pH-dependent conformational change of West Nile flavivirus E protein and virus-mediated membrane fusion, *J. Gen. Virol.*, **69 (Pt 6)**, pp. 1247–1254.

42. Wyman, T.B., Nicol, F., Zelphati, O., Scaria, P.V., Plank, C., Szoka, F.C., Jr. (1997). Design, synthesis, and characterization of a cationic peptide that binds to nucleic acids and permeabilizes bilayers, *Biochemistry*, **36**, pp. 3008–3017.

43. Zhu, S., Lansakara, P.D., Li, X., Cui, Z. (2012). Lysosomal delivery of a lipophilic gemcitabine prodrug using novel acid-sensitive micelles improved its antitumor activity, *Bioconjug. Chem.*, **23**, pp. 966–980.

44. Lee, C.H., Cheng, S.H., Huang, I.P., Souris, J.S., Yang, C.S., Mou, C.Y., Lo, L.W. (2010). Intracellular pH-responsive mesoporous silica nanoparticles for the controlled release of anticancer chemotherapeutics, *Angew. Chem. Int. Ed. Engl.*, **49**, pp. 8214–8219.

45. Parrott, M.C., Finniss, M., Luft, J.C., Pandya, A., Gullapalli, A., Napier, M.E., DeSimone, J.M. (2012). Incorporation and controlled release of silyl ether prodrugs from PRINT nanoparticles, *J. Am. Chem. Soc.*, **134**, pp. 7978–7982.

46. Bernardos, A., Mondragon, L., Aznar, E., Marcos, M.D., Martinez-Manez, R., Sancenon, F., Soto, J., Barat, J.M., Perez-Paya, E., Guillem, C., Amoros, P. (2010). Enzyme-responsive intracellular controlled release using nanometric silica mesoporous supports capped with "saccharides", *ACS Nano.*, **4**, pp. 6353–6368.

47. Lee, J.S., Groothuis, T., Cusan, C., Mink, D., Feijen, J. (2011). Lysosomally cleavable peptide-containing polymersomes modified with anti-EGFR antibody for systemic cancer chemotherapy, *Biomaterials*, **32**, pp. 9144–9153.

48. Griset, A.P., Walpole, J., Liu, R., Gaffey, A., Colson, Y.L., Grinstaff, M.W. (2009). Expansile nanoparticles: Synthesis, characterization, and *in vivo* efficacy of an acid-responsive polymeric drug delivery system, *J. Am. Chem. Soc.*, **131**, pp. 2469–2471.

49. You, J.O., Auguste, D.T. (2009). Nanocarrier cross-linking density and pH sensitivity regulate intracellular gene transfer, *Nano. Lett.*, **9**, pp. 4467–4473.

50. Lin, J.H., Lu, A.Y. (1997). Role of pharmacokinetics and metabolism in drug discovery and development, *Pharmacol. Rev.*, **49**, pp. 403–449.

51. Kroemer, G., Jaattela, M. (2005). Lysosomes and autophagy in cell death control, *Nat. Rev. Cancer*, **5**, pp. 886–897.

52. Bareford, L.M., Swaan, P.W. (2007). Endocytic mechanisms for targeted drug delivery, *Adv. Drug Deliv. Rev.*, **59**, pp. 748–758.

53. Rajendran, L., Knolker, H.J., Simons, K. (2010). Subcellular targeting strategies for drug design and delivery, *Nat. Rev. Drug Discov.*, **9**, pp. 29–42.

54. Johannes, L., Romer, W. (2010). Shiga toxins–from cell biology to biomedical applications, *Nat. Rev. Microbiol.*, **8**, pp. 105–116.

55. Malik, A. (2009). Targeting endothelial cell surface receptors: Novel mechanisms of microvascular endothelial barrier transport, *J. Med. Sci.*, **2**, pp. 13–17.

56. Xu, Q., Liu, Y., Su, S., Li, W., Chen, C., Wu, Y. (2012). Anti-tumor activity of paclitaxel through dual-targeting carrier of cyclic RGD and transferrin conjugated hyperbranched copolymer nanoparticles, *Biomaterials*, **33**, pp. 1627–1639.

57. Saul, J.M., Annapragada, A.V., Bellamkonda, R.V. (2006). A dual-ligand approach for enhancing targeting selectivity of therapeutic nanocarriers, *J. Control. Release*, **114**, pp. 277–287.
58. Papademetriou, I.T., Garnacho, C., Schuchman, E.H., Muro, S. (2013). *In vivo* performance of polymer nanocarriers dually-targeted to epitopes of the same or different receptors, *Biomaterials*, **34**, pp. 3459–3466.
59. Tesar, D.B., Tiangco, N.E., Bjorkman, P.J. (2006). Ligand valency affects transcytosis, recycling and intracellular trafficking mediated by the neonatal Fc receptor, *Traffic*, **7**, pp. 1127–1142.
60. Marsh, E., Leopold, P., Jones, N., Maxfield, F. (1995). Oligomerized transferrin receptors are selectively retained by a lumenal sorting signal in a long-lived endocytic recycling compartment, *J. Cell Biol.*, **129**, pp. 1509–1522.
61. Ding, B.S., Dziubla, T., Shuvaev, V.V., Muro, S., Muzykantov, V.R. (2006). Advanced drug delivery systems that target the vascular endothelium, *Mol. Interv.*, **6**, pp. 98–112.
62. Wang, J., Tian, S., Petros, R.A., Napier, M.E., Desimone, J.M. (2010). The complex role of multivalency in nanoparticles targeting the transferrin receptor for cancer therapies, *J. Am. Chem. Soc.*, **132**, pp. 11306–11313.
63. Muzykantov, V. (2013). Targeted drug delivery to endothelial adhesion molecules, *ISRN Vasc. Med.*, **2013**.
64. Garnacho, C., Albelda, S.M., Muzykantov, V.R., Muro, S. (2008). Differential intra-endothelial delivery of polymer nanocarriers targeted to distinct PECAM-1 epitopes, *J. Control. Release.*, **130**, pp. 226–233.
65. Chacko, A.M., Nayak, M., Greineder, C.F., Delisser, H.M., Muzykantov, V.R. (2012). Collaborative enhancement of antibody binding to distinct PECAM-1 epitopes modulates endothelial targeting, *PLoS One*, **7**, pp. e34958.
66. Frey, A., Giannasca, K.T., Weltzin, R., Giannasca, P.J., Reggio, H., Lencer, W.I., Neutra, M.R. (1996). Role of the glycocalyx in regulating access of microparticles to apical plasma membranes of intestinal epithelial cells: Implications for microbial attachment and oral vaccine targeting, *J. Exp. Med.*, **184**, pp. 1045–1059.
67. Hatakeyama, H., Akita, H., Maruyama, K., Suhara, T., Harashima, H. (2004). Factors governing the *in vivo* tissue uptake of transferrin-coupled polyethylene glycol liposomes *in vivo*, *Int. J. Pharm.*, **281**, pp. 25–33.

68. Koval, M., Preiter, K., Adles, C., Stahl, P.D., Steinberg, T.H. (1998). Size of IgG-opsonized particles determines macrophage response during internalization, *Exp. Cell Res.*, **242**, pp. 265–273.
69. Champion, J.A., Mitragotri, S. (2006). Role of target geometry in phagocytosis, *Proc. Natl. Acad. Sci. U.S.A.*, **103**, pp. 4930–4934.
70. Beningo, K.A., Wang, Y.L. (2002). Fc-receptor-mediated phagocytosis is regulated by mechanical properties of the target, *J. Cell Sci.*, **115**, pp. 849–856.
71. Vasir, J.K., Labhasetwar, V. (2008). Quantification of the force of nanoparticle-cell membrane interactions and its influence on intracellular trafficking of nanoparticles, *Biomaterials*, **29**, pp. 4244–4252.
72. Yue, Z.G., Wei, W., Lv, P.P., Yue, H., Wang, L.Y., Su, Z.G., Ma, G.H. (2011). Surface charge affects cellular uptake and intracellular trafficking of chitosan-based nanoparticles, *Biomacromolecules*, **12**, pp. 2440–2446.
73. Basu, S.K., Goldstein, J.L., Anderson, R.G., Brown, M.S. (1981). Monensin interrupts the recycling of low density lipoprotein receptors in human fibroblasts, *Cell*, **24**, pp. 493–502.
74. Liang, W., Curran, P.K., Hoang, Q., Moreland, R.T., Fishman, P.H. (2004). Differences in endosomal targeting of human (beta)1- and (beta)2-adrenergic receptors following clathrin-mediated endocytosis, *J. Cell Sci.*, **117**, pp. 723–734.
75. Stein, B.S., Sussman, H.H. (1986). Demonstration of two distinct transferrin receptor recycling pathways and transferrin-independent receptor internalization in K562 cells, *J. Biol. Chem.*, **261**, pp. 10319–10331.
76. Bhat, M., Hickey, A.J. (2000). Effect of chloroquine on phagolysosomal fusion in cultured guinea pig alveolar macrophages: Implications in drug delivery, *AAPS PharmSci.*, **2**, pp. E34.
77. Carpentier, S., N'Kuli, F., Grieco, G., Van Der Smissen, P., Janssens, V., Emonard, H., Bilanges, B., Vanhaesebroeck, B., Gaide Chevronnay, H.P., Pierreux, C.E., Tyteca, D., Courtoy, P.J. (2013). Class III phosphoinositide 3-kinase/VPS34 and dynamin are critical for apical endocytic recycling, *Traffic*, **14**, pp. 933–948.
78. Futerman, A.H., van Meer, G. (2004). The cell biology of lysosomal storage disorders, *Nat. Rev. Mol. Cell Biol.*, **5**, pp. 554–565.
79. Muro, S. (2012). Strategies for delivery of therapeutics into the central nervous system for treatment of lysosomal storage disorders, *Drug Deliv. Transl. Res.*, **2**, pp. 169–186.

80. Lieberman, A.P., Puertollano, R., Raben, N., Slaugenhaupt, S., Walkley, S.U., Ballabio, A. (2012). Autophagy in lysosomal storage disorders, *Autophagy*, **8**, pp. 719–730.

81. Sanchez-Perez, A.M., Claramonte-Clausell, B., Sanchez-Andres, J.V., Herrero, M.T. (2012). Parkinson's disease and autophagy, *Parkinsons Dis.*, **2012**, pp. 429524.

82. Fraldi, A., Annunziata, F., Lombardi, A., Kaiser, H.J., Medina, D.L., Spampanato, C., Fedele, A.O., Polishchuk, R., Sorrentino, N.C., Simons, K., Ballabio, A. (2010). Lysosomal fusion and SNARE function are impaired by cholesterol accumulation in lysosomal storage disorders, *Embo J.*, **29**, pp. 3607–3620.

83. Xu, M., Liu, K., Swaroop, M., Porter, F.D., Sidhu, R., Firnkes, S., Ory, D.S., Marugan, J.J., Xiao, J., Southall, N., Pavan, W.J., Davidson, C., Walkley, S.U., Remaley, A.T., Baxa, U., Sun, W., McKew, J.C., Austin, C.P., Zheng, W. (2012). delta-Tocopherol reduces lipid accumulation in Niemann–Pick type C1 and Wolman cholesterol storage disorders, *J. Biol. Chem.*, **287**, pp. 39349–39360.

84. Seregin, S.S., Amalfitano, A. (2011). Gene therapy for lysosomal storage diseases: Progress, challenges and future prospects, *Curr. Pharm. Des.*, **17**, pp. 2558–2574.

85. Desnick, R.J., Schuchman, E.H. (2002). Enzyme replacement and enhancement therapies: Lessons from lysosomal disorders, *Nat. Rev. Genet.*, **3**, pp. 954–966.

86. Boado, R.J., Zhang, Y., Xia, C.F., Wang, Y., Pardridge, W.M. (2008). Genetic engineering of a lysosomal enzyme fusion protein for targeted delivery across the human blood–brain barrier, *Biotechnol. Bioeng.*, **99**, pp. 475–484.

87. Rappaport, J., Garnacho, C., Muro, S. (2014). Clathrin-mediated endocytosis is impaired in type A-B Niemann–Pick disease model cells and can be restored by ICAM-1-mediated enzyme replacement, *Mol. Pharm.*, **11**, pp. 2887–2895.

88. Garnacho, C., Dhami, R., Simone, E., Dziubla, T., Leferovich, J., Schuchman, E.H., Muzykantov, V., Muro, S. (2008). Delivery of acid sphingomyelinase in normal and Niemann–Pick disease mice using intercellular adhesion molecule-1-targeted polymer nanocarriers, *J. Pharmacol. Exp. Ther.*, **325**, pp. 400–408.

89. Hsu, J., Northrup, L., Bhowmick, T., Muro, S. (2012). Enhanced delivery of α-glucosidase for Pompe disease by ICAM-1-targeted nanocarriers: Comparative performance of a strategy for three distinct lysosomal storage disorders, *Nanomedicine*, **8**, pp. 731–739.

Chapter 13

Endolysosomal Escape into Cytosol

Manju Saraswathy and David Oupický
Center for Drug Delivery and Nanomedicine, Department of Pharmaceutical Sciences, University of Nebraska Medical Center, 985830 Nebraska Medical Center, Omaha, NE 68198-5830, USA
david.oupicky@unmc.edu

Despite tremendous progress in the development of nanosized drug delivery systems, the ability to deliver drugs to their intracellular site of action still remains a major challenge. Majority of nanocarriers used for intracellular delivery of therapeutic agents enter cells by endocytosis and are often routed for degradation into lysosomes. Thus, escape from the endolysosomal pathway is required for successful delivery of drugs that exert their activity in the cytosol, nucleus, and organelles other than lysosomes. Multiple approaches have been explored to facilitate endosomal escape of nanocarriers and subsequent cytosolic delivery of the delivered drugs. This chapter details the main and most successful strategies to achieve endosomal escape of nanocarriers.

13.1 Introduction

Nanocarriers are excellent platforms to enhance therapeutic efficacy of small-molecule drugs as well as macromolecular drugs such

Drug Delivery Across Physiological Barriers
Edited by Silvia Muro

ISBN 978-981-4669-40-5 (Hardcover), 978-981-4669-41-2 (eBook)
www.panstanford.com

as proteins, peptides, and nucleic acids by improved delivery to the designated sites of action with minimum systemic toxicity [1–6]. Nanocarriers based on various materials, including polymers, lipids, and inorganic materials, have been proposed and tested for drug delivery in multiple diseases, including cancer, diabetes, AIDS, malaria, and tuberculosis [7–10]. It is well recognized that physicochemical properties, such as size, size distribution, shape, surface charge, and surface hydrophilicity, play a major role in the ability of nanocarriers to systemically deliver drugs to the site of action [11–13]. For example, nanocarriers with sizes in the range of 10 to 200 nm exhibit most favorable pharmacokinetic properties and are best suited for systemic drug delivery thanks to prolonged plasma circulation, ability to penetrate deep into tissues, and ability to cross fenestrations present in the epithelial lining of some organs and tissues (e.g., liver and tumors). The *in vivo* design aspects of nanocarrier-mediated drug delivery have been reviewed extensively [14–16].

Following systemic delivery, nanocarriers have to enter cells and navigate through intracellular trafficking pathways to deliver their cargo to the desired intracellular organelles and to maximize the therapeutic effect [17, 18]. Appropriate nanocarriers should be selected depending on the location of the intracellular target. The intracellular targets may be located in the cytoplasm, nucleus, mitochondria or other subcellular compartments. Multiple factors control the ability of nanocarriers to deliver drugs to their intracellular targets, including binding to the cell surface, traversing plasma membrane, escaping lysosomal degradation, and overcoming the nuclear envelope [19–21]. This chapter details the importance of endolysosomal escape of nanocarriers and the strategies that were developed to facilitate the endolysosomal escape.

13.2 Importance of Endosomal Escape of Nanocarriers

Endocytosis has been established as a major cellular uptake mechanism of nanocarriers. Different mechanisms of endocytosis that are distinguished by specific molecular regulators, such as clathrin-mediated endocytosis, caveolae-mediated endocytosis, macropinocytosis, and other clathrin- and caveolae-independent endocytosis [e.g., flotillin-dependent endocytosis and RhoA (Ras homolog gene family, member A)-dependent endocytosis], are well described in the literature and are discussed in other chapters of this book [22–24]. The multitude of endocytic uptake mechanisms depends on the formation of intracellular vesicles following invagination of the cell membrane. The intracellular vesicles with entrapped nanocarriers eventually fuse with early or sorting endosomes. Large fraction of the nanocarriers may be recycled back to the membrane and transported out of the cell. Often, the nanocarriers are trafficked to late endosomes, which undergo rapid acidification to pH 6–4.9 by the action of membrane-bound ATPase proton pumps. Eventually, late endosomes fuse with acidic lysosomes (pH ~4.5) that contain numerous degradative enzymes. The routing of nanocarriers into lysosomes results in a potential degradation of drugs susceptible to easy enzymatic degradation (nucleic acids, proteins, peptides). Even if not degraded, drugs that remain sequestered in the lysosomes are unable to exert their therapeutic effect, unless lysosomes are their site of action. Therefore, it is believed that enzymatically sensitive therapeutics, such as nucleic acids and proteins, should escape from the endosomal compartment at an early stage to avoid lysosomal degradation [4, 25]. On the other hand, some types of nanocarriers require exposure to lysosomal enzymes or acidic pH to trigger drug release and in those cases, early release from the endosomes might be counterproductive. Thus, the timing of endosomal release has to be considered within the context of each nanocarrier design individually. Numerous strategies have been devised to facilitate and improve the escape of nanocarriers from the endolysosomal trafficking and to enhance the therapeutic effect of the delivered therapeutic agents [26–28]. The most common strategies are discussed in the following sections.

13.3 Endosomal Escape Strategies

As discussed earlier, it is important to facilitate escape of nanocarriers that carry sensitive therapeutics from the endosomes at an early stage to improve the therapeutic efficacy. Multiple mechanisms

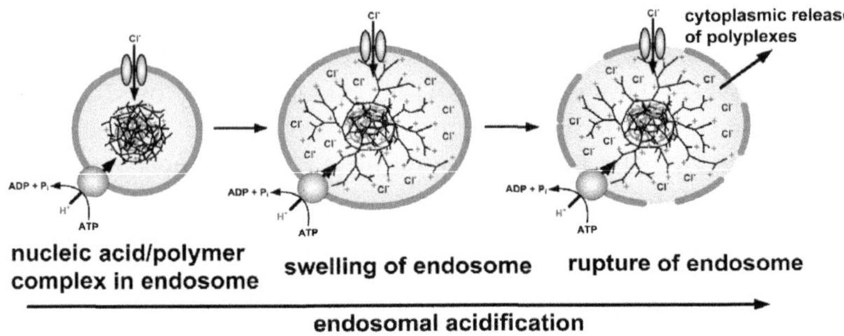

Figure 13.1 Schematic representation of the proton sponge effect, leading to late endolysosomal burst release of nanocarriers into the cytoplasm. Reprinted from Ref. 34, Copyright 2012, with permission from American Chemical Society.

of endosomal escape of nanocarriers have been put forward in the last two decades. Many of the successful approaches have resulted from improved understanding of viral and bacterial mechanisms of endosomal escape [29–31]. Other methods, such as proton sponge effect, have no counterpart in nature and are unique to the synthetic nanocarriers.

13.3.1 *Proton Sponge Effect*

The "proton sponge" effect or pH-buffering effect has been among the most widely exploited mechanisms of endosomal escape of polycation-based systems for delivery of nucleic acids, such as DNA and siRNA [32, 33]. The proton sponge effect utilizes the fact that during endosomal maturation, membrane-bound V-ATPase actively pumps protons from cytoplasm to the endosomes, which leads to the acidification of the endosomal compartment. Under such acidic conditions, polycations with pK_a close to the endosomal pH become protonated and that induces extensive inflow of ions and water into the endosomes. The subsequent osmotic swelling leads to the rupture of the endosomal membrane (endosomolysis) and results in the release of the nanocarriers into the cytoplasm (Fig. 13.1) [34].

Despite the widespread use of the proton sponge effect as the guiding rationale for the development of new polycations, multiple reports suggest that proton sponge may not be the dominant effect behind enhanced endosomal escape of such carriers [33, 35]. Benjaminsen and colleagues used nanoparticle-based pH sensor to measure lysosomal pH in cells treated with the most successful proton sponge polycation poly(ethyleneimine) (PEI) [33]. Contrary to expectations, the results showed that PEI caused no pH change in lysosomes. It was reasoned that even though PEI has buffering capability, the capacity of the V-ATPase pump is large enough to compensate and maintain lysosomal pH by increasing the influx of protons. A complementary hypothesis has been proposed to better explain endosomal escape of polycationic nanocarriers such as dendrimers. According to this umbrella hypothesis, polycations expand volumetrically during protonation at low pH. The electrostatic repulsion between the neighboring positively charged amine groups facilitates the unfolding of the terminal branches of the polymer, which then assumes a fully extended conformation. It was demonstrated that increase in volume caused by polymer swelling within the constrained space of endosome triggered the disruption of endosomal membrane and facilitated the release of polyplexes into the cytoplasm. The umbrella hypothesis was confirmed using a number of polyamidoamine (PAMAM) dendrimers with different degrees of flexibility and ability to expand volumetrically at the endolysosomal pH range. It was found that dendrimers with optimal flexibility of their branches exhibited an increased apparent volume and mediated superior transfection efficiency compared to sterically constrained intact dendrimers [34, 36]. Additional support for this hypothesis was provided by studies with covalently crosslinked polyplexes that suggested that constraining molecular movement within polyplexes impaired endosomal escape [37].

Nanocarriers with high density of primary, secondary, and tertiary amines possess a greater buffering potential than those having only primary or secondary amines. This is believed to explain the enhanced endosomal escape and increased cytosolic delivery by polycations such as PEI. A number of cationic polymers with high buffering capacity at the endosomal pH have been developed as successful vectors for nucleic acid delivery [34, 38–40]. Among them, PEI has been the most successful example of this class of polycations and transfection reagents. PEI has been demonstrated to successfully transfect nucleic acids into a broad range of cells both *in vitro* and *in vivo* [41, 42]. However, clinical application of PEI has been limited due to its toxicity [43]. Because toxicity of PEI is closely associated with the number of ionizable amines, which are required to facilitate endosomal escape, many studies have attempted to reduce the toxicity of PEI without negatively affecting its endosomal escape ability. For example, a recent study by Seow and colleagues demonstrated reduced toxicity of PEI/DNA complexes by *in situ* oxidation of PEI amines without any adverse effect on its buffering capacity [44]. In another study, Gabrielson and colleagues reported that partial acetylation of branched PEI can decrease toxicity and enhance DNA transfection. It was found that, despite the decreased number of protonizable amines, the weakened polymer/DNA interactions by the acetylation of the PEI amines were beneficial for the release of DNA from the polyplexes and that this was responsible for the enhanced transfection [45].

Guided by the proton sponge hypothesis, multiple reports showed that endosomal escape of cationic polymers with low transfection efficacy can be improved by modification with imidazole moieties (e.g., histidine) [46]. It was argued that, similar to PEI, endosomal protonation of the histidyl imidazole ring (pK_a ~6) can trigger endosomal membrane rupture and release of nanocarriers into the cytoplasm. Polylysine (PLL) was one of the first polycations used for gene delivery [47]. However, it exhibited low transfection efficiency believed to be due to the lack of buffering capacity. In an attempt to improve PLL activity, the polymer was modified with imidazole and the resulting polyplexes showed significant enhancement in transfection [48, 49]. A similar effect of imidazole modification on increased transfection was reported for naturally occurring cationic polysaccharide chitosan. For example, histidine-modified chitosan facilitated more efficient endosomal escape as documented by widespread cytoplasmic distribution of DNA in a study by Chang and colleagues [50]. Similarly, incorporation of histamine into the structure of nanoparticles with cationic shell enhanced cytoplasmic delivery of siRNA [51].

Dendrimers are another important class of successful polycations in gene delivery that take advantage of the proton sponge effect [52]. In particular, PAMAM dendrimers have a large number of tertiary and primary amino groups and show significant pH buffering effect [53, 54]. In the case of dendrimers, the primary amines on the surface enable efficient condensation of nucleic acids, while the inner amines act as proton sponge, which enhances the endosomal escape. Chemical modification of PAMAM dendrimers has been reported to enhance the endosomal escape of nucleic acids. For example, modification with histidine significantly enhanced DNA transfection of PAMAM dendrimers [55].

Recent advances in high-throughput methods of polymer synthesis resulted in accelerated discovery of biodegradable polycations with high transfection activity and low toxicity [56]. In particular, poly(β-amino ester)s emerged as important class of pH-buffering polycations with high transfection activity [57–59].

A novel approach to enhancing endosomal escape by increasing the proton-sponge-mediated osmotic pressure in the endosomes was reported by Lim and colleagues [60]. The authors synthesized polycations degradable in the acidic pH of the endolysosomal pathway and used them to deliver siRNA. The results suggested that the pH-buffering ability of the polycations, combined with the endosomal degradation of the polycations, led to increased osmotic pressure in the endosomes and more efficient endosomal release of the siRNA.

13.3.2 *Photochemical Disruption of Endosomal Membrane*

Another method to disrupt endosomes and facilitate cytoplasmic delivery of nanocarriers is photochemical internalization (PCI). In PCI, nanocarriers are internalized along with photosensitizers that integrate into the plasma membrane and distribute into endolysosomal membranes during endocytosis. Exposure of the treated cells to light of a photosensitizer-specific wavelength results in the disruption of the endolysosomal membranes [61–63]. The mechanism of membrane disruption relies on the controlled generation of reactive oxygen species in the membranes and subsequent oxidative damage. This approach has been used successfully in the delivery of small-molecule drugs (e.g., doxorubicin, bleomycine), DNA, siRNA, and proteins (e.g., gelonin) [62, 64–67].

Photosensitizers used in clinical applications are usually highly reactive substances with short range of action (10–20 nm) and

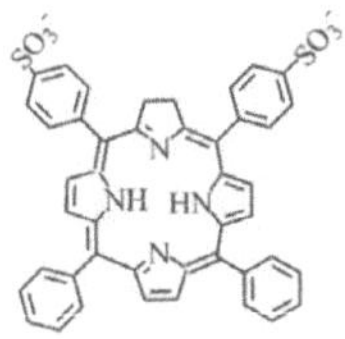

Disulfonated meso-tetraphenylporphin (TPPS2a)

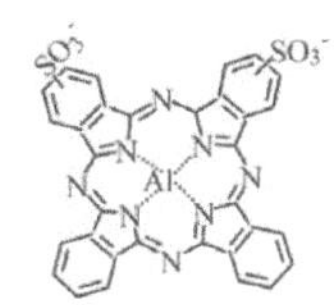

Disulfonated aluminum phthalocyanine (AlPcS2a)

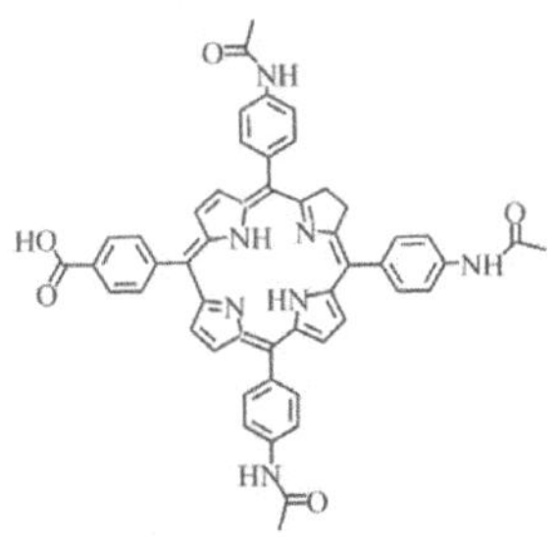

5,10,15-tri(4-acetamidophenyl)-20-mono(4-carboxyl-phenyl)porphyrin (TAMCPP)

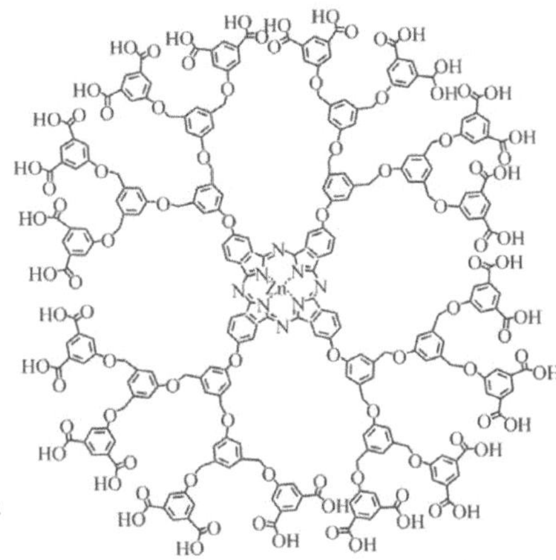

Dendrimer phthalocyanine (DPc)

Figure 13.2 Examples of common photosensitizers for photochemical internalization.

short lifetime (0.01–0.04 μs) to restrict the damaging effect to the endosomal membrane and to avoid damage to other cellular components [68, 69]. Photosensitizers used in PCI are typically amphiphilic compounds that can bind to the plasma membrane and concentrate in the endosomal membranes after endocytosis. Examples of common photosensitizers are shown in Fig. 13.2 [63, 70].

PCI is capable of strongly enhancing the transfection activity of DNA delivered by polyplexes. Available studies demonstrate the positive effect of PCI for enhanced transfection of PLL polyplexes *in vitro* [63]. Similar to PLL, protamine sulfate is a relatively weak transfection agent due to its limited ability to facilitate endosomal escape. The use of PCI in combination with protamine sulfate/DNA polyplexes increased the transfection almost 10-

fold [71]. The difference in transfection was mainly due to the increased endosomal escape of the polyplexes. PCI can even enhance endosomal escape of proton sponge polycations like PEI, suggesting low efficacy of the proton sponge mechanism. For example, PCI improved the delivery of PEI/DNA polyplexes as documented by a 600-fold enhancement of luciferase gene expression [72].

Effectiveness of PCI in enhancing the delivery of small-molecule drugs and proteins has been demonstrated with the dendrimer conjugates of doxorubicin, camptothecin, and saporin [66, 73, 74]. In another study, Mathews and colleagues demonstrated improved delivery of the antibiotic bleomycin to glioma cells by PCI, which resulted in increased antitumor activity [75]. Applicability of PCI to improve cytoplasmic delivery of inorganic nanoparticles has been demonstrated by Mackowiak and colleagues on mesoporous silica nanoparticles. The nanoparticles were covalently modified with a photosensitizer activated by red light, and proof-of-principle studies confirmed enhanced cytosolic delivery of fluorescent dye calcein, which was encapsulated in the silica nanoparticles [76].

PCI is clearly a highly effective method of enhancing endosomal release. The main problem that prevents its widespread applicability is limited penetration depth of light in tissues. Advances in fiber optics and laser technology, however, appear to extend the possibility of controlled illumination of tissues deep in the body in places such as the gastrointestinal tract, lungs, and pancreas [77].

13.3.3 *Temperature-Triggered Endosomal Escape*

Another externally triggered approach used to improve the endosomal escape relies on temperature-responsive nanocarriers. Although not investigated as much as PCI or proton sponge, available evidence suggests that endosomal escape triggered by temperature changes (hypothermia, hyperthermia) is a viable option. Such nanocarriers are characterized by a critical phase-transition temperature at which they undergo nonlinear change in their volume [15, 78]. The abrupt change in nanocarrier volume in a constrained space of the endosome is believed to trigger disruption of the endosomal membrane and release of the nanocarriers into the cytoplasm. For example, Pluronic block copolymers exhibit a thermally reversible swelling/deswelling behavior, which has been shown to trigger endosomal escape of siRNA [79]. The rapid volume expansion of the Pluronic nanocarriers induced by a short cold-shock treatment could burst endosomal membrane and deliver siRNA into the cytoplasm. Similar mechanism of action was used to improve endosomal escape of nanogels based on copolymers of lactic acid with Pluronics and PLL. The reported nanogels underwent reversible volume transition from nano- to microscale particles in response to a temperature decrease (150 nm at 37°C versus 1.4 μm at 15°C) [80]. In addition to rupturing endosomal membrane, the nanogels also induced necrotic cell death by destabilizing cytoskeleton and structural integrity of the cellular membranes by increased internal hydrostatic pressure.

13.3.4 *Endosomal Escape by Membrane-Active Peptides*

Membrane-active peptides have been extensively used as part of drug delivery systems because of their ability to destabilize endosomal membranes and facilitate cytoplasmic delivery. The mechanism of action of membrane-active peptides relies on endosomal membrane fusion or pore formation. Most of the reported peptides are either cationic and/or amphiphilic in nature, and depending on their physicochemical characteristics, they exhibit different levels of selectivity for endosomal membranes (Table 13.1).

Endosomal membrane fusion is one of the main escape mechanisms of nanocarriers driven by fusogenic peptides or cell-penetrating peptides [81]. This strategy is inspired by the escape mechanism of enveloped viruses, which induce endosomal membrane disruption by fusion between the virus and endosomal membranes at acidic pH [82]. For example, the HA2 subunit of hemagglutinin glycoprotein present in the coat of the influenza virus acts as a fusogenic agent and helps the virus to efficiently disrupt endosomal membrane to enable cytosolic translocation [83]. During the process, hemagglutinin undergoes conformational change from an anionic hydrophilic coil at pH 7.4 to a hydrophobic and membrane-active α-helical structure at acidic endosomal pH. The α-helical peptide structures lead to the fusion of viral and endosomal

Table 13.1 Endosomolytic proteins and peptides

Proteins/Peptides	Source	Endosomal escape mechanism
Hemagglutinin (HA2)	Virus	Membrane fusion
INF-7	Virus	Membrane fusion
Penton base	Virus	Pore formation
gp41	Virus	Pore formation and membrane fusion
TAT	Virus	Unclear
L2 peptide	Virus	Membrane fusion
Diphtheria toxin (DT)	Bacteria	Membrane fusion
Pneumococcal Pneumolysin (PLO)	Bacteria	Pore formation
Pseudomonas aeruginosa exotoxin A (ETA)	Bacteria	Pore formation
Gelonin	Plant	Unclear
Melittin	Bee venom	Pore formation
(R-Ahx-R)(4) AhxB	Synthetic	Unclear
Glycoprotein H (gpH) from herpes simplex	Synthetic	Membrane fusion/flip-flop of membrane lipid
GALA	Synthetic	Membrane fusion
KALA	Synthetic	Membrane fusion
EB1	Synthetic	Unclear
CADY	Synthetic	Bypass endosome through nonendosomal uptake
HEL peptide and analogue	Synthetic	Membrane fusion
Penetratin	Synthetic	Unclear

Source: Adapted from Refs. [30, 77].

membranes and in subsequent membrane destabilization and release of the viral genome into the cytoplasm (Fig. 13.3) [84, 85].

The HA2 subunit of hemagglutinin protein of the influenza virus was used successfully to impart endosomolytic activity to transferrin-targeted PLL/DNA complexes as early as 1992 [82]. Several synthetic peptides were developed based on the N-terminal sequence of the HA2 subunit of hemagglutinin. Among the most successful ones has been a peptide named GALA. GALA is a synthetic amphiphilic peptide with a Gly-to-Glu substitution at position 4 of the mutant sequence [30, 86]. When the pH of the surrounding environment decreases from 6.5 to 5.0, GALA undergoes conformational change from a random coil to an amphipathic α-helix. In one of the early studies, GALA-modified dendrimers demonstrated enhanced

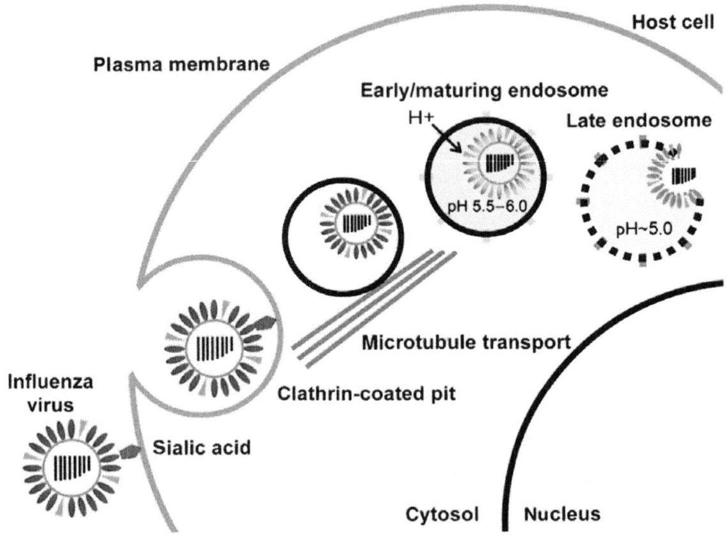

Figure 13.3 Schematic representation of influenza A virus endocytosis followed by the endosomal membrane fusion at pH ~5. Reprinted from Ref. 85, Copyright 2010, with permission from Nature Publishing Group.

transfection efficacy due to improved endosomal escape [87]. Similarly, GALA-modified nanoparticles demonstrated enhanced gene silencing by siRNA compared with unmodified nanoparticles [27]. Even though negatively charged GALA effectively facilitates endosomal escape of nanocarriers, it could not be used alone for nucleic acid delivery. In order to overcome these limitations, cationic amphipathic peptide (KALA) was developed. KALA is a modified version of GALA in which Glu is partially replaced with Lys. KALA adopts α-helix conformation in a wider pH range and undergoes a pH-dependent conformational change from amphipathic α-helical to a mixture of α-helix and random coil as the pH is lowered [88]. Guo and colleagues studied a number of KALA-based α-helical peptides as nonviral gene carriers [89]. The authors varied the content of cationic amino acids (Lys and His) in the KALA peptides to identify optimal composition for maximum transfection efficacy. In order to overcome problems associated with low molecular weight, the KALA peptide was oligomerized via disulfides formed by Cys residues introduced into the peptide structure. The oligomerization of the peptides did not negatively affect the endosomolytic activity of the systems, which demonstrated greater transfection efficiency, but lower cytotoxicity as compared to PEI.

Synthetic pH-sensitive polymers that can mimic the function of fusogenic peptides were developed as excellent safer alternatives to peptides [90, 91]. In particular, it was the pioneering work by Tirrell and colleagues on pH-dependent membrane disruption by poly(ethylacrylic acid) that paved the way for these synthetic alternatives [92]. Similar to GALA peptides, these polyacids are designed to be inactive at physiological pH, while becoming membrane active at lower pH of endosome. Stayton and colleagues demonstrated that pH-sensitive copolymers based on propylacrylic acid can enhance the transfection efficacy of various gene-delivery systems [93, 94].

Pore formation is another endosomal escape mechanism utilized in the design of nanocarriers that was inspired by nature. For example, nonenveloped viruses may display capsid-associated penetration proteins that induce the formation of pores in the endosomal membrane and facilitate viral transport into the cytosol. The pore formation is based on the interplay between membrane tension that enlarges the pore and line tension that closes the pore [95, 96]. Pore formation is initiated by the binding of the peptides onto the lipid bilayer of the endosomal membrane. The endosomal membrane tension induced by these peptides then leads to the formation of structural defects called pores.

A variety of theoretical structural models (Fig. 13.4) have been used to describe the mechanism of pore formation [97–99]. In the barrel-stave model, for example, peptides form a barrel from stave-like peptide cluster and orient themselves perpendicular to the plane of the membrane to form transmembrane pores. Pore formation described by toroidal pore model is also mediated by aggregates of peptides that enter into the membrane in a perpendicular orientation. This model differs from the barrel-stave model in that the peptides are always associated with the lipid headgroups. In forming such a pore, the lipid monolayer bends continuously from the top to the bottom like a "toroidal hole" [100]. In the carpet model, endosomolytic peptides accumulate on the membrane surface until its integrity is ruptured and transient

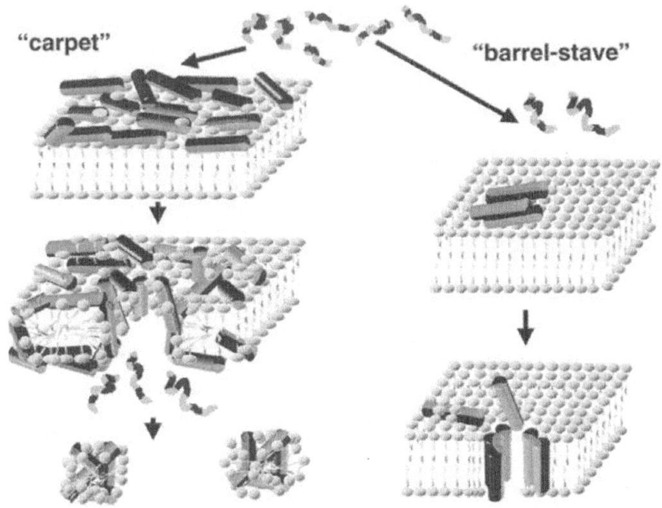

Figure 13.4 Schematic representation of pore formation: barrel-stave model and carpet model. Reprinted from Ref. 100, Copyright 2001, with permission from Elsevier.

holes are formed. Most often, high concentrations of endosomolytic peptides result in the complete collapse of the membrane [101]. Mellitin, from honeybee venom, is one of the most widely studied pore-forming peptides. For example, cytoplasmic delivery of siRNA could be enhanced by conjugating mellitin to nanocarriers [102].

Rationally designed pore-forming polymers have been developed recently, which provide similar endosomal escape functionality as natural peptides. For example, Yin and colleagues showed that complexes formulated with helix-forming copolymers based on poly(L-glutamate) and chitosan derivatives can enhance transfection efficacy of DNA via pore formation in the endosomal membrane [103].

13.3.5 *Endosomal Escape by Lipid-Mediated Mechanisms*

Lipid-based nanocarriers have been extensively investigated for intracellular targeting of various drugs. In particular, cationic lipids have received significant attention because of their ability to facilitate delivery of DNA and siRNA into the nucleus and cytoplasm, respectively. There are several basic mechanisms by which cationic lipids allow nucleic acids to escape from the endolysosomes.

Flip-flop mechanism explains endosomal escape of complexes of cationic lipids with nucleic acids (lipoplexes). This mechanism of action is driven by the ability of membrane phospholipids to flip to opposing membrane bilayer leaflets [104, 105]. During the process, the endosomal membrane becomes destabilized by entrapped cationic lipoplexes, which triggers the anionic lipids facing the cytoplasmic side of the membrane to flip and face the endosomal lumen. The anionic lipids of the endosomal membrane laterally diffuse into the lipoplexes and form charge-neutralized ion pairs. As a result, the nucleic acids dissociate from the lipoplexes and are released into the cytoplasm (Fig. 13.5) [106].

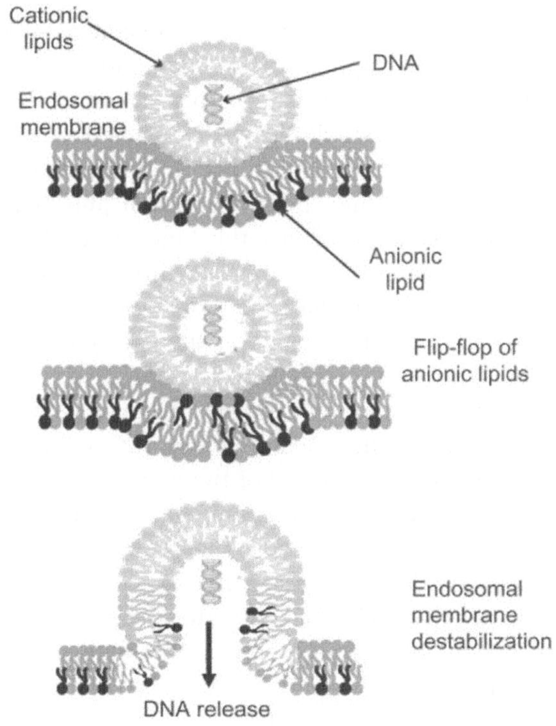

Figure 13.5 Schematic representation of endosomal escape of lipoplexes via flip-flop mechanism. Reprinted from Ref. 106, Copyright 2008, with permission from Elsevier.

Another widely explored mechanism of endosomal escape associated with the use of lipid nanocarriers relies on the incorporation of pH-sensitive moieties or fusogenic lipids. Fusogenic lipids undergo a phase transition in acidic conditions that further enhances the destabilization of endolysosomal membrane via flip-flop mechanism. DOPE (1,2-dioleoyl-sn-glycero-3-phosphoethanolamine) is an excellent example of a fusogenic lipid that can be easily incorporated into lipid nanocarriers [2, 107]. The transformation of DOPE from lipid bilayer structure at pH 7.4 to inverted hexagonal structure (fusogenic phase) at acidic pH contributes to the destabilization of endosomal membrane.

Various strategies have been proposed for the formulation of pH-sensitive lipid nanocarriers. For example, pH-sensitive liposomes can be prepared that are stable at physiological pH but become protonated in the acidic environment of the endosome to facilitate endosomal membrane rupture [108]. Allon and colleagues studied gene delivery to the lung using lipid nanocarriers prepared with DOPS (1,2-dioleoyl-sn-glycero-3-phospho-L-serine), DOPE, and cholesterol [109]. The low endosomal pH protonates the amino groups and neutralizes the carboxyl group of phosphatidylserines (PS). The positive charge brings the lipids into close proximity to the negatively charged endosomal membrane, which facilitates membrane destabilization via the flip-flop mechanism and drug release into the cytoplasm. In general, increasing the PS content in the lipid nanocarriers improves endosomal escape.

PEGylation is a widely employed method to enhance the stability of lipid nanocarriers in the bloodstream. However, PEGylation interferes with the endosomal escape of the nanocarriers by inhibiting fusion of the lipids with endosomal membrane. Multiple strategies have been developed to overcome the problems associated with the use of PEGylation. Chan and colleagues studied pH-sensitive hydrazone linker to attach PEG to lipoplexes as a way of enhancing endosomal escape [8]. In this approach, PEG is selectively removed from the lipoplexes in the acidic environment of late endosomes. The removal of PEG chains enables the electrostatic interactions between the cationic lipids and endosomal membrane, leading to membrane fusion and release of the encapsulated cargo into the cytoplasm. In a similar fashion, Guo and colleagues designed a biocompatible lipid nanocarrier with good serum stability using pH-sensitive orthoester linker [110]. The nanocarrier hydrolyzed rapidly during endocytosis, which resulted in endosomal escape via the flip-flop mechanism.

13.4 Conclusions and Perspectives

Despite tremendous progress in the development of various nanocarriers for delivery of drugs, intracellular targeting remains a major challenge. Among the main hurdles that limit our ability to target drug delivery to most intracellular locations is inefficient endosomal escape of the nanocarriers. As discussed in this chapter, different approaches have been proposed and significant progress has been made in our ability to release nanocarriers from the endolysosomal pathway. However, the efficiency of endosomal escape remains low for most nanocarriers. Improved understanding of intracellular trafficking of nanocarriers will be required to promote design and development of the next generation of approaches that can safely deliver therapeutic agents to the desired intracellular locations.

References

1. Couvreur, P. (2013). Nanoparticles in drug delivery: Past, present and future, *Adv. Drug Deliv. Rev.*, **65**, pp. 21–23.
2. Farhood, H., Serbina, N., and Huang, L. (1995). The role of dioleoyl phosphatidylethanolamine in cationic liposome mediated gene transfer *BBA-Biomembranes*, **1235**, pp. 289–295.
3. Gu, W., et al. (2013). Polymer nanocarrier system for endosome escape and timed release of siRNA with complete gene silencing and cell death in cancer cells, *Biomacromolecules*, **14**, pp. 3386–3389.
4. Gu, Z., et al. (2011). Tailoring nanocarriers for intracellular protein delivery, *Chem. Soc. Rev.*, **40**, pp. 3638–3655.
5. Panyam, J., and Labhasetwar, V. (2012). Biodegradable nanoparticles for drug and gene delivery to cells and tissue, *Adv. Drug Deliv. Rev.*, **55**, pp. 329–347.

6. Parveen, S., Misra, R., and Sahoo, S.K. (2012). Nanoparticles: A boon to drug delivery, therapeutics, diagnostics and imaging, *Nanomedicine,* **8**, pp. 147–166.
7. Almeida, A.J., and Souto, E. (2007). Solid lipid nanoparticles as a drug delivery system for peptides and proteins, *Adv. Drug Deliv. Rev.* **59**, pp. 478–490.
8. Chan, C.-L., et al. (2012). Endosomal escape and transfection efficiency of PEGylated cationic liposome–DNA complexes prepared with an acid-labile PEG-lipid, *Biomaterials,* **33**, pp. 4928–4935.
9. Ojea-Jimenez, I., et al. (2013). Engineered inorganic nanoparticles for drug delivery applications *Curr. Drug Metab.,* **14**, pp. 518–530.
10. Pusic, K., et al. (2013). Iron oxide nanoparticles as a clinically acceptable delivery platform for a recombinant blood-stage human malaria vaccine, *FASEB J.* **27**, pp. 1153–1166.
11. Dobrovolskaia, M.A., and McNeil, S.E. (2007). Immunological properties of engineered nanomaterials *Nat. Nanotechnol.,* **2**, pp. 469–478.
12. Ernsting, M.J., et al. (2013). Factors controlling the pharmacokinetics, biodistribution and intratumoral penetration of nanoparticles, *J Control Release,* **172**, pp. 782–794.
13. Kulkarni, S.A., and Feng, S.S. (2013). Effects of particle size and surface modification on cellular uptake and biodistribution of polymeric nanoparticles for drug delivery, *Pharm Res.* **30** pp. 12–22
14. Chacko, A.-M., et al. (2011). Targeted nanocarriers for imaging and therapy of vascular inflammation, *Curr. Opin. Colloid Interface Sci,* **16**, pp. 215–227.
15. Ganta, S., et al. (2008). A review of stimuli-responsive nanocarriers for drug and gene delivery *J. Control. Release,* **126**, pp. 187–204.
16. Zhang, J., and Ma, P.X. (2013). Cyclodextrin-based supramolecular systems for drug delivery: Recent progress and future perspective, *Adv. Drug Deliv. Rev.* **65** pp. 1215–1233.
17. Torchilin, V.P. (2006). Recent approaches to intracellular delivery of drugs and DNA and organelle targeting, *Annu. Rev. Biomed. Eng.,* **8**, pp. 343–375.
18. Kocbek, P., et al. (2013). Targeting intracellular compartments by magnetic polymeric nanoparticles, *Eur. J. Pharm. Sci.,* **50**, pp. 130–138.
19. Bally, M.B., et al. (1999). Biological barriers to cellular delivery of lipid-based DNA carriers, *Adv. Drug Deliv. Rev.,* **38**, pp. 291–315.

20. Gottfried, L.F., and Dean, D.A. (2013). Extracellular and intracellular barriers to non-viral gene transfer. In *Novel Gene Therapy Approaches* Wei, M., and Good, D. (Eds.), InTech, pp. 75–88.
21. Schneider, M., et al. (2013). Crossing biological barriers for advanced drug delivery, *Eur. J. Pharm. Biopharm.* **85**, pp. 521–530.
22. Duncan, R., and Richardson, S.C. (2012). Endocytosis and intracellular trafficking as gateways for nanomedicine delivery: Opportunities and challenges, *Mol Pharm,* **9**, pp. 2380–2402.
23. Hillaireau, H., and Couvreur, P. (2009). Nanocarriers' entry into the cell: Relevance to drug delivery, *Cell Mol Life Sci.,* **66**, pp. 2873–2896.
24. Zaki, N.M., and Tirelli, N. (2010). Gateways for the intracellular access of nanocarriers: A review of receptor-mediated endocytosis mechanisms and of strategies in receptor targeting, *Exp. Opin. Drug Deliv.,* **7**, pp. 895–913.
25. Guo, X., and Huang, L. (2011). Recent advances in nonviral vectors for gene delivery, *Acc. Chem. Res.* **45**, pp. 971–979.
26. Godbey, W., Wu, K.K., and Mikos, A.G. (1999). Poly(ethylenimine) and its role in gene delivery, *J. Control. Release,* **60**, pp. 149–160.
27. Hatakeyama, H., et al. (2009). A pH-sensitive fusogenic peptide facilitates endosomal escape and greatly enhances the gene silencing of siRNA-containing nanoparticles *in vitro* and *in vivo*, *J. Control. Release,* **139**, pp. 127–132.
28. Khalil, I., et al. (2007). Octaarginine-modified multifunctional envelope-type nanoparticles for gene delivery, *Gene Ther.,* **14**, pp. 682–689.
29. Sawant, R.R., Patel, N.R., and Torchilin., V.P. (2013). Therapeutic delivery using cell-penetrating peptides, *Eur. J. Nanomedicine,* **5**, pp. 141–158.
30. Varkouhi, A.K., et al. (2011). Endosomal escape pathways for delivery of biologicals, *J. Control. Release,* **151**, pp. 220–228.
31. Vasir, J.K., and Labhasetwar, V. (2007). Biodegradable nanoparticles for cytosolic delivery of therapeutics, *Adv Drug Deliv Rev,* **59**, pp. 718–728.
32. Akinc, A., et al. (2005). Exploring polyethyleniminemediated DNA transfection and the proton sponge hypothesis, *J. Gene Med.* **7**, pp. 657–663.
33. Benjaminsen, R.V., et al. (2012). The possible "proton sponge" effect of polyethylenimine (PEI) does not include change in lysosomal pH, *Mol. Ther.* **21**, pp. 149–157.

34. Nguyen, J., and Szoka, F.C. (2012). Nucleic acid delivery: The missing pieces of the puzzle? *Acc Chem. Res.* **45**, pp. 1153–1162.
35. Funhoff, A.M., et al. (2004). Endosomal escape of polymeric gene delivery complexes is not always enhanced by polymers buffering at low pH, *Biomacromolecules*, **5**, pp. 32–39.
36. Tang, M.X., Redemann, C.T., and Szoka, F.C. (1996). *In vitro* gene delivery by degraded polyamidoamine dendrimers, *Bioconjugate Chem*, **7**, pp. 703–714.
37. Oupicky, D., Carlisle, R.C., and Seymour, L.W. (2001). Triggered intracellular activation of disulfide crosslinked polyelectrolyte gene delivery complexes with extended systemic circulation *in vivo*, *Gene Ther.*, **8**, pp. 713–724.
38. Oba, M. (2013). Study on development of polymeric micellar gene carrier and evaluation of its functionality, *Biol. Pharm. Bull.* **36**, pp. 1045–1051.
39. Thomas, M., and Klibanov, A. (2003). Non-viral gene therapy: Polycation-mediated DNA delivery, *Appl. Microbiol. Biotechnol.*, **62**, pp. 27–34.
40. Chen, J., Wu, C., and Oupický, D. (2009). Bioreducible hyperbranched poly(amido amine)s for gene delivery, *Biomacromolecules*, **10**, pp. 2921–2927.
41. Pfeifer, C., et al. (2011). Dry powder aerosols of polyethylenimine (PEI)-based gene vectors mediate efficient gene delivery to the lung, *J. Control. Release*, **154**, pp. 69–76.
42. Urban-Klein, B., et al. (2005). RNAi-mediated gene-targeting through systemic application of polyethylenimine (PEI)-complexed siRNA *in vivo*, *Gene Ther*, **12**, pp. 461–466.
43. Chollet, P., et al. (2002). Side-effects of a systemic injection of linear polyethylenimine-DNA complexes, *J. Gene Med.*, **4**, pp. 84–91.
44. Seow, W.Y., et al. (2013). Oxidation as a facile strategy to reduce the surface charge and toxicity of polyethyleneimine gene carriers, *Biomacromolecules*, **14**, pp. 2340–2346.
45. Gabrielson N.P., and Pack, D.W. (2006). Acetylation of polyethylenimine enhances gene delivery via weakened polymer/DNA interactions, *Biomacromolecules*, **7**, pp. 2427–2435.
46. Midoux, P., et al. (2009). Chemical vectors for gene delivery: A current review on polymers, peptides and lipids containing histidine or imidazole as nucleic acids carriers, *Br. J. Pharmacol*, **157**, pp. 166–178.

47. Pack, D.W., et al. (2005). Design and development of polymers for gene delivery, *Nat. Rev. Drug Discov.*, **4**, pp. 581–593.

48. Benns, J.M., et al. (2000). pH-sensitive cationic polymer gene delivery vehicle: N-Ac-poly(L-histidine)-graft-poly(L-lysine) comb shaped polymer, *Bioconjugate Chem.* **11**, pp. 637–645.

49. Midoux, P., and Monsigny, M. (1999). Efficient gene transfer by histidylated polylysine/pDNA complexes, *Bioconjugate Chem.* **10**, pp. 406–411.

50. Chang, K.-L., et al. (2010). Efficient gene transfection by histidine-modified chitosan through enhancement of endosomal escape, *Bioconjugate Chem.* **21**, pp. 1087–1095.

51. Shrestha, R., et al. (2012). Endosomal escape and siRNA delivery with cationic shell crosslinked knedel-like nanoparticles with tunable buffering capacities, *Biomaterials,* **33**, pp. 8557–8568.

52. Dufès, C., Uchegbu, I.F., and Schätzlein, A.G. (2005). Dendrimers in gene delivery, *Adv. Drug Deliv. Rev.* **57**, pp. 2177–2202.

53. Pichon, C., Billiet, L., and Midoux, P. (2010). Chemical vectors for gene delivery: Uptake and intracellular trafficking, *Curr. Opin. Biotechnol.,* **21**, pp. 640–645.

54. Patil, M.L., Zhang, M., and Minko, T. (2011). Multifunctional triblock nanocarrier (PAMAM-PEG-PLL) for the efficient intracellular siRNA delivery and gene silencing, *ACS Nano,* **5**, pp. 1877–1887.

55. Wen, Y., et al. (2012). Serum tolerance and endosomal escape capacity of histidine-modified pDNA-loaded complexes based on polyamidoamine dendrimer derivatives, *Biomaterials,* **33**, pp. 8111–8121.

56. Lynn, D.M., and Langer, R. (2000). Degradable poly(ß-amino esters): Synthesis, characterization, and self-assembly with plasmid DNA, *J. Am. Chem. Soc.,* **122**, pp. 10761–10768.

57. Zugates, G.T., et al. (2006). Synthesis of poly (ß-amino ester)s with thiol-reactive side chains for DNA delivery, *J. Am. Chem. Soc.,* **128**, pp. 12726–12734.

58. Anderson, D.G., et al. (2005). Structure/property studies of polymeric gene delivery using a library of poly(ß-amino esters), *Mol. Ther.* **11**, pp. 426–434.

59. Lee, J.-S., et al. (2009). Gold, poly(ß-amino ester) nanoparticles for small interfering RNA delivery, *Nano Lett.* **9**, pp. 2402–2406.

60. Lim, H., et al. (2013). Acid-degradable cationic poly(ketal amidoamine) for enhanced RNA interference *in vitro* and *in vivo*, *Biomacromolecules,* **14**, pp. 240–247.

61. Gnanasammandhan, M.K., Bansal, A., and Zhang Y. (2013). Light-activated endosomal escape using upconversion nanoparticles for enhanced delivery of drugs, *SPIE BiOS,* **8594**.
62. Lai, P.-S., et al. (2007). Doxorubicin delivery by polyamidoamine dendrimer conjugation and photochemical internalization for cancer therapy, *J. Control Release,* **122**, pp. 39–46.
63. Prasmickaite, L., Høgset, A., and Berg, K. (2001). Evaluation of different photosensitizers for use in photochemical gene transfection, *Photochem. Photobiol.* **73**, pp. 388–395.
64. Berg, K., et al. (2005). Site-specific drug delivery by photochemical internalization enhances the antitumor effect of bleomycin, *Clin. Cancer Res.* **11**, pp. 8476–8485.
65. Dietze, A., Engesaeter, B., and Berg, K. (2005). Transgene delivery and gelonin cytotoxicity enhanced by photochemical internalization in fibroblast-like synoviocytes (FLS) from rheumatoid arthritis patients, *Photochem. Photobiol. Sci.,* **4**, pp. 341–347.
66. Nishiyama, N., et al. (2005). Light-induced gene transfer from packaged DNA enveloped in a dendrimeric photosensitizer, *Nat. Mater.* **4**, pp. 934–941.
67. Oliveira, S., Hogset, A., and Storm, G. (2008). Delivery of siRNA to the target cell cytoplasm: Photochemical internalization facilitates endosomal escape and improves silencing efficiency, *in vitro* and *in vivo*, *Curr. Pharm. Des.* **14**, pp. 3686–3697.
68. Weissig, V., and D'Souza, G.G. (2010). *Organelle-Specific Pharmaceutical Nanotechnology* (John Wiley & Sons, Inc., Hoboken, NJ).
69. Moan, J., and Berg, K. (1991). The photodegradation of porphyrins in cells can be used to estimate the lifetime of singlet oxygen, *Photochem Photobiol.,* **53**, pp. 549–553.
70. Allison, R.R., et al. (2004). Photosensitizers in clinical PDT, *Photodiagnosis Photodyn. Ther.,* **1**, pp. 27–42.
71. Hirschberg, H., et al. (2012). Enhanced gene transfection by photochemical internalization of protomine sulfate/DNA complexes, *SPIE BiOS,* **8**.
72. Kloeckner, J., et al. (2004). Photochemically enhanced gene delivery of EGF receptor-targeted DNA polyplexes, *J. Drug Target.* **12**, pp. 205–213.
73. Shieh, M.J., et al. (2008). Non-toxic phototriggered gene transfection by PAMAM-porphyrin conjugates, *J. Control. Release,* **129**, pp. 200–206.

74. Selbo, P.K., et al. (2010). Photochemical internalization provides time- and space-controlled endolysosomal escape of therapeutic molecules, *J. Control. Release,* **148**, pp. 2–12.

75. Mathews, M.S., et al. (2012). Photochemical internalization of bleomycin for glioma treatment, *J. Biomed. Opt.,* **17**, pp. 058001.

76. Mackowiak, S.A., et al. (2013). Targeted drug delivery in cancer cells with red-light photoactivated mesoporous silica nanoparticles, *Nano Lett.* **13**, pp. 2576–2583.

77. Liang, W., and Lam, J.K. (2012). Endosomal escape pathways for non-viral nucleic acid delivery systems. In *Molecular Regulation of Endocytosis* Ceresa, B. (Ed.), Intech, Rijeka, Croatia, pp. 421–467.

78. Choi, S.H., Lee, S.H., and Park, T.G. (2006). Temperature-sensitive pluronic/poly(ethylenimine) nanocapsules for thermally triggered disruption of intracellular endosomal compartment, *Biomacromolecules,* **7**, pp. 1864–1870.

79. Lee, S.H., et al. (2008). Thermally sensitive cationic polymer nanocapsules for specific cytosolic delivery and efficient gene silencing of siRNA: Swelling induced physical disruption of endosome by cold shock, *J. Control. Release,* **125**, pp. 25–32.

80. Lee, Y., et al. (2009). Thermally triggered intracellular explosion of volume transition nanogels for necrotic cell death, *J. Control. Release,* **135**, pp. 89–95.

81. Nakase, I., Kobayashi, S., and Futaki, S. (2010). Endosomedisruptive peptides for improving cytosolic delivery of bioactive macromolecules, *Pept. Sci.* **94**, pp. 763–770.

82. Skehel, J.J., and Wiley, D.C. (2000). Receptor binding and membrane fusion in virus entry: The influenza hemagglutinin, *Annu. Rev. Biochem.* **69**, pp. 531–569.

83. Wagner, E., et al. (1992). Coupling of adenovirus to transferrin-polylysine/DNA complexes greatly enhances receptor-mediated gene delivery and expression of transfected genes, *Proc. Natl. Acad. Sci. U.S.A.,* **89**, pp. 6099–6103.

84. Martin, I., Ruysschaert, J.M., and Epand, R.M. (1999). Role of the N-terminal peptides of viral envelope proteins in membrane fusion, *Adv. Drug Deliv. Rev.,* **38**, pp. 233–255.

85. Lee, K.K. (2010). Architecture of a nascent viral fusion pore, *EMBO J.,* **29**, pp. 1299–1311.

86. Sasaki, K., et al. (2008). An artificial virus-like nano carrier system: Enhanced endosomal escape of nanoparticles via synergistic action of

pH-sensitive fusogenic peptide derivatives, *Anal. Bioanal. Chem.* **391**, pp. 2717–2727.

87. Haensler, J., and Szoka, F.C. Jr. (1993). Polyamidoamine cascade polymers mediate efficient transfection of cells in culture, *Bioconjugate Chem.* **4**, pp. 372–379.
88. Wyman, T.B., et al. (1997). Design, synthesis, and characterization of a cationic peptide that binds to nucleic acids and permeabilizes bilayers, *Biochemistry,* **36**, pp. 3008–3017.
89. Guo, X.D., et al. (2012). Oligomerized alpha-helical KALA peptides with pendant arms bearing cell-adhesion, DNA-binding and endosome-buffering domains as efficient gene transfection vectors, *Biomaterials,* **33**, pp. 6284–6291.
90. Murthy, N., et al. (2003). Design and synthesis of pH-responsive polymeric carriers that target uptake and enhance the intracellular delivery of oligonucleotides *J. Control. Release,* **89**, pp. 365–374.
91. Cheung, C.Y., et al. (2001). A pH-sensitive polymer that enhances cationic lipid-mediated gene transfer, *Bioconjugate Chem.* **12**, pp. 906–910.
92. Thomas, J.L., Barton, S.W., and Tirrell, D.A. (1994). Membrane solubilization by a hydrophobic polyelectrolyte: Surface activity and membrane binding, *Biophys. J.,* **67**, pp. 1101–1106.
93. Lackey, C.A., et al. (2002). A biomimetic pH-responsive polymer directs endosomal release and intracellular delivery of an endocytosed antibody complex, *Bioconjugate Chem,* **13**, pp. 996–1001.
94. Tai, H., et al. (2012). pH-responsive hyperbranched copolymers from OnePot RAFT copolymerization, *Macromol. Mater. Eng.,* **297**, pp. 1175–1183.
95. Huang, H.W., Chen, F.-Y., and Lee, M.-T. (2004). Molecular mechanism of peptide-induced pores in membranes, *Phys. Rev. Lett.* **92**, pp. 198304.
96. Lee, M.-T., et al. (2008). Mechanism and kinetics of pore formation in membranes by water-soluble amphipathic peptides, *Proc. Natl. Acad. Sci. U.S.A.,* **105**, pp. 5087–5092.
97. Arnusch, C.J., Pieters, R.J., and Breukink, E. (2012). Enhanced membrane pore formation through high-affinity targeted antimicrobial peptides, *PloS One,* **7**, pp. e39768.
98. Huang, H.W. (2000). Action of antimicrobial peptides: Two-state model, *Biochemistry,* **39**, pp. 8347–8352.

99. Powers, J.-P.S., and Hancock, R.E. (2003). The relationship between peptide structure and antibacterial activity, *Peptides*, **24**, pp. 1681–1691.

100. Shai, Y., and Oren, Z. (2001). From "carpet" mechanism to de-novo designed diastereomeric cell-selective antimicrobial peptides, *Peptides*, **22**, pp. 1629–1641.

101. Yang, L., et al. (2001). Barrel-stave model or toroidal model? A case study on melittin pores, *Biophys. J.*, **81**, pp. 1475–1485.

102. Hou, K.K., et al. (2013). Melittin derived peptides for nanoparticle based siRNA transfection, *Biomaterials*, **34**, pp. 3110–3119.

103. Yin, L., et al. (2013). Nonviral gene delivery via membrane-penetrating, mannose-targeting supramolecular self-assembled nanocomplexes, *Adv. Mater.*, **22**, pp. 3063–3070.

104. Xu, Y., and Szoka, F.C. (1996). Mechanism of DNA release from cationic liposome/DNA complexes used in cell transfection, *Biochemistry*, **35**, pp. 5616–5623.

105. Zelphati, O., and Szoka, F.C. (1996). Mechanism of oligonucleotide release from cationic liposomes, *Proc. Natl. Acad. Sci. U.S.A.*, **93**, pp. 11493–11498.

106. Morille, M., et al. (2008). Progress in developing cationic vectors for non-viral systemic gene therapy against cancer, *Biomaterials*, **29**, pp. 3477–3496.

107. Heyes, J., et al. (2005). Cationic lipid saturation influences intracellular delivery of encapsulated nucleic acids, *J. Control. Release*, **107**, pp. 276–287.

108. Dominska, M., and Dykxhoorn, D.M. (2010). Breaking down the barriers: siRNA delivery and endosome escape, *J. Cell Sci.* **123**, pp. 1183–1189.

109. Allon, N., et al. (2012). A new liposome-based gene delivery system targeting lung epithelial cells using endothelin antagonist, *J. Control. Release*, **160**, pp. 217–224.

110. Guo, X., MacKay, J. Andrew, and Szoka, F.C. Jr. (2003). Mechanism of pH-triggered collapse of phosphatidylethanolamine liposomes stabilized by an orthoester polyethyleneglycol lipid, *Biophys. J.*, **84**, pp. 1784–1795.

Chapter 14

Intracellular Transport to Mitochondria and Other Organelles

Gerard G. M. D'Souza,[a] **Paul H. Nguyen,**[a] **and Volkmar Weissig**[b]

[a] *Department of Pharmaceutical Sciences, MCPHS University, 179 Longwood Avenue, Boston, MA 02115, USA*

[b] *Department of Pharmaceutical Sciences, Midwestern University College of Pharmacy Glendale, 19555 N. 59th Avenue, Glendale, AZ 8530, USA*

gerard.dsouza@mcphs.edu

The efficacy of a drug depends on its ability to interact with its specific molecular target. For drug molecules that act on intracellular targets, the need to mediate efficient intracellular transport to the site of the molecular target is being increasingly recognized. This chapter discusses current strategies for intracellular delivery to mitochondria as examples that might be applied to other organelles as well.

14.1 Introduction

Biodistribution is a term that most pharmaceutical scientists associate with the distribution of drug molecules in the body. Generally speaking, such approaches to describing biodistribution are based on the assumption that the drug has some sort of

Drug Delivery Across Physiological Barriers
Edited by Silvia Muro

ISBN 978-981-4669-40-5 (Hardcover), 978-981-4669-41-2 (eBook)
www.panstanford.com

predisposition for certain compartments. The blood is usually the central compartment with other organs representing peripheral compartments of distribution. However, a growing number of researchers are taking that concept further and looking at the same biodistribution principles as they apply to drugs in a body at a different scale. This body is the cell. The cell has various compartments with arguably the cytosol representing the central compartment and the various other membrane-bound organelles serving as peripheral compartments. Drug transport into cells, particularly to lysosomes and the cytosol, are topics that have been discussed in other chapters of this book; therefore, this chapter will focus on a discussion of drug transport primarily to the mitochondria.

The recognition of the role that mitochondria play in human health and disease is evidenced by the emergence in recent decades of a whole new field of "mitochondrial medicine" [1, 2]. Mitochondria are once again the focus of basic as well as applied research in the biomedical sciences with the result that significant advances have been made in understanding mitochondrial biology and disease progression. The discovery that might have the greatest repercussions for how delivery strategies are designed is interestingly not associated with breakthroughs in understanding the molecular mechanisms of transport but with the very nature of mitochondria in live cells. Unlike other cell organelles, mitochondria were long thought to be a population of discrete entities as evidenced by the pervasive use of mitochondria and mitochondrion to describe the collective and singular states, respectively. This model presented a rather complicated situation when approaching mitochondria-specific delivery. How could delivery to every single mitochondrion be achieved? However, it is becoming widely accepted that rather than a collection of individual organelles that we call mitochondria, a living cell possesses one large interconnected mitochondrial network and that what were previously described as individual mitochondria are in fact just fragments of the larger network with the level of fragmentation being related to cell viability. The greater the level of mitochondrial fragmentation, the more likely that the cell has lost viability. Conveniently, the mitochondrial network concept negates the idea that mitochondria-specific delivery approaches must mediate delivery to all mitochondria or at the very least to a significantly large subset of the total cellular population of mitochondria. Successful delivery to any part of the mitochondrial network should mean that the delivered cargo is accessible by the entire mitochondrial network. However, this scenario raises the question of whether and how the delivered cargo is then made available to the rest of the mitochondrial network [3]. It, therefore, stands to reason that a comprehensive approach to mitochondrial delivery requires an understanding of not just mechanisms of transport to mitochondria and into the mitochondrial compartment but also mechanisms of transport within the mitochondrial network.

14.2 Major Concepts of Subcellular Drug Transport

Our current understanding of drug transport is based on two major mechanisms. A partition/charge-mediated passive mechanism and a chaperone/transporter-mediated active mechanism. With this in mind, it should be clear that molecules that enter the intracellular space can distribute within the cell based on their relative solubilities in various subcellular regions as well as electrostatic interactions within these regions. Moreover, the molecule may also be subject to prevalent modes of active transport within the cell. An additional level of complexity is associated with the fact that the cytosol is a region with a high level of molecular crowding due to the presence of the cytoskeletal network, organelles, and a concentration of macromolecules between 50 and 400 g/L [4, 5]. Transport or diffusion events in such a crowded milieu, especially for higher molecular weight drugs, is very different from the type of diffusion that is observed in dilute solutions. Generally, intracellular diffusion is described as hindered diffusion [6, 7]. Additionally, the fluid-phase viscosity of the cytoplasm and binding to intracellular components are believed to influence the diffusion of solutes inside a cell [8–10]. Therefore, it is very likely that the observed predispositions of various molecules for subcellular compartments, including the mitochondria, might be the net result of diffusion, partition, electrostatic, and active-transport-based phenomena.

14.2.1 *Partition/Charge-Mediated Passive Mechanisms*

For a large number of small molecules, it is possible to explain and even predict subcellular disposition primarily on the basis of the physicochemical properties of the molecule. This concept of quantitative structure–activity relationships (QSAR) is arguably a common starting point for the development of a subcellular targeting strategy [11]. The QSAR approach has been used to study the chemical features of molecules that are known to selectively accumulate at or inside mammalian mitochondria within living cells [12]. A nonbiased sample of reported mitochondriotropic compounds was generated from the literature and then examined using physicochemical classifications, a Fick–Nernst–Planck physicochemical model, and QSAR models. It has been postulated that selective mitochondrial accumulation involves electric potential, ion-trapping, and complex formation with cardiolipin, while nonspecific accumulation involves membrane partitioning and nonspecific access requires only low lipophilicity. Using QSAR and Fick–Nernst approaches, it was possible to predict the mitochondriotropic behavior of more than 80% of the selected compounds. Therefore, the QSAR approach represents a useful tool to assess the mitochondriotropism of potential drug molecules, and in theory, this approach should also be applicable for predicting subcellular localization of small molecules in other organelle compartments [13]. Table 14.1 lists criteria that could be used to identify accumulation in some subcellular compartments based on the QSAR approach. The ability to predict the influence of the likely site of accumulation within the cell could prove to be a powerful tool in drug design to either select molecules

Table 14.1 QSAR criteria for accumulation of molecules in various subcellular compartments

	log P	AI	Z	pK_a
Mitochondria	0–5	—	>0	—
Endoplasmic reticulum	0–6	3.5–6	>0	—
Lysosome	−5 to 0	—	>0	<10
Nonspecific uptake into biomembranes	8–5	—	—	—

AI: amphiphilicity index; log P: partition coefficient; pK_a: acidity constant; Z: charge with a desired subcellular accumulation or identify molecules that would benefit from subcellular targeting strategies.

14.2.2 *Chaperone/Transporter-Mediated Active Mechanisms*

While it is usually appropriate to assume that low-molecular-weight drugs are transported to a large extent by passive mechanisms, the role of active transport mechanisms cannot be ignored. For high-molecular-weight drugs, active transport mechanisms are far more likely to be the primary determinants of drug disposition within the cell. In addition to specific mechanisms of molecular transport, it stands to reason that vesicular transport as well as normal organellar uptake mechanisms could also contribute to the transport of drugs within the cell and into the various organelles. The mitochondria, for example, have a very elegant transport system for importing all of the mitochondrial proteins encoded by the nuclear genome as well as various other metabolites and nucleic acids. Both mitochondrial membranes contain an elaborate network of protein translocases together with a variety of chaperones and processing enzymes in the matrix and intermembrane space to mediate protein import. Although the mitochondrial protein import machinery is still under intense investigation, a relatively clear picture of this highly complex mechanism has emerged over the past years [14–26]. Generally, proteins synthesized in the cytosol and to be transported into mitochondria possess an amino-terminal targeting sequence [mitochondrial leader sequence (MLS) peptide]. Although no amino acid sequence homology has been observed among MLS peptides of different mitochondrial pre-proteins, two common attributes have been noted. MLS peptides display usually a net positive charge, and they have the ability to form amphiphilic α-helices [27]. Proteins bearing an MLS peptide are escorted through the cytosol by chaperones to the translocase of the outer mitochondrial membrane complex. After crossing the outer membrane, a so-called translocase of the inner mitochondrial membrane complexes mediates the further transport into the mitochondrial matrix or insertion into the inner mitochondrial membrane. Finally, inside the mitochondrial matrix, the leader peptide is cleaved off by a matrix processing protease. Thus, MLSs of the various mitochondrial proteins represent a second subset of potential mitochondrial targeting ligands for drug delivery to mitochondria.

In addition to proteins, the mitochondria are also able to transport nucleotides and nucleic acids. The adenine nucleotide transporter, in conjunction with the voltage-dependent anion channel and the mitochondrial permeability transition pore complex, could also mediate the transport of molecules into mitochondria. Additionally, mitochondria of lower organisms like yeast and *Leishmania* exhibit nucleic acid import capabilities that can be adapted to mammalian mitochondria [28–31]. More intriguing is the recent discovery that mammalian mitochondria are capable of importing linear DNA [32]. Taken together, the aforementioned transport mechanisms offer a starting point for designing approaches to achieving the specific transport of drug molecules to mitochondria. Further, many of these approaches could be adapted to achieve subcellular transport of drugs to other organelles as well.

14.3 Major Concepts of Organelle-Specific Drug Delivery

When developing drug molecules, a potential drug molecule might not have the desired subcellular accumulation. Strategies for dealing with such a situation fall into two major categories: direct molecular modification and carrier-mediated delivery.

14.3.1 *Direct Molecular Modification*

Most molecular modification approaches to subcellular delivery are based on molecular conjugation. This concept involves conjugating a drug molecule with suboptimal subcellular localization to a molecule with a high affinity for the desired subcellular compartment. Several well-known mitochondriotropic molecules could potentially serve as mitochondriotropic ligands. Rhodamine 123, which has been used extensively as a relatively nontoxic stain for mitochondria in living cells since its introduction in 1982, is probably the best known of the mitochondriotropics [33]. Other examples of mitochondriotropic cations are cyanine dyes such as N,N′-bis(2-ethyl-1,3-dioxolane) kryptocyanine [34], Victoria blue BO [35], dequalinium chloride [36], and phosphonium salts such as methyltriphenylphosphonium bromide [37].

Over the last two decades, a variety of conjugates composed of biologically active molecules and the mitochondriotropic triphenylphosphonium (TPP) cation have been synthesized [38, 39]. The bioactive cargos include mostly antioxidants [40] and anticancer agents, but peptide nucleic acids [41] have also been subject to TPP conjugation. The resulting conjugates have been found to have high mitochondria-specific accumulation and not surprisingly the TPP cation has emerged as the first choice of targeting ligand in most current mitochondria-specific delivery approaches. Recently, TPP has been conjugated to porphyrins for potential mitochondria-targeted photodynamic therapy. Meso-tetraphenylporphyrin derivatives bearing either the triphenylphosphonium ion or the triethylammonium ion were shown to be mitochondria-targeted and to exhibit light-dependent toxicity in MCF-7 human breast cancer cell line, suggesting their application potential in photodynamic therapy. Additionally, the triphenyl phosphonium cation and various other mitochondriotropic residues were used to synthesize several examples of a class of molecules described as geldanamycin mitochondrial matrix inhibitors (Gamitrinibs). Gamitrinibs are designed to selectively target and inhibit Hsp90 activity in the mitochondria of human carcinoma cells by acting as ATPase antagonists. TEMPOL is a piperidine nitroxide that acts as a prodrug and is reduced to a hydroxylamine in mitochondria by ubiquinol. The metabolite induces apoptosis in cancer cells by action on complex I [42]. Conjugation of a TPP residue to TEMPOL improved its mitochondria-specific accumulation and its reaction with ubiquinol in mitochondrial membranes [43].

As previously discussed, molecules in the cell can reach the mitochondria by two major mechanisms. Molecules [44] such as TPP, rhodamine, and various other mitochondria-specific stains utilize the charge/partition mechanism. However, the active transport mechanism could also be exploited to deliver drug molecules to the mitochondria. The various mitochondrial matrix proteins that are synthesized in the cytosol and imported into the matrix by the mitochondrial import machinery rely on short leader sequences to mediate this specific import process. Just like the TPP cation, the leader sequences of various matrix proteins can be used to mediate the mitochondria-specific accumulation of a variety of bioactive molecules [45]. This approach is particularly useful for proteins and peptides but has also been used for nucleic acids and even for small molecules. An antioxidant, consisting of manganese (Mn) porphyrin and signal peptide for mitochondrial targeting, was synthesized in an attempt to improve the action of the antioxidant on mitochondria [46]. The resulting Mn-porphyrin-oligopeptide conjugate exhibited significant superoxide dismutase activity and decomposed peroxynitrite ($ONOO^-$). Following intracellular delivery, the antioxidant conjugate recovered the viability of lipopolysaccharide-stimulated cells. Using a similar approach, the pro-apoptotic and potential anticancer peptide, $(KLAKLAK)_2$, was engineered to improve mitochondrial localization, which resulted in improved action of the peptide [47].

The approaches described so far utilize mechanisms that are found naturally in mammalian mitochondria. However, it has even been possible to transfer mitochondria-specific transport mechanisms from lower organisms to mammalian mitochondria resulting in improved mitochondria-specific transport. During the end of the 1990s, a natural tRNA import pathway was identified in yeast and subsequently adapted to drive the import of tRNA from the cytosol into the mitochondria of mammalian cells [28]. It was first shown that yeast mitochondria were able to import cytoplasmic tRNAs based on the recognition of specific aminoacylated motifs on the anticodon arm [29, 30]. Cytoplasmic tRNAs modified with the correct amino acylation identity could then be targeted to the mitochondrial matrix, where they participated in mitochondrial translation. Later, it was demonstrated that human mitochondria, which did not normally import tRNAs, were able to internalize yeast tRNA derivatives *in vitro* in the presence of an essential yeast import factor [28]. Finally, it was shown that synthetic tRNA molecules engineered with the appropriate yeast aminoacylation identity could be imported into human mitochondria [31].

14.3.2 *Carrier-Mediated Delivery*

The strategies discussed so far involve direct chemical modification of drug molecules. The primary concern with the molecular conjugation approach is that the conjugation might affect the pharmacological action of the drug molecule or that the mitochondriotropic ligand loses its affinity for the mitochondria after conjugation. An early attempt at conjugating rhodamine 123 to *cis*-diamminedichloroplatinum (II) serves as a good example of the latter [48]. Platinum rhodamine 123, a tight ion pair composed of one platinum (II) tetrachlorodianion and two molecules of rhodamine 123, showed increased accumulation of platinum, about 70-fold compared to non-complexed *cis*-diamminedichloroplatinum (II). However, platinum rhodamine complex did not exhibit mitochondria-targeting capability. Roughly 50% of platinum rhodamine 123 was found in the nuclei and about 30% was detected in the cytosol of SCC-25 or SCC-25/CP cells [48]. In such cases, pharmaceutical nanocarriers such as liposomes, micelles, and solid nanoparticles offer a nonchemical approach to modify the disposition of drug molecules. All chemistry can be performed on the components of the nanocarrier system that can then be loaded with the drug to afford targeted delivery [49–52]. Most pharmaceutical nanocarriers can be modified to allow tissue-specific and cell-specific accumulation [53–57]. However, it is only recently that nanocarrier technology has reached the point where control over subcellular delivery appears to be really feasible. Current efforts toward the design of nanocarriers with the potential for subcellular targeting have generally employed one of two major schemes. The first is based on the inherent predisposition of the nanocarrier for a particular compartment, and the second is based on attaching subcellular targeting ligands to the surface of nanocarriers to redirect their accumulation to the desired compartment. However, nanocarriers can help on an even simpler level as illustrated by an approach toward the delivery of mastoparan into cells [58]. Liposomes were modified with transferrin and a fusogenic peptide Chol-GALA. The transferrin modification enhances liposomal uptake into cells via endocytosis after which the peptide facilitates release from the endosomes into the cytosol. Thus, by just increasing the intracellular content of the drug, the delivery approach achieved a higher concentration of the drug that was potentially available to interact with the subcellular target. At this point, it is appropriate to consider that endosomal escape strategies also can be considered to be subcellular delivery approaches with the cytosol being the intended target compartment for delivery. Given that this book has chapters devoted to endosomal trafficking and endosomal escape, these will not be discussed further here.

Micelles have also been proposed for the delivery of hydrophobic drugs to various subcellular organelles, including mitochondria [59]. The fluorescently labeled micelles used in the study were found to be distributed through several cytoplasmic organelles, including a majority of them being associated with the mitochondria. Arguably DQAsomes are the earliest example of pharmaceutical carriers designed for organelle-specific delivery. Prepared from the mitochondriotropic molecule dequalinium chloride, these vesicular nanocarriers were first studied for mitochondria-specific DNA delivery. DQAsomes were also shown to be capable of changing the subcellular distribution of paclitaxel to increase the accumulation of the drug in mitochondria. The mitochondria-specific delivery led to improved apoptotic activity at paclitaxel concentrations, at which the free drug does not have a significant cytotoxic effect [60]. Paclitaxel-loaded DQAsomes have also been tested for their ability to inhibit the growth of human colon cancer tumors in a mouse model [61]. The antitumor efficiency of DQAsomal encapsulated paclitaxel was also further enhanced by modifying the DQAsomal surface with folic acid (FA) [62]. The folate receptor is a folate high-affinity membrane-binding protein, which is overexpressed in a large variety of human tumors [63–65]. FA conjugates are internalized in a tumor-cell-specific manner by receptor-mediated endocytosis resulting in an increased toxicity of the corresponding drug [66–68]. Cell cytotoxicity studies using folate receptor expressing HeLa cells suggested that FA-conjugated DQAsomes possessed better antitumor activity as compared to plain paclitaxel-loaded DQAsomes, FA-conjugated paclitaxel-loaded liposomes, and the free drug. Based on the data, it was concluded that FA-conjugated DQAsomes delivered the drug not only to the cytosol but also to mitochondria, whereas FA-conjugated liposomes delivered the drug into the cytosol only [62].

Another approach to the design of mitochondria-specific nanocarriers is to modify existing nanocarriers with mitochondriotropic ligands. In this regard, TPP again served as the mitochondriotropic ligand in liposomal and polymer-based nanocarriers. Liposomes have been well characterized as delivery systems and are a popular choice due to their biocompatibility, ease of surface modification, capacity to encapsulate hydrophilic or hydrophobic drugs [69]. The first indication that liposomes could be rendered mitochondriotropic by surface modification with a mitochondriotropic residue came from a report that so-called proteoliposomes prepared by incorporating a crude mitochondrial membrane fraction into liposomes colocalized with endogenous mitochondria in pre-implantation embryos [70]. Further investigation of the concept of using ligands to alter the subcellular distribution of liposomes was based on the synthesis of stearyl triphenylphosphonium (STPP) [71]. The stearyl residue of STPP serves as a lipid anchor to modify the surface of liposomes with the TPP residue and resulted in a liposomal preparation with a marked predisposition for mitochondria [71]. STPP-liposomes were shown to effectively direct the accumulation of rhodamine-labeled phosphatidylethanolamine (Rh-PE) to mitochondria in live cells [72]. Based on flow cytometry, STPP-liposomes exhibited the same level of cell association as liposomes with the same cationic charge. However, the subsequent subcellular localization analyzed by confocal microscopy was markedly different, indicating that the mitochondriotropic ligand and not the surface charge is what determines mitochondria-specific association of the nanocarrier [72]. It was also found that the TPP ligand did not change the *in vivo* distribution and tumor accumulation of long-circulating PEGylated liposomes. STPP-liposomes did, however, improve both the *in vitro* and *in vivo* efficacy of ceramide. Taken together, these data suggest that the tendency of long-circulating liposomes to passively accumulate (via the enhanced permeability and retention effect) in solid tumors can be combined with organelle-specific tropism conferred by modification with an appropriate ligand to potentiate the effect of an encapsulated antitumor agent.

An alternative approach to developing mitochondria-specific liposomes has focused on the concept that liposomes with a tendency to selectively fuse with mitochondrial membranes are more likely to associate with mitochondria upon cell entry. Referred to as MITO-Porter, these liposomes are surface modified with octaarginine residues to facilitate their entry into cells via macropinocytosis [73]. The lipid composition was selected on the basis of high levels of fusion with the mitochondrial membrane and the release of its cargo to the intramitochondrial compartment in living cells [73]. Based on confocal microscopy data, MITO-Porter liposomes have been used to deliver green fluorescent protein [73] as well as propidium iodide [74] to mitochondria suggesting that they can be used to deliver macromolecules as well as small molecules to mitochondria. In an approach aimed at exploring earlier work suggesting the utility of an isolated mitochondrial fraction as a crude source of mitochondriotropic residues for liposomal targeting, it was recently shown that such liposomes could be used to deliver bioactives to mitochondria [75]. In addition to showing for the first time that the approach could be utilized to deliver DNA to mitochondria, it was found that the lipid-enriched mitochondrial fraction was significantly better at directing the mitochondria-specific accumulation of liposomes compared to the crude mitochondrial fraction [75].

In addition to the vesicular systems discussed so far, polymeric nanoparticles have also been explored as mitochondria-specific delivery systems [76–78]. The overall approach has been to use a mitochondriotropic ligand conjugated to the polymer. While there is at least one report of the use of a MLS as the mitochondriotropic ligand, most of the polymeric systems are based on the use of the TPP ligand. TPP modification was first reported to result in a mitochondriotropic *N*-(2-hydroxylpropyl)methacrylamide (HPMA) copolymer-based nanoparticle [77, 78]. TPP has more recently been used to create a poly-D,L-lactic-co-glycolic acid (PLGA) and polyethylene glycol (PEG) mitochondriotropic nanoparticle platform [79]. The approach is based on blending a PLGA-block-PEG-TPP polymer (PLGA-b-PEG-TPP) with PLGA-b-PEG-OH or PLGA-COOH. *In vitro* screening was used to identify a size- and charge-optimized mitochondriotropic formulation that was then used to deliver

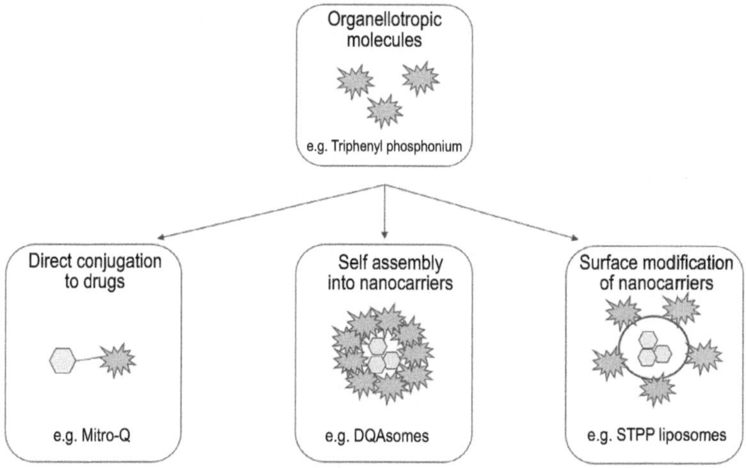

Figure 14.1 Schematic representation of approaches for organelle-specific delivery.

various classes of mitochondria-acting therapeutics. The PLGA-b-PEG-TPP platform is reported to have significantly improved the therapeutic effect of lonidamine, α-tocopheryl succinate, curcumin, and 2,4-dinitrophenol [79].

Inorganic nanoparticles have also been shown to be capable of mitochondria-specific delivery. Hypocrellin A, a photodynamic drug, was encapsulated in a water-soluble amorphous silica nanocage [80]. These drug-loaded nanocages are reportedly able to specifically accumulate in the mitochondria of cancer cells and improve the photosensitizing effect of hypocrelin A [80]. It is, however, unclear what mediates the mitochondria-specific accumulation of the nanocage system.

It is also interesting to note that changes in nanoparticle architecture result in changes in subcellular disposition [81]. Fluorescein isothiocyanate labeled layered double hydroxide nanoparticles were prepared from Mg_2Al under conditions that yielded either hexagonal sheets (50–150 nm wide and 10–20 nm thick) or nanorods (30–60 nm wide and 100–200 nm long). A comparison of the subcellular distribution of these two types of preparations revealed that the nanorods trafficked to the nucleus but the hexagonal sheets remained in the cytoplasm [81].

Taken together, the various studies described so far strongly support the hypothesis that nanocarriers can indeed control the subcellular accumulation of bioactive molecules and as such represent a useful tool in the development of organelle-targeted drug delivery strategies. One must, however, be cautious in the knowledge that the approaches described so far are based only on current understanding of subcellular trafficking. Figure 14.1 summarizes the current strategies for achieving organelle-specific delivery.

14.4 Conclusions and Perspectives

Approaches to organelle-specific delivery have been increasingly explored over the past few years, and the proof of principle of using such approaches to improve the action of drug molecules has been established. Some molecular modification approaches have even been investigated in clinical studies, but most of the approaches, especially nanocarrier-based systems, require further study before clinical application becomes commonplace. The detailed mechanisms by which specific subcellular localization is achieved are still unclear. It is, however, possible to speculate based on some of the results reported so far. For example, taking the results of all the liposome-based approaches together, it seems reasonable to hypothesize that these liposomal systems might utilize a common mechanism for achieving mitochondria-specific accumulation. Upon interaction with the cell membrane and the subsequent endocytic processes, the mitochondriotropic lipids become incorporated into the endosomal membrane, thereby changing the subcellular trafficking of the endosomal vesicle. The endosomal vesicle is transported to the mitochondrial network and fuses with the mitochondrial membranes. If this were indeed the mechanism, it would require that the mitochondriotropic ligand used in any carrier-based strategy for mitochondrial delivery be capable of redistribution to the endosomal membrane after endocytosis. Results from the first study with STPP-modified HPMA seem to support this notion. It was observed that while the STPP-modified polymers did exhibit association with isolated mitochondria, experiments with cells revealed predominantly lysosomal association of the polymer [77]. A subsequent study reported that microinjection using fluorescently labeled constructs suggested improved mitochondrial targeting ability based on microscopic analysis [78]. It is, therefore, interesting that the PLGA-based system showed excellent mitochondriotropism. However, given that the PLGA system uses a polymer blend approach, it might be possible that the STPP-modified polymer fraction is able to redistribute to the endosomal membrane resulting in the retrafficking of the endosomal vesicle to mitochondria. Unfortunately, the ability to test this and other hypotheses about subcellular trafficking mechanisms is still limited by current technology. Most knowledge of subcellular trafficking processes is based on studies with solid particles [82–88]. Extending these experimental approaches to quasi-stable vesicular systems such as liposomes and micelles is dependent on the availability of suitable imaging technology especially for real-time fluorescence confocal imaging of live cells [89–92]. Emerging technologies such as total internal reflection microscopy [93] and label-free approaches such as Raman microscopy [94–96] also hold promise for elucidating subcellular transport mechanisms.

References

1. Murphy, M.P., and Smith, R.A. (2000). Drug delivery to mitochondria: The key to mitochondrial medicine, *Adv. Drug Deliv. Rev.*, **2**, pp. 235–250.
2. Larson, N.-G., and Luft, R. (1999). Revolution in mitochondrial medicine, *FEBS Lett.*, pp. 199–202.
3. Partikian, A., Olveczky, B., Swaminathan, R., Li, Y., and Verkman, A.S. (1998). Rapid diffusion of green fluorescent protein in the mitochondrial matrix, *J. Cell Biol.*, **4**, pp. 821–829.
4. Minton, A.P. (2006). How can biochemical reactions within cells differ from those in test tubes, *J. Cell Sci.*, Pt **14**, pp. 2863–2869.
5. Ellis, R.J., and Minton, A.P. (2003). Cell biology: Join the crowd, *Nature*, **6953**, pp. 27–28.
6. Sanabria, H., Kubota, Y., and Waxham, M.N. (2007). Multiple diffusion mechanisms due to nanostructuring in crowded environments, *Biophys. J.*, **1**, pp. 313–322.

7. Goins, A.B., Sanabria, H., and Waxham, M.N. (2008). Macromolecular crowding and size effects on probe microviscosity, *Biophys. J.*, **11**, pp. 5362–5373.
8. Lukacs, G.L., Haggie, P., Seksek, O., Lechardeur, D., Freedman, N., and Verkman, A.S. (2000). Size-dependent DNA mobility in cytoplasm and nucleus, *J. Biol. Chem.*, **3**, pp. 1625–1629.
9. Seksek, O., Biwersi, J., and Verkman, A.S. (1997). Translational diffusion of macromolecule-sized solutes in cytoplasm and nucleus, *J. Cell Biol.*, **1**, pp. 131–142.
10. Weiss, M. (2008). Probing the interior of living cells with fluorescence correlation spectroscopy, *Ann. N.Y. Acad. Sci.*, **1130**, pp. 21–27.
11. Horobin, R.W. (2001). Uptake, distribution and accumulation of dyes and fluorescent probes within living cells: A structure-activity modelling approach, *Adv. Colour Sci. Technol.*, pp. 101–107.
12. Horobin, R.W., Trapp, S., and Weissig, V. (2007). Mitochondriotropics: A review of their mode of action, and their applications for drug and DNA delivery to mammalian mitochondria, *J. Control. Release*, **3**, pp. 125–136.
13. Horobin, R.W., and Weissig, V. (2005). A QSAR-modeling perspective on cationic transfection lipids. 1. Predicting efficiency and understanding mechanisms, *J. Gene Med.*, **8**, pp. 1023–1034.
14. Cavadini, P., Gakh, O., and Isaya, G. (2002). Protein import and processing reconstituted with isolated rat liver mitochondria and recombinant mitochondrial processing peptidase, *Methods*, **4**, pp. 298–306.
15. Endo, T., and Kohda, D. (2002). Functions of outer membrane receptors in mitochondrial protein import, *Biochim. Biophys. Acta.*, **1**, pp. 3–14.
16. Gordon, D.M., Dancis, A., and Pain, D. (2000). Mechanisms of mitochondrial protein import, *Essays Biochem.*, **2000**, pp. 61–73.
17. Hood, D.A., Adhihetty, P.J., Colavecchia, M., Gordon, J.W., Irrcher, I., Joseph, A.M., Lowe, S.T., and Rungi, A.A. (2003). Mitochondrial biogenesis and the role of the protein import pathway, *Med. Sci. Sports Exerc.*, **1**, pp. 86–94.
18. Jensen, R.E., and Johnson, A.E. (2001). Opening the door to mitochondrial protein import, *Nat. Struct. Biol.*, **12**, pp. 1008–1010.
19. Moro, F., Okamoto, K., Donzeau, M., Neupert, W., and Brunner, M. (2002). Mitochondrial protein import: Molecular basis of the ATP-dependent interaction of MtHsp70 with Tim44, *J. Biol. Chem.*, **9**, pp. 6874–6880.

20. Pfanner, N., and Wiedemann, N. (2002). Mitochondrial protein import: Two membranes, three translocases, *Curr. Opin. Cell Biol.*, **4**, pp. 400–411.
21. Pfanner, N., and Truscott, K.N. (2002). Powering mitochondrial protein import, *Nat. Struct. Biol.*, **4**, pp. 234–236.
22. Koehler, C.M., Jarosch, E., Tokatlidis, K., Schmid, K., Schweyen, R.J., and Schatz, G. (1998). Import of mitochondrial carriers mediated by essential proteins of the intermembrane space, *Science*, **5349**, pp. 369–373.
23. Koehler, C.M., Merchant, S., and Schatz, G. (1999). How membrane proteins travel across the mitochondrial intermembrane space, *Trends Biochem. Sci.*, **11**, pp. 428–432.
24. Koehler, C.M. (2000). Protein translocation pathways of the mitochondrion, *FEBS Lett.*, **1-2**, pp. 27–31.
25. Murphy, M.P., Leuenberger, D., Curran, S.P., Oppliger, W., and Koehler, C.M. (2001). The essential function of the small Tim proteins in the TIM22 import pathway does not depend on formation of the soluble 70-kilodalton complex, *Mol. Cell. Biol.*, **18**, pp. 6132–6138.
26. Lithgow, T. (2000). Targeting of proteins to mitochondria, *FEBS Lett.*, **1–2**, pp. 22–26.
27. Scheffler, I.E. (2001). Mitochondria make a come back, *Adv. Drug Deliv. Rev.*, **1–2**, pp. 3–26.
28. Kolesnikova, O.A., Entelis, N.S., Mireau, H., Fox, T.D., Martin, R.P., and Tarassov, I.A. (2000). Suppression of mutations in mitochondrial DNA by tRNAs imported from the cytoplasm, *Science*, **5486**, pp. 1931–1933.
29. Kazakova, H.A., Entelis, N.S., Martin, R.P., and Tarassov, I.A. (1999). The aminoacceptor stem of the yeast tRNA(Lys) contains determinants of mitochondrial import selectivity, *FEBS Lett.*, **2–3**, pp. 193–197.
30. Entelis, N.S., Kieffer, S., Kolesnikova, O.A., Martin, R.P., and Tarassov, I.A. (1998). Structural requirements of tRNALys for its import into yeast mitochondria, *Proc. Natl. Acad. Sci. U.S.A.*, **6**, pp. 2838–2843.
31. Entelis, N.S., Kolesnikova, O.A., Dogan, S., Martin, R.P., and Tarassov, I.A. (2001). 5 S rRNA and tRNA import into human mitochondria. Comparison of *in vitro* requirements, *J. Biol. Chem.*, **49**, pp. 45642–4553.
32. Koulintchenko, M., Temperley, R.J., Mason, P.A., Dietrich, A., and Lightowlers, R.N. (2006). Natural competence of mammalian mitochondria allows the molecular investigation of mitochondrial gene expression, *Hum. Mol. Genet.*, **1**, pp. 143–154.

33. Chen, L.B., Summerhayes, I.C., Johnson, L.V., Walsh, M.L., Bernal, S.D., and Lampidis, T.J. (1982). Probing mitochondria in living cells with rhodamine 123, *Cold Spring Harb. Symp. Quant. Biol.*, **1982**, pp. 141–155.
34. Oseroff, A.R., Ohuoha, D., Ara, G., McAuliffe, D., Foley, J., and Cincotta, L. (1986). Intramitochondrial dyes allow selective *in vitro* photolysis of carcinoma cells, *Proc. Natl. Acad. Sci. U.S.A*, **24**, pp. 9729–9733.
35. Morgan, J., Whitaker, J.E., and Oseroff, A.R. (1998). GRP78 induction by calcium ionophore potentiates photodynamic therapy using the mitochondrial targeting dye Victoria blue BO, *Photochem. Photobiol.*, **1**, pp. 155–164.
36. Weiss, M.J., Wong, J.R., Ha, C.S., Bleday, R., Salem, R.R., Steele, G.D., Jr., and Chen, L.B. (1987). Dequalinium, a topical antimicrobial agent, displays anticarcinoma activity based on selective mitochondrial accumulation, *Proc. Natl. Acad. Sci. U.S.A.*, **15**, pp. 5444–5448.
37. Liberman, E.A., Topaly, V.P., Tsofina, L.M., Jasaitis, A.A., and Skulachev, V.P. (1969). Mechanism of coupling of oxidative phosphorylation and the membrane potential of mitochondria, *Nature*, **198**, pp. 1076–1078.
38. Murphy, M.P. (2001). Development of lipophilic cations as therapies for disorders due to mitochondrial dysfunction, *Expert Opin. Biol. Ther.*, **5**, pp. 753–764.
39. Muratovska, A., Lightowlers, R.N., Taylor, R.W., Wilce, J.A., and Murphy, M.P. (2001). Targeting large molecules to mitochondria, *Adv. Drug Deliv. Rev.*, **1–2**, pp. 189–198.
40. Ross, M.F., Kelso, G.F., Blaikie, F.H., James, A.M., Cocheme, H.M., Filipovska, A., Da Ros, T., Hurd, T.R., Smith, R.A., and Murphy, M.P. (2005). Lipophilic triphenylphosphonium cations as tools in mitochondrial bioenergetics and free radical biology, *Biochemistry (Mosc)*, **2**, pp. 222–230.
41. Muratovska, A., Lightowlers, R.N., Taylor, R.W., Turnbull, D.M., Smith, R.A., Wilce, J.A., Martin, S.W., and Murphy, M.P. (2001). Targeting peptide nucleic acid (PNA) oligomers to mitochondria within cells by conjugation to lipophilic cations: Implications for mitochondrial DNA replication, expression and disease, *Nucleic Acids Res.*, **9**, pp. 1852–1863.
42. Monti, E., Supino, R., Colleoni, M., Costa, B., Ravizza, R., and Gariboldi, M.B. (2001). Nitroxide TEMPOL impairs mitochondrial function and induces apoptosis in HL60 cells, *J. Cell. Biochem.*, **2**, pp. 271–276.
43. Trnka, J., Blaikie, F.H., Smith, R.A., and Murphy, M.P. (2008). A mitochondria-targeted nitroxide is reduced to its hydroxylamine by ubiquinol in mitochondria, *Free Radic. Biol. Med.*, **7**, pp. 1406–1419.

44. Modica-Napolitano, J.S., and Aprille, J.R. (2001). Delocalized lipophilic cations selectively target the mitochondria of carcinoma cells, *Adv. Drug Deliv. Rev.*, **1–2**, pp. 63–70.

45. Yousif, L.F., Stewart, K.M., Horton, K.L., and Kelley, S.O. (2009). Mitochondria-penetrating peptides: Sequence effects and model cargo transport, *ChemBioChem*, **12**, pp. 2081–2088.

46. Asayama, S., Kawamura, E., Nagaoka, S., and Kawakami, H. (2006). Design of manganese porphyrin modified with mitochondrial signal peptide for a new antioxidant, *Mol. Pharm.*, **4**, pp. 468–470.

47. Horton, K.L., and Kelley, S.O. (2009). Engineered apoptosis-inducing peptides with enhanced mitochondrial localization and potency, *J. Med. Chem.*, **10**, pp. 3293–3299.

48. Teicher, B.A., Holden, S.A., Jacobs, J.L., Abrams, M.J., and Jones, A.G. (1986). Intracellular distribution of a platinum-rhodamine 123 complex in cis-platinum sensitive and resistant human squamous carcinoma cell lines, *Biochem. Pharmacol.*, **19**, pp. 3365–3369.

49. Ganta, S., Devalapally, H., Shahiwala, A., and Amiji, M. (2008). A review of stimuli-responsive nanocarriers for drug and gene delivery, *J. Control. Release*, **3**, pp. 187–204.

50. Jabr-Milane, L., van Vlerken, L., Devalapally, H., Shenoy, D., Komareddy, S., Bhavsar, M., and Amiji, M. (2008). Multi-functional nanocarriers for targeted delivery of drugs and genes, *J. Control. Release*, **2**, pp. 121–128.

51. Mishra, B., Patel, B.B., and Tiwari, S. (2009). Colloidal nanocarriers: A review on formulation technology, types and applications toward targeted drug delivery, *Nanomedicine*, **6**, pp. 9–24.

52. Torchilin, V.P. (2007). Targeted pharmaceutical nanocarriers for cancer therapy and imaging, *AAPS J.*, **2**, pp. E128–147.

53. Ma, L., Liu, Y.L., Ma, Z.Z., Dou, H.L., Xu, J.H., Wang, J.C., Zhang, X., and Zhang, Q. (2009). Targeted treatment of choroidal neovascularization using integrin-mediated sterically stabilized liposomes loaded with combretastatin A4, *J. Ocul. Pharmacol. Ther.*, **3**, pp. 195–200.

54. Hatakeyama, H., Akita, H., Ishida, E., Hashimoto, K., Kobayashi, H., Aoki, T., Yasuda, J., Obata, K., Kikuchi, H., Ishida, T., Kiwada, H., and Harashima, H. (2007). Tumor targeting of doxorubicin by anti-MT1-MMP antibody-modified PEG liposomes, *Int. J. Pharm.*, **1–2**, pp. 194–200.

55. Nallamothu, R., Wood, G.C., Kiani, M.F., Moore, B.M., Horton, F.P., and Thoma, L.A. (2006). A targeted liposome delivery system for combretastatin A4: Formulation optimization through drug loading and *in vitro* release studies, *PDA J. Pharm. Sci. Technol.*, **3**, pp. 144–155.

56. Frisch, B., Carriere, M., Largeau, C., Mathey, F., Masson, C., Schuber, F., Scherman, D., and Escriou, V. (2004). A new triantennary galactose-targeted PEGylated gene carrier, characterization of its complex with DNA, and transfection of hepatoma cells, *Bioconjug. Chem.*, **4**, pp. 754–764.

57. Torchilin, V.P., Levchenko, T.S., Lukyanov, A.N., Khaw, B.A., Klibanov, A.L., Rammohan, R., Samokhin, G.P., and Whiteman, K.R. (2001). p-Nitrophenylcarbonyl-PEG-PE-liposomes: Fast and simple attachment of specific ligands, including monoclonal antibodies, to distal ends of PEG chains via p-nitrophenylcarbonyl groups, *Biochim. Biophys. Acta.*, **2**, pp. 397–411.

58. Yamada, Y., Shinohara, Y., Kakudo, T., Chaki, S., Futaki, S., Kamiya, H., and Harashima, H. (2005). Mitochondrial delivery of mastoparan with transferrin liposomes equipped with a pH-sensitive fusogenic peptide for selective cancer therapy, *Int. J. Pharm.*, **1–2**, pp. 1–7.

59. Savic, R., Luo, L., Eisenberg, A., and Maysinger, D. (2003). Micellar nanocontainers distribute to defined cytoplasmic organelles, *Science*, **5619**, pp. 615–618.

60. D'Souza, G.G., Cheng, S.M., Boddapati, S.V., Horobin, R.W., and Weissig, V. (2008). Nanocarrier-assisted subcellular targeting to the site of mitochondria improves the pro-apoptotic activity of paclitaxel, *J. Drug Target.*, **7**, pp. 578–585.

61. Cheng, S.M., Pabba, S., Torchilin, V.P., Fowle, W., Kimpfler, A., Schubert, R., and Weissig, V. (2005). Towards mitochondria-specific delivery of apoptosis-inducing agents: DQAsomal incorporated paclitaxel, *J. Drug Deliv. Sci. Tec.*, **15**, pp. 81–86.

62. Vaidya, B., Paliwal, R., Rai, S., Khatri, K., Goyal, A.K., Mishra, N., and Vyas, S.P. (2009). Cell-selective mitochondrial targeting: A new approach for cancer therapy, *Cancer Therapy*, pp. 141–148.

63. Sega, E.I., and Low, P.S. (2008). Tumor detection using folate receptor-targeted imaging agents, *Cancer Metastasis Rev.*, **4**, pp. 655–664.

64. Zhao, X.B., and Lee, R.J. (2004). Tumor-selective targeted delivery of genes and antisense oligodeoxyribonucleotides via the folate receptor, *Adv. Drug Deliv. Rev.*, **8**, pp. 1193–1204.

65. Ke, C.Y., Mathias, C.J., and Green, M.A. (2003). The folate receptor as a molecular target for tumor-selective radionuclide delivery, *Nucl. Med. Biol.*, **8**, pp. 811–817.

66. Esmaeili, F., Ghahremani, M.H., Ostad, S.N., Atyabi, F., Seyedabadi, M., Malekshahi, M.R., Amini, M., and Dinarvand, R. (2008). Folate-receptor-

targeted delivery of docetaxel nanoparticles prepared by PLGA-PEG-folate conjugate, *J. Drug Target*, **5**, pp. 415–423.

67. Kim, S.H., Jeong, J.H., Mok, H., Lee, S.H., Kim, S.W., and Park, T.G. (2007). Folate receptor targeted delivery of polyelectrolyte complex micelles prepared from ODN-PEG-folate conjugate and cationic lipids, *Biotechnol. Prog.*, **1**, pp. 232–237.
68. Leamon, C.P., Reddy, J.A., Vlahov, I.R., Westrick, E., Dawson, A., Dorton, R., Vetzel, M., Santhapuram, H.K., and Wang, Y. (2007). Preclinical antitumor activity of a novel folate-targeted dual drug conjugate, *Mol. Pharm.*, **5**, pp. 659–667.
69. Torchilin, V.P. (2005). Recent advances with liposomes as pharmaceutical carriers, *Nat. Rev. Drug Discov.*, **2**, pp. 145–160.
70. Inoki, Y., Hakamata, Y., Hamamoto, T., Kinouchi, T., Yamazaki, S., Kagawa, Y., and Endo, H. (2000). Proteoliposomes colocalized with endogenous mitochondria in mouse fertilized egg, *Biochem. Biophys. Res. Commun.*, **1**, pp. 183–191.
71. Boddapati, S.V., Tongcharoensirikul, P., Hanson, R.N., D'Souza, G.G., Torchilin, V.P., and Weissig, V. (2005). Mitochondriotropic liposomes, *J. Liposome Res.*, **1–2**, pp. 49–58.
72. Boddapati, S.V., D'Souza, G.G., Erdogan, S., Torchilin, V.P., and Weissig, V. (2008). Organelle-targeted nanocarriers: Specific delivery of liposomal ceramide to mitochondria enhances its cytotoxicity *in vitro* and *in vivo*, *Nano Lett.*, ***8***, pp. 2559–2563.
73. Yamada, Y., Akita, H., Kamiya, H., Kogure, K., Yamamoto, T., Shinohara, Y., Yamashita, K., Kobayashi, H., Kikuchi, H., and Harashima, H. (2008). MITO-Porter: A liposome-based carrier system for delivery of macromolecules into mitochondria via membrane fusion, *Biochim. Biophys. Acta.*, **2**, pp. 423–432.
74. Yasuzaki, Y., Yamada, Y., and Harashima, H. (2010). Mitochondrial matrix delivery using a MITO-Porter, a liposome-based carrier that specifies fusion with mitochondrial membranes, *Biochem. Biophys. Res. Commun.*, 397, pp. 181–186.
75. Wagle, M.A., Martinville, L.E., and D'Souza, G.G. (2011). The utility of an isolated mitochondrial fraction in the preparation of liposomes for the specific delivery of bioactives to mitochondria in live mammalian cells, *Pharm. Res.*, **11**, pp. 2790–2796.
76. Lee, M., Choi, J.S., Choi, M.J., Pak, Y.K., Rhee, B.D., and Ko, K.S. (2007). DNA delivery to the mitochondria sites using mitochondrial leader peptide conjugated polyethylenimine, *J. Drug Target.*, **2**, pp. 115–122.

77. Callahan, J., and Kopecek, J. (2006). Semitelechelic HPMA copolymers functionalized with triphenylphosphonium as drug carriers for membrane transduction and mitochondrial localization, *Biomacromolecules*, **8**, pp. 2347–2356.

78. Cuchelkar, V., Kopeckova, P., and Kopecek, J. (2008). Novel HPMA copolymer-bound constructs for combined tumor and mitochondrial targeting, *Mol. Pharm.*, **5**, pp. 776–786.

79. Marrache, S., and Dhar, S. (2012). Engineering of blended nanoparticle platform for delivery of mitochondria-acting therapeutics, *Proc. Natl. Acad. Sci. U.S.A.*, **40**, pp. 16288–16293.

80. Zhou, L., Liu, J.H., Ma, F., Wei, S.H., Feng, Y.Y., Zhou, J.H., Yu, B.Y., and Shen, J. (2010). Mitochondria-targeting photosensitizer-encapsulated amorphous nanocage as a bimodal reagent for drug delivery and biodiagnose *in vitro*, *Biomed. Microdevices*, **12**, pp. 655–663.

81. Xu, Z.P., Niebert, M., Porazik, K., Walker, T.L., Cooper, H.M., Middelberg, A.P., Gray, P.P., Bartlett, P.F., and Lu, G.Q. (2008). Subcellular compartment targeting of layered double hydroxide nanoparticles, *J. Control. Release*, **1**, pp. 86–94.

82. Yacobi, N.R., Malmstadt, N., Fazlollahi, F., Demaio, L., Marchelletta, R., Hamm-Alvarez, S.F., Borok, Z., Kim, K.J., and Crandall, E.D. (2009). Mechanisms of alveolar epithelial translocation of a defined population of nanoparticles, *Am. J. Respir. Cell Mol. Biol.*, **42**, pp. 604–614.

83. Hillaireau, H., and Couvreur, P. (2009). Nanocarriers' entry into the cell: Relevance to drug delivery, *Cell. Mol. Life Sci.*, **66**, pp. 2873–2896.

84. Smirnov, P. (2009). Cellular magnetic resonance imaging using superparamagnetic anionic iron oxide nanoparticles: Applications to *in vivo* trafficking of lymphocytes and cell-based anticancer therapy, *Methods Mol. Biol.*, **512**, pp. 333–353.

85. Harush-Frenkel, O., Altschuler, Y., and Benita, S. (2008). Nanoparticle-cell interactions: Drug delivery implications, *Crit. Rev. Ther. Drug Carrier Syst.*, **6**, pp. 485–544.

86. Huser, T. (2008). Nano-biophotonics: New tools for chemical nano-analytics, *Curr. Opin. Chem. Biol.*, **5**, pp. 497–504.

87. Vasir, J.K., and Labhasetwar, V. (2008). Quantification of the force of nanoparticle-cell membrane interactions and its influence on intracellular trafficking of nanoparticles, *Biomaterials*, **31**, pp. 4244–4252.

88. Rajan, S.S., Liu, H.Y., and Vu, T.Q. (2008). Ligand-bound quantum dot probes for studying the molecular scale dynamics of receptor endocytic trafficking in live cells, *ACS Nano*, **6**, pp. 1153–1166.

89. Perrine, K.A., Lamarche, B.L., Hopkins, D.F., Budge, S.E., Opresko, L.K., Wiley, H.S., and Sowa, M.B. (2007). High speed method for *in situ* multispectral image registration, *Microsc. Res. Tech.*, **4**, pp. 382–389.

90. Rabut, G., and Ellenberg, J. (2004). Automatic real-time three-dimensional cell tracking by fluorescence microscopy, *J. Microsc.*, **Pt 2**, pp. 131–137.

91. Sunaguchi, M., Nishi, M., Mizobe, T., and Kawata, M., (2003). Real-time imaging of green fluorescent protein-tagged beta 2-adrenergic receptor distribution in living cells, *Brain Res.*, **1–2**, pp. 21–32.

92. Jester, J.V., Andrews, P.M., Petroll, W.M., Lemp, M.A., and Cavanagh, H.D. (1991). *In vivo*, real-time confocal imaging, *J. Electron Microsc. Tech.*, **1**, pp. 50–60.

93. Byrne, G.D., Pitter, M.C., Zhang, J., Falcone, F.H., Stolnik, S., and Somekh, M.G. (2008). Total internal reflection microscopy for live imaging of cellular uptake of sub-micron non-fluorescent particles, *J. Microsc.*, Pt **1**, pp. 168–179.

94. Matthaus, C., Kale, A., Chernenko, T., Torchilin, V., and Diem, M. (2008). New ways of imaging uptake and intracellular fate of liposomal drug carrier systems inside individual cells, based on Raman microscopy, *Mol. Pharm.*, **2**, pp. 287–293.

95. Fujita, K., and Smith, N.I. (2008). Label-free molecular imaging of living cells, *Mol. Cells*, **6**, pp. 530–535.

96. Freudiger, C.W., Min, W., Saar, B.G., Lu, S., Holtom, G.R., He, C., Tsai, J.C., Kang, J.X., and Xie, X.S. (2008). Label-free biomedical imaging with high sensitivity by stimulated Raman scattering microscopy, *Science*, **5909**, pp. 1857–1861.

Index

An environmentally friendly book printed and bound in England by www.printondemand-worldwide.com

PEFC Certified

This product is from sustainably managed forests and controlled sources

www.pefc.org

PEFC/16-33-415

This book is made of chain-of-custody materials; FSC materials for the cover and PEFC materials for the text pages.